[redacted]

S0-CLL-064

Compendium of Wheat Diseases

SECOND EDITION

M. V. Wiese

WITHDRAWN

[redacted]

APS PRESS

The American Phytopathological Society

Financial Contributors

Ciba-Geigy Inc.
Greensboro, NC
(representative Karen Ludlow)

E. I. du Pont de Nemours & Company, Inc.
Agricultural Products Department
Newark, DE
(representative William C. Reische)

Gustafson, Inc.
Dallas, TX
(representative Kyle W. Rushing)

Illinois Foundation Seeds, Inc.
Champaign, IL
(representative Dale Cochran)

Merck and Company, Inc.
Summerville, OR
(representative Ken Brown)

Mobay Corporation
Agricultural Chemicals Division
Kansas City, MO
(representative Susan A. York)

North Idaho Foundation Seed Association
Lewiston, ID
(representative Loren Kambitsch)

Pennwalt Corporation, AgChem Division
Fresno, CA
(representative Don Gargano)

Pioneer Hi-Bred International, Inc.
Johnston, IA
(representative Jeffrey R. Beard)

United AgriSeeds, Inc.
Champaign, IL
(representative Jim Nelson)

Front cover photograph by Steve Kronmiller, APS Press, St. Paul, MN
Back cover photograph by M. V. Wiese, College of Agriculture,
University of Idaho, Moscow, ID

Library of Congress Catalog Card Number: 87-070237
International Standard Book Number: 0-89054-076-4

©1977, 1987 by The American Phytopathological Society
Published 1977. Second Edition 1987

All rights reserved.
No part of this book may be reproduced in any form
by photocopy, microfilm, retrieval system, or any other means,
without written permission from the publisher.

Printed in the United States of America

The American Phytopathological Society
3340 Pilot Knob Road
St. Paul, Minnesota 55121, USA

Preface

The first edition of the Compendium of Wheat Diseases, published by the American Phytopathological Society in 1977, proved to be a popular and useful document. Currently, 18,000 copies are in circulation among scientists, students, wheat producers, and wheat industry personnel.

Like the first edition, this second edition is a thorough and authoritative treatment of wheat diseases and is international in scope. It has been written to serve as a practical reference for both plant pathologists and general audiences. It is amply illustrated, uses descriptive terminology, and includes a glossary, index, and selected references. In developing this second edition, special attention was given to updating symptom and pathogen descriptions, to adding post-1976 literature citations, to bringing disease and pathogen nomenclature up to date, and to improving and supplementing the more than 160 photographs included in the first edition.

Any abnormality of the wheat plant or its seed is considered herein to be a disease. Damage caused by insects, however, is discussed only as it relates to, resembles, or confounds other disease diagnoses. The first and largest section of the Compendium describes over 100 infectious diseases caused by bacteria, fungi, nematodes, viruses and viruslike organisms, and witchweeds. The numerous diseases caused by fungi are subdivided according to the part of the plant on which the abnormality usually occurs. Physiological (nonparasitic) disorders caused by environmental factors are discussed in the second section on noninfectious diseases. Frequent cross-references in the text point out similarities and contrasts between diseases, symptoms, and host-parasite interactions.

Symptoms, pathogens, and disease cycles are given special attention and, as much as possible, the history, distribution, and economic importance of each disease are included. Control measures are presented as principles because specific controls for individual diseases often vary greatly with time and location. Detailed information about currently recommended controls should be obtained from competent area specialists.

Like the first edition, this second edition of the Compendium of Wheat Diseases has benefited from the input of numerous scientists. Each entry in the first edition was sent to one or more experts for technical updating. Comments from these initial reviews were incorporated, and the revised text was returned to the same and to additional scientists for final approval. In all, over 70 scientists contributed to the second edition.

Approximately one-fourth of the photographs in the first edition have been replaced or supplemented with improved illustrations. This second edition contains six new disease entries and significantly more illustrations of disease symptoms, pathogens, and disease cycles.

APS Press, the publishing arm of the American Phytopathological Society, approved the development of this second edition of the Compendium of Wheat Diseases in 1985. Dr. R. James Cook, USDA-ARS, Root Disease and Biological Control Laboratory, Pullman, Washington, served as the Associate Editor and as the liaison between the publisher and author. In addition, Dr. Cook reviewed and revised sections of the second edition and suggested many of the technical contributors on the following pages.

Dr. Barry Jacobsen, Department of Plant Pathology, University of Illinois, Urbana, reviewed parts of the manuscript and successfully solicited financial support for this project from several organizations.

In addition to the assistance of Drs. Cook and Jacobsen, the second edition has benefited from the input of more than 70 technical contributors from around the world, who reviewed and updated entries in the first edition, contributed new photographs, volunteered helpful comments, supplied current literature citations, and reviewed segments of the final manuscript. Their knowledgeable and timely input helped bring this project efficiently to completion and has made this second edition a truly comprehensive and authoritative resource. Drs. Patrick E. Lipps, Saad L. Hafez, and Robert H. Callihan contributed original drafts of the entries Snow Rot, Stem Nematode, and Chemical Injuries, respectively.

Financial resources for developing this second edition were provided primarily by the American Phytopathological Society and the University of Idaho, College of Agriculture. Generous contributions from the organizations listed on page ii covered the cost of clerical help, supplies, and photographic and postal services. Office and library facilities were provided by the University of Idaho and its College of Agriculture. I am grateful to Dr. Larry Branen, Dean of the College of Agriculture, University of Idaho, and Dr. Gary Lee, Director of the Idaho Agricultural Experiment Station, for the release time that was essential to develop this second edition.

Jacki Strope assisted with literature and library searches, typed and proofread each revision of the manuscript, and served as corresponding secretary. Her careful attention to detail and accurate reproduction of literature citations and scientific names and terms were most helpful.

I am grateful to the American Phytopathological Society and the APS Press for the unique opportunity to develop this second edition. Just as in 1977, when it was my privilege to write the first edition of this Compendium, I now am grateful to the many cooperating institutions and scientists who supported this project and to the many scientists who generously contributed expert assistance. My reward has been the knowledge I've gained, the acquaintances I've made, and the satisfaction I've felt in the last 10 years from seeing the Compendium of Wheat Diseases in use by scientists and students of wheat diseases the world over.

I share this accomplishment also with my wife Susie and my children Pat, Kris, and Steve, and I thank them for their understanding and support.

Technical Contributors

Kenneth R. Barker
Department of Plant Pathology
North Carolina State University
Raleigh, NC 27695

Donald J. S. Barr
Biosystematics Research Institute
Central Experimental Farm
Ottawa, Ontario K1A 0C6, Canada

William W. Bockus
Department of Plant Pathology
Kansas State University
Manhattan, KS 66506

Myron K. Brakke
USDA-ARS
Department of Plant Pathology
University of Nebraska
Lincoln, NE 68583

Lewis E. Browder
USDA-ARS
Department of Plant Pathology
Kansas State University
Manhattan, KS 66506

Rob H. Brown
Plant Research Institute
Victoria Department of Agriculture
Burnley, Victoria 3121, Australia

George W. Bruehl
Department of Plant Pathology
Washington State University
Pullman, WA 99164

George W. Buchenau
Department of Plant Science
South Dakota State University
Brookings, SD 57007

Robert H. Callihan
Department of Plant, Soil and Entomological Sciences
University of Idaho
Moscow, ID 83843

Thomas W. Carroll
Department of Plant Pathology
Montana State University
Bozeman, MT 59717

Carlos Castro
EMBRAPA
Caixa Postal 403-96.100
Pelotas - RS - Brazil

Lloyd N. Chiykowski
Plant Research Center
Agriculture Canada Research Branch
Ottawa, Ontario, Canada K1A 0C6

Neil W. Christensen
Department of Soil Science
Oregon State University
Corvallis, OR 97331

Joseph L. Clayton
Department of Botany and Plant Pathology
Michigan State University
East Lansing, MI 48824

Gilbert E. Cook
E. I. du Pont de Nemours & Co., Inc.
Greenacres, WA 99016

R. James Cook
USDA-ARS
Root Disease and Biological Control Laboratory
Department of Plant Pathology
Washington State University
Pullman, WA 99164

Barry M. Cunfer
Department of Plant Pathology
Georgia Experiment Station
Experiment, GA 30212

Harry S. Fenwick
Department of Plant, Soil and Entomological Sciences
University of Idaho
Moscow, ID 83843

Robert L. Forster
University of Idaho
Research and Extension Center
Kimberly, ID 83341

Richard A. Frederiksen
Department of Plant Pathology and Microbiology
Texas A&M University
College Station, TX 77843

Dennis W. Fulbright
Department of Botany and Plant Pathology
Michigan State University
East Lansing, MI 48824

Dennis C. Gross
Department of Plant Pathology
Washington State University
Pullman, WA 99164

Saad L. Hafez
University of Idaho
Southwestern Idaho Research and Extension Center
Parma, ID 83660

L. Pat Hart
Department of Botany and Plant Pathology
Michigan State University
East Lansing, MI 48824

Allen S. Heagle
USDA-ARS
Air Quality Research Laboratory
North Carolina State University
Raleigh, NC 27606

James A. Hoffmann
USDA-ARS
Crops Research Laboratory
Utah State University
Logan, UT 83422

High W. Homan
Department of Plant, Soil and Entomological Sciences
University of Idaho
Moscow, ID 83843

Robert M. Hosford, Jr.
Department of Plant Pathology
North Dakota State University
Fargo, ND 58105

Don M. Huber
Department of Botany and Plant Pathology
Purdue University
West Lafayette, IN 47907

Andrew O. Jackson
Department of Plant Pathology
University of California
Berkeley, CA 94720

Barry J. Jacobsen
Department of Plant Pathology
University of Illinois
Urbana, IL 61801

Stanley G. Jensen
USDA-ARS
Department of Plant Pathology
University of Nebraska
Lincoln, NE 68583

James B. Johnson
Department of Plant, Soil and Entomological Sciences
University of Idaho
Moscow, ID 83843

Brian R. Kerry
Department of Nematology
Rothamsted Experimental Station
Harpenden, Herts. AL5 2JQ, England

Elizabeth S. Klepper
USDA-ARS
Columbia Basin Agricultural Research Center
Pendleton, OR 97801

Cal F. Konzak
Department of Agronomy and Soils
Washington State University
Pullman, WA 99164

Lance W. Kress
Environmental Research Division
Argonne National Laboratory
Argonne, IL 60439

Willem G. Langenberg
USDA-ARS
Department of Plant Pathology
University of Nebraska
Lincoln, NE 68583

Roland F. Line
USDA-ARS
Department of Plant Pathology
Washington State University
Pullman, WA 99164

Patrick E. Lipps
Department of Plant Pathology
Ohio Agricultural Research and Development Center
Wooster, OH 44691

Richard M. Lister
Department of Botany and Plant Pathology
Purdue University
West Lafayette, IN 47907

Don E. Mathre
Department of Plant Pathology
Montana State University
Bozeman, MT 59717

Robert McDole
Department of Plant, Soil and Entomological Sciences
University of Idaho
Moscow, ID 83843

Glen A. Murray
Department of Plant, Soil and Entomological Sciences
University of Idaho
Moscow, ID 83843

Timothy D. Murray
Department of Plant Pathology
Washington State University
Pullman, WA 99164

Jens J. Nielsen
Agriculture Canada Research Station
Winnipeg, Manitoba R3T 2M9, Canada

Gary N. Odvody
Texas A&M University
Agricultural Experiment Station
Corpus Christi, TX 78410

Donald Penner
Department of Crop and Soil Sciences
Michigan State University
East Lansing, MI 48824

Keith S. Pike
Washington State University
Irrigated Agriculture Research and Extension Center
Prosser, WA 99350

Roger T. Plumb
Department of Plant Pathology
Rothamsted Experimental Station
Harpenden, Herts. AL5 2JQ, England

Anne S. Prabhu
EMBRAPA
Caixa Postal 179
Goias, Brazil

Laurence H. Purdy
Department of Plant Pathology
University of Florida
Gainesville, FL 32611

A. S. Rao
Department of Botany
Nagarjuna University
Nagarjuna Nagar
Andhra Pradesh, India

Richard A. Reinert
USDA-ARS
Air Quality Research
North Carolina State University
Raleigh, NC 27650

John H. Riesselman
Department of Plant Pathology
Montana State University
Bozeman, MT 59717

Alan P. Roelfs
USDA-ARS
Cereal Rust Laboratory
University of Minnesota
St. Paul, MN 55108

Gerald S. Santo
Washington State University
Irrigated Agriculture Research and Extension Center
Prosser, WA 99350

David B. Sauer
USDA-ARS
U.S. Grain Marketing Research Laboratory
Kansas State University
Manhattan, KS 66506

Norman W. Schaad
Department of Plant, Soil and Entomological Sciences
University of Idaho
Moscow, ID 83843

Albert L. Scharen
USDA-ARS
Department of Plant Pathology
Montana State University
Bozeman, MT 59717

George Semeniuk
109 Spring Street
Richland, WA 99352

Gregory E. Shaner
Department of Botany and Plant Pathology
Purdue University
West Lafayette, IN 47907

Eugene L. Sharp
Department of Plant Pathology
Montana State University
Bozeman, MT 59717

Pierre A. Signoret
Ecole Nationale Supérieure Agronomique
Montpellier, Cedex, France

Richard W. Smiley
Oregon State University
Columbia Basin Agricultural Research Center
Pendleton, OR 97801

Roland G. Timian
USDA-ARS
Department of Plant Pathology
North Dakota State University
Fargo, ND 58105

Robert D. Tinline
Agriculture Canada Research Station
Saskatoon, Saskatchewan S7N 0X2, Canada

John F. Tuite
Department of Botany and Plant Pathology
Purdue University
West Lafayette, IN 74907

Anne K. Vidaver
Department of Plant Pathology
University of Nebraska
Lincoln, NE 68583

Maurice L. Vitosh
Department of Crop and Soil Sciences
Michigan State University
East Lansing, MI 48824

James A. Vomocil
Department of Soil Science
Oregon State University
Corvallis, OR 97331

J. M. Waller
Commonwealth Mycological Institute
Kew, Surrey TW9 3AF, England

David M. Weller
USDA-ARS
Root Disease and Biological Control Laboratory
Department of Plant Pathology
Washington State University
Pullman, WA 99164

Ralph E. Whitesides
Department of Agronomy and Soils
Washington State University
Pullman, WA 99164

Richard A. Wiese
Department of Agronomy
University of Nebraska
Lincoln, NE 68583

Contents

Part II. Noninfectious Diseases

Introduction

Wheat is a nutritious, convenient, and economical source of food. It provides about 20% of the world's food calories and is a staple for nearly 40% of the world's population. Only rice, corn, and perhaps potatoes are as important. In many countries more wheat is consumed per capita than any other food.

Wheat grain provides important carbohydrates, proteins, vitamins, and minerals for growth and maintenance. It is consumed principally as bread but is a basic ingredient of numerous other food products. Wheat kernels contain gluten, a unique protein that is elastic and entraps carbon dioxide during fermentation of leavened dough.

Although wheat is grown primarily as a food crop, the plant and seed also find use in industrial products and as feed for livestock. Virtually all parts of the plant have value. Wheat straw is used as fuel, animal bedding, and organic matter for soil. Wheat germ oil is used as a food supplement. Wheat grain and bran are important livestock feeds, and young wheat plants serve as livestock forage.

Fossil records date wheatlike kernels to 15,000 B.C. Early humans in the Near and Middle East (along the Fertile Crescent) found seeds of native wheatlike grasses a reliable and palatable food source. These plants were the first to be cultivated, and they eventually influenced human longevity, distribution, and life-style. The baking and brewing industries evolved from the long association of human beings with cereal grains.

Today wheat occupies approximately 20% of the world's cultivated land and is the most important agricultural commodity in international trade. Most wheat is grown in the Northern Hemisphere; North America, Europe, China, and the Soviet Union account for over 80% of the world supply.

The term "staff of life" well conveys our past and present dependence on bread and the wheat from which it is made. The following passage from *Wheat and Wheat Improvement* (Agronomy Monograph 13, 1967, page ix, quoted by permission of the American Society of Agronomy) well describes the many facets of wheat in our culture:

> To a botanist, wheat is a grass. To a chemist, it is organic compounds, and to a geneticist, a challenging organism. To a farmer, it means a cash crop, and to a hauler, freight. To a laborer, it means employment, to a merchant, it is produce. To a miller it is grist, and to a baker, flour. The banker sees it as chattel, and the politician as a problem. Animals browse and feed on it and it sustains parasites. The conservationist uses it as ground cover. In religion, it is used as a symbol. The artist and photographer see it as a model. To millions it provides a livelihood, and to millions more a lifegiving food.

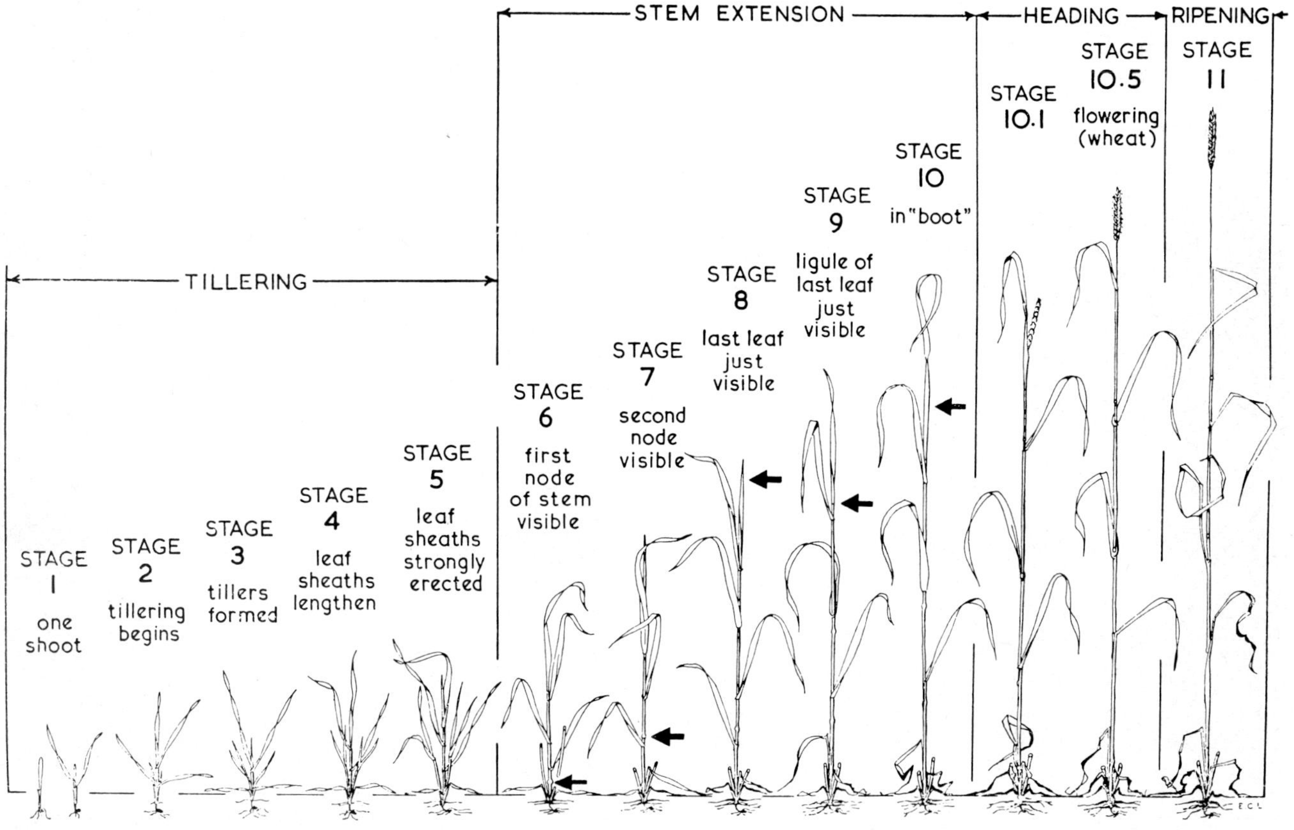

Fig. 1. Growth stages in cereals. (©Crown Copyright 1986. Reprinted, by permission, from Large, 1954, Fig. 2)

The Wheat Plant

Wheat is an annual grass that is widely cultivated as a small-grain cereal crop from sea level to altitudes over 3,000 m. It is best adapted to well-drained, clay-loam soils and to arid or semiarid environments in the temperate zone. Wheat plants may reach over 2 m in height, but most cultivated types are about 1 m tall. Mature wheat roots are fibrous and typically 1–2 m deep.

Like many other grasses, wheat plants emerge from seed as a coleoptile with seminal roots. Subsequently, tillers and roots develop from a crown that normally forms just beneath the soil surface. The crown is a compacted series of nodes that separate when internodal tissue expands and gives rise to the stem or culm. The upper culm internode (the peduncle) bears the spike or head. Culm length and strength are important because lodging (leaning or falling) can be a serious yield and harvest problem.

Unstressed wheat plants produce leaves, tillers, and root axes in a definite pattern. The number of heat units (growing degree-days) after seeding is primarily responsible for the rate of plant development. Stress causes tillers to elongate later, slower, or not at all. By observing which leaves and tillers are affected and by using a schedule of degree-days and anatomic growth stages, the time and severity of the stress can be estimated.

The developmental stages of wheat and certain other grass plants are usually described by the Feekes growth scale (Fig. 1). Other systems for naming and enumerating the growth stages of wheat have been developed by Romig (see Calpouzos et al, 1976), Zadoks et al (1974), and Klepper et al (1982). The latter systems were designed to have greater resolution, especially at the seedling and seed maturation stages, than the Feekes scale. For example, the node-naming system of Klepper et al (1982) identifies all leaves, tillers, and roots on a wheat plant. The plant shown in Fig. 2, for example, has 5.4 leaves on the main stem; tillers at the 0, 1, and 2 nodes; a well-branched seminal root system; and a few young, unbranched roots at the crown node. The coleoptilar tiller (T0) has 3.4 leaves, the first tiller (T1) has 2.6 leaves, and the second tiller (T2) has 1.9 leaves. Under this system, wheat roots can be described as being associated with node 1, 2, or 3 (crown roots) or with node 0, the scutellar or epiblast node, in the seed (seminal roots).

The wheat culm develops as an elongation of upper crown internodes. Wheat leaves form at culm nodes; they are alternately arranged and parallel-veined and consist of a blade and a culm-clasping sheath (Fig. 3). The head or spike emerges at the top of the plant during the last stages of culm elongation but forms at the culm apex, which initially is sheathed by leaves not far from soil level. The uppermost group of culm nodes that

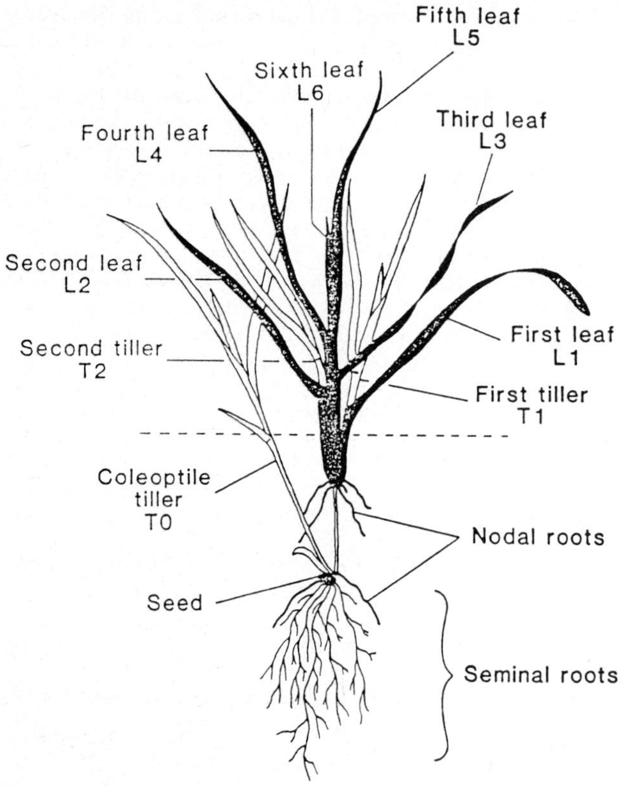

Fig. 2. Major anatomic features of a young wheat plant. (Courtesy E. S. Klepper)

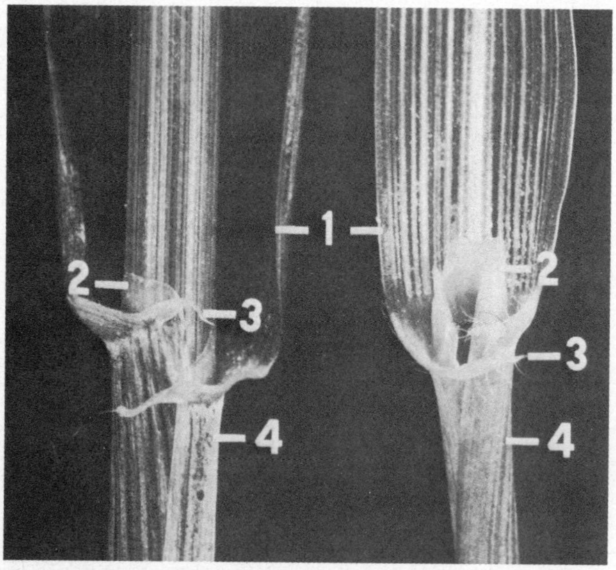

Fig. 3. Wheat leaves, showing blade (1), ligule (2), auricle (3), and sheath (4). The culm has been removed from the specimen on the right. (Reprinted, by permission of the American Society of Agronomy, from Quisenberry and Reitz, 1967)

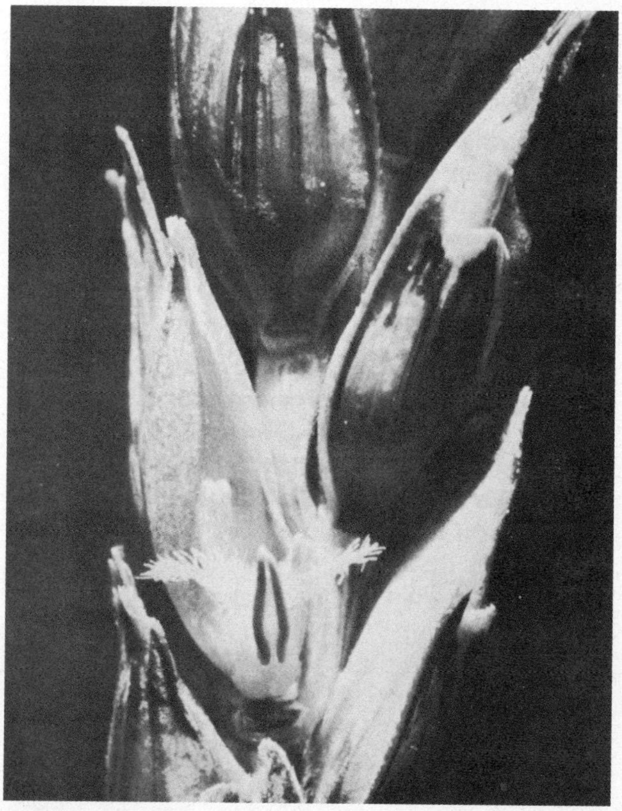

Fig. 4. Portion of wheat spike, showing the interior of one floret. (Reprinted, by permission of the American Society of Agronomy, from Quisenberry and Reitz, 1967)

form on the head are compacted into a rachis, on which the flower parts and seed differentiate. Each node on the rachis gives rise to a spikelet, which contains up to six florets set within glumes (Fig. 4). Each floret when fertilized can give rise to a kernel, which is attached at its embryo end and bears a brush of persistent hairlike epidermal cells at its terminus (Fig. 5). Wheat is typically self-pollinated (via anthers within each enclosed floret). Cross-pollination under field conditions normally involves less than 2% of all florets.

Botanically, wheat belongs to the genus *Triticum* L. in the Hordeae tribe of the grass family Gramineae. This genus is distinguished by one to several spikelets separately and alternately attached on opposite sides of the rachis. Modern *Triticum* species probably originated in the Near East. A chance mating between a wild einkorn and a grass could have produced einkorn (*T. monococcum* L.), a diploid European wheat ally with seven chromosome pairs ($n = 7$). Einkorn seeds trace to 6,700 B.C. and the river valleys between the Mediterranean and Red seas. Modern wheats appear to be descendants of crosses between einkorn and emmer (*T. dicoccum* Schrank), a tetraploid with 14 chromosome pairs ($n = 14$).

Modern *Triticum* species fall into three natural groups based on chromosome number: diploids ($n = 7$), tetraploids ($n = 14$), and hexaploids ($n = 21$). Each group of seven chromosome pairs (genome) apparently was contributed to modern wheat by different ancestral parents. Although usually less than 2% of all seed is cross-pollinated in wheat, natural outcrossing between wheatlike grasses is presumed to have initiated the modern polyploid wheats. *T. aestivum* L. (common wheat) and *T. compactum* Host (club wheat) are hexaploid (contain three genomes), and *T. durum* Desf. (durum wheat) is diploid (contains two genomes). These three species account for about 90% of the cultivated crop. None of the ancestral parents of these wheat species are grown commercially today.

In addition to taxonomic relationships, wheat classification also hinges on agronomic criteria and on kernel texture and color. Hard wheat, for example, has vitreous kernels with 11–17% protein and is used primarily for bread and rolls. Soft wheat is generally sown in the autumn and contains 6–11% protein. The weaker gluten of soft wheat makes it more desirable for cakes, cookies, pastries, and crackers. Most club wheats are soft white winter types. Most durum wheats are white and spring-sown. Because of the coarse nature of their ground product (semolina), durum wheats are used in macaroni, spaghetti, and other pasta products. True winter wheats require a cold period to induce erect growth and reproductive development, whereas spring wheats do not require a cold period to set seed.

Wheat breeders making conventional crosses between selected male and female wheat parents have produced improved cultivars that have been of great significance locally, nationally, and internationally. New genetic engineering techniques will likely lead to further cultivar improvements. Cultivars with increased yield potential, uniformity, and disease resistance are highly desirable, as are those adapted for specific environments, regions, and uses. Thus, wheat today is a heterogeneous crop that can be differentiated by chromosome number, spring or winter growth habit, response to day length, seed color, winter and drought hardiness, resistance to pests and diseases, plant height, straw strength, and quality for food, feed, and forage.

Selected References

Calpouzos, L., Roelfs, A. P., Madison, M. E., Martin, F. B., Welsh, J. R., and Wilcoxson, R. D. 1976. A new model to measure yield losses caused by stem rust in spring wheat. Minn. Agric. Exp. Stn. Tech. Bull. 307. 23 pp.

Klepper, B., Rickman, R. W., and Peterson, C. M. 1982. Quantitative characterization of vegetative development in small grain cereals. Agron. J. 74:789–792.

Klepper, B., Belford, R. K., and Rickman, R. W. 1984. Root and shoot development in winter wheat. Agron. J. 76:117–122.

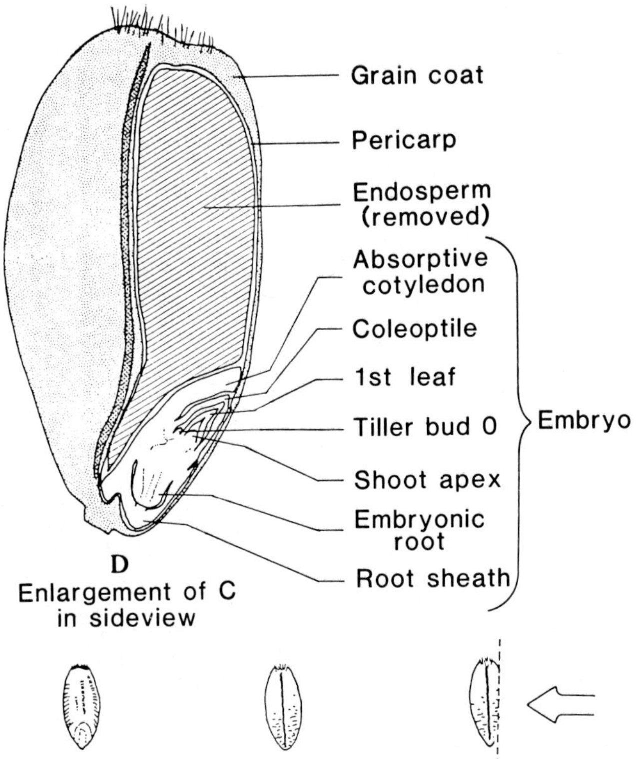

Fig. 5. Major anatomic features of a mature wheat kernel. (Courtesy E. S. Klepper)

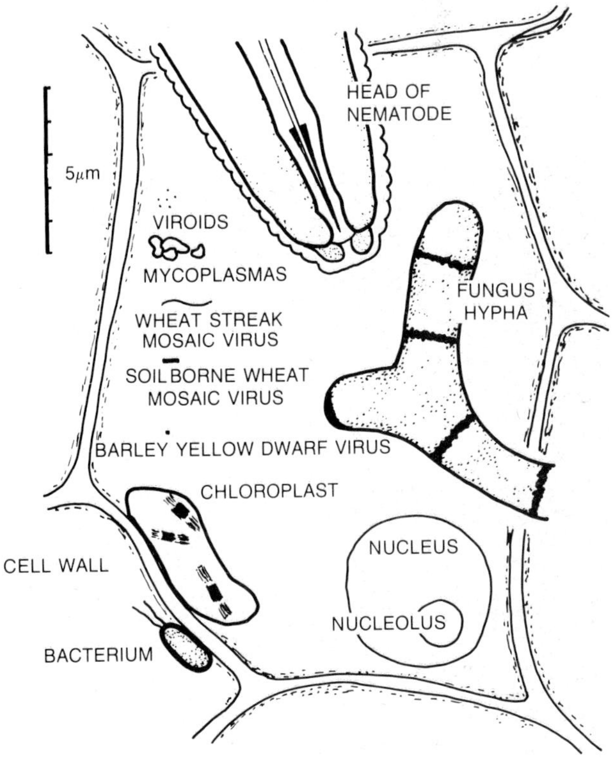

Fig. 6. Relative sizes of plant pathogens and a host plant cell. (Modified from G. N. Agrios, Plant Pathology, Fig. 2, 2nd ed., copyright 1978, Academic Press, New York)

Large, E. C. 1954. Growth stages in cereals. Illustration of the Feekes scale. Plant Pathol. 3:128–129.

Quisenberry, K. S., and Reitz, L. P. 1967. Wheat and Wheat Improvement. Agronomy Monograph 13. American Society of Agronomy, Madison, WI.

Zadoks, J. C., Chang, T. T., and Konzak, C. F. 1974. A decimal code for the growth stages of cereals. Weed Res. 14:415–421.

Wheat Diseases

Wheat plants in all stages of growth and in all natural environments are subject to various mechanical, physiologic, and biological stresses that interfere with their normal growth and development. Weather, toxicants, pollutants, insects, viruses, fungi, nematodes, bacteria, and weeds are primary hazards to wheat production. The principle biological agents that cause wheat diseases are fungi, viruses, bacteria, and nematodes. These agents are parasitic and cause infectious diseases that are transmissible from plant to plant (Fig. 6). Herein, however, any abnormality in the wheat plant is considered a disease.

Each year about 20% of the wheat that otherwise would be available for food and feed worldwide is lost to disease either in the field or in storage. Successful wheat production hinges on sound crop, soil, insect, disease, weed, and water management and use of improved cultivars, new technology, and mass-produced nutrients. Among all constraints to wheat yield and quality, temperature, available nutrients, and moisture are perhaps the most important.

The actual number of wheat diseases is unknown. About 50 are routinely important economically, and nearly 200 have been described. All diseases, noninfectious as well as infectious, are injurious in some areas, in some years, and on some plant part. All draw attention because of symptoms or signs and generate concern because of their detrimental effects on the quality and/or quantity of plants, straw, or grain.

All parts of the wheat plant are subject to disease, and one or more diseases can occur on virtually every plant and in every field. The magnitude of disease damage depends on environment and cultivar. With infectious diseases, the type, population, and virulence of the pathogens and their vectors also influence disease severity.

Wheat diseases are introduced individually in this Compendium. However, the diagnostician must be prepared to recognize combinations of symptoms, signs, and pathogens and to deal with symptoms common to more than one disease syndrome or as modified by prevailing or changing environmental conditions.

Part I. Infectious Diseases

Diseases Caused by Bacteria and Mycoplasmas

Bacteria are prokaryotic and mostly unicellular organisms. Over 2,000 species are recognized, most of which lack chlorophyll and are saprophytic. A few are parasitic and pathogenic to plants and/or animals. Bacteria display remarkable variability and adaptability. They persist in numerous environments, often in heterogeneous populations. Although most bacteria can be isolated and readily cultured, they can be difficult to classify and identify to species.

Bacterial cells divide by binary fission, snapping division, or pinching off. In favorable environments, large numbers of new cells can be produced within minutes or hours. Motility is common and is achieved primarily via flagella. Plant-pathogenic bacteria are usually unicellular nonspore-forming rods about 1–3 μm long (Figs. 6 and 7).

Symptoms of bacterial diseases of wheat include water-soaking, chlorosis, necrosis, bleaching, spotting, rots, and deformations. Occasionally, slime or mucus is visible. In addition to specific and general host symptoms, bacterial wheat pathogens are identified by host specificity, cell and colony morphology, and reactions in serological and biochemical tests.

Bacteria enter wheat through wounds and natural openings and inhabit vascular tissues and intercellular spaces. Many are pathogenic primarily through the action of enzymes and toxins, some cause metabolic imbalances in their hosts, and the pathogenic action of others is unknown. Free moisture and moderate-to-warm temperatures are general requirements for pathogen and disease development. Dispersal of plant-pathogenic bacteria is largely mechanical and mediated by people, seed, insects, air currents, and water.

Mycoplasmas are bacteria in the class Mollicutes. They are prokaryotic but smaller than commonly known bacteria and lack a rigid cell wall. The heterogeneous protoplasm of each cell is held within a delicate membrane that can assume various morphological shapes that range from 0.08 to 1 μm wide (Fig. 8). Some mycoplasmas are the smallest organisms capable of replication on artificial media.

Because mycoplasmas cause symptoms in plants that resemble those of virus diseases and still cannot be cultured (except for a few spiroplasmas), they often escape detection as

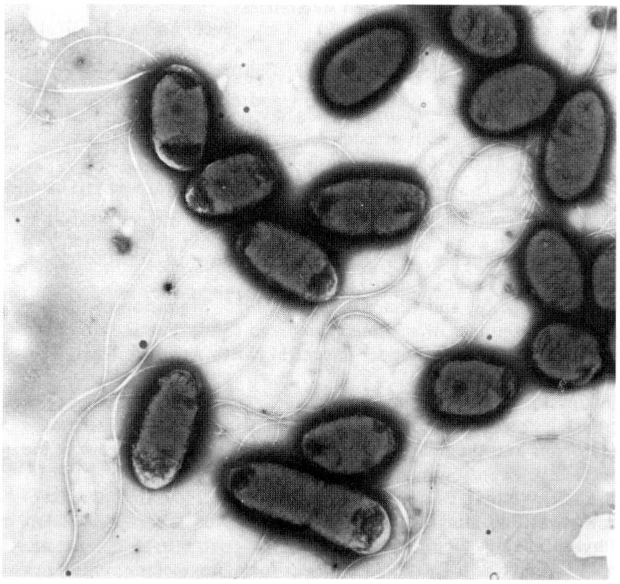

Fig. 7. Flagellated cells of an *Erwinia* sp. bacterium (×10,000). (Courtesy C. F. Gonzalez)

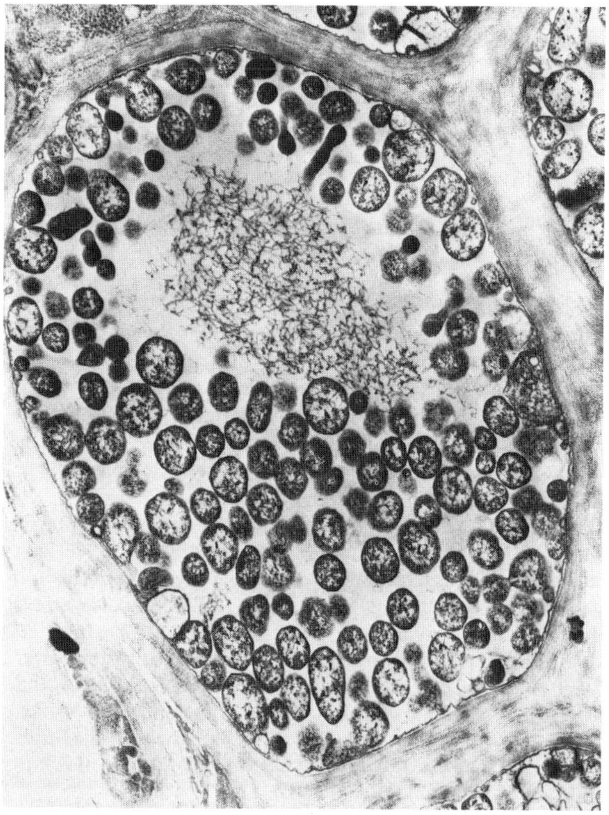

Fig. 8. Aster yellows mycoplasma in a host phloem cell. (Courtesy E. E. Banttari)

and produce a diffusible green fluorescent pigment on King's medium B or similar media. On beef-peptone agar, colonies are circular, shining, and translucent. The bacterium is aerobic and catalase-positive but negative for arginine dihydrolase and oxidase. It grows optimally between 25 and 28° C, with limits at 2 and 37° C.

Disease Cycle and Control

P. syringae pv. *atrofaciens* is seedborne and also may persist in soil. During wet weather, water in the creases of spikelets traps the bacterium disseminated in windblown dust or on insects. The pathogen multiplies near glume joints and lies dormant when moisture is limiting. Epidemics related to the dispersal of *P. syringae* pv. *atrofaciens* on seed from diseased plants have not been verified. Apparently bacterial populations in soil and on seed and plant debris are less important to disease development than is moisture. Nevertheless, clean or chemically disinfected seed should be selected for planting.

Selected References

Boewe, G. H. 1960. Diseases of wheat, oats, barley and rye. Ill. Nat. Hist. Surv. Circ. 48.

McCulloch, L. 1920. Basal glume rot of wheat. J. Agric. Res. 18:543–551.

Noble, R. J. 1933. Basal glume rot. Agric. Gaz. N.S.W. 44:107–109.

Sands, D. C., and Rovira, A. D. 1970. Isolation of fluorescent pseudomonads with a selective medium. Appl. Microbiol. 20:513–514.

Wilkie, J. P. 1973. Basal glume rot of wheat in New Zealand. N.Z. J. Agric. Res. 16:155–160.

Black Chaff

Black chaff (also called bacterial stripe) occurs worldwide on wheat, rye, triticale, oats, barley, and numerous grasses. The first descriptions of the disease appeared at the end of the 19th century. Each early account cited darkened chaff (glumes) often accompanied by chlorotic and necrotic blotches on leaves. The disease was first described on wheat in 1902. The causal agent was identified as a bacterium in 1916 and was shown to be seedborne in 1919.

Damage from black chaff varies from negligible to severe. Destructive epidemics on wheat and other cereals have been reported sporadically since 1917. The disease often goes unnoticed in dryland wheat in temperate regions. However, in subtropical regions like the southeastern United States and under sprinkler irrigation, yield losses in wheat and barley can be significant.

Symptoms

Black chaff symptoms are most obvious after heading. They include brown-black, water-soaked, and necrotic streaks and blotches, especially on glumes and/or leaves (Plates 4 and 5). Typically, upper portions of glumes are discolored, but darkening can be total. Awns also are darkened irregularly after first being discolored at their base. Bands of healthy and necrotic tissue on awns are indicative of black chaff. Diagnosis must be done carefully because symptoms of brown necrosis and glume blotch (melanism) are similar. Isolating the causal bacterium on yeast extract-dextrose $CaCO_3$ (YDC) or other semiselective agar may be necessary to confirm black chaff.

In wet weather, abundant bacterial growth may appear as slime or viscous droplets on diseased tissues (Fig. 10). When dry, such exudates appear fragile, light-colored, and scalelike. As with bacterial leaf blight, water-soaked lesions may develop on young leaves. Occasionally a lime green zone surrounds such lesions. Culms may show dark longitudinal streaks, and kernels may become shrunken at their base and nongerminable (see Basal Glume Rot). Diseased heads mature late and may be sterile if infected before flowering.

Causal Organism

Xanthomonas campestris pv. *translucens* (J. J. & R.) Dye (syns. *X. campestris* (Pam.) Dows., *X. translucens* (J. J. & R.) Dows. var. *undulosa* (J. J. & R.)), the causal bacterium of black chaff, is an aerobic, gram-negative rod. Cells measure 0.5×1.0 μm and are motile by one polar flagellum. A yellow (brominated arylpolyene ester) pigment, xanthomonadin, produced in culture is diagnostic. The bacterium is oxidase-negative and catalase-positive. Colonies on YDC agar are slow-growing, pale yellow, mucoid, convex, and smooth, with entire margins. Their growth is optimal at 28–30° C.

Etiology and Control

X. campestris pv. *translucens* is seedborne. Its persistence in host residues and soil is suspected but not well documented. It tolerates relatively wide ranges of temperature and moisture conditions. Wheat is invaded through stomata and wounds and supports the bacterium in tissue creases and intercellular spaces that harbor free water. The bacterium is spread by splashing rain, sprinkler irrigation, plant contact, and spike-visiting insects such as aphids. The bacterium also is dispersed on seed and residues from infected plants. Free water aids but may not be necessary for infection and disease development.

The best known control for black chaff is to use seed that is free of the pathogen. Thus, seed assay and certification programs are beneficial. If possible, frequent and especially sprinkler irrigation should be avoided. The use of resistant or tolerant cultivars also can reduce or eliminate black chaff.

Selected References

Cunfer, B. M., and Scolari, B. L. 1982. *Xanthomonas campestris* pv. *translucens* on triticale and other small grains. Phytopathology 72:683–686.

Fig. 10. Exudate of bacterial cells (*Xanthomonas campestris* pv. *translucens*) from leaves with black chaff. (Courtesy M. G. Boosalis)

Hagborg, W. A. F. 1946. The diagnosis of bacterial black chaff of wheat. Sci. Agric. 26:140–146.

Kim, H. K. 1982. Epidemiological, genetical, and physiological studies of the bacterial leaf streak pathogen, *Xanthomonas campestris* pv. *translucens* (J.J.R.) Dowson. Ph.D. thesis. Montana State University, Bozeman. 94 pp.

Schaad, N. W., and Forster, R. L. 1985. A semi-selective agar medium for isolating *Xanthomonas campestris* pv. *translucens* from wheat seeds. Phytopathology 75:260–263.

Wallin, J. R. 1946. Seed and seedling infection of barley, bromegrass and wheat by *Xanthomonas translucens* var. *cerealis*. Phytopathology 36:446–457.

Bacterial Leaf Blight

Bacterial leaf blight is caused by *Pseudomonas syringae* pv. *syringae*, a strain of an extremely common and widely dispersed bacterium. The organism was described previously as a pathogen on corn and sorghum but was not characterized as a wheat pathogen until 1972. Bacterial leaf blight resembles and often occurs in conjunction with other leaf spot diseases and physiologic disorders. Yield losses caused by bacterial leaf blight have not been thoroughly assessed, but foliage destruction may exceed 50%. The disease frequently occurs in the north central United States. Common wheats, durums, and triticales are susceptible, as are certain other cereals, including oats and rye.

Symptoms

Bacterial leaf blight develops on uppermost leaves after plants reach the boot stage. Initial water-soaked spots less than 1 mm in diameter expand, become necrotic, and turn from gray-green to tan-white. The spots may coalesce into irregular streaks or blotches within two to three days (Plate 6). Entire leaves may become necrotic, while heads and glumes typically are symptomless. During wet periods, droplets of bacterial cells may develop within lesions.

Causal Organism

P. syringae pv. *syringae* van Hall is a unicellular, gram-negative, blunt rod, $0.5–0.7 \times 0.8–2.2\ \mu m$. Its cells are motile by one or more polar flagella and produce a green, fluorescent pigment in culture (especially on King's medium B). Colonies on beef extract agar are gray-white to bluish. The bacterium liquefies gelatin and is oxidase-negative and catalase-positive. It grows optimally in culture at 28° C. Its temperature limits for growth are 1 and 35° C.

P. syringae is a composite species that includes numerous saprophytic strains and many pathogenic strains differentiated by host specificity. Many *P. syringae* isolates do not parasitize wheat but may invade woody or other herbaceous hosts.

Etiology and Control

P. syringae pv. *syringae* is found in soil and water and on plant surfaces the world over. Physiologic forms virulent on wheat require moisture for infection. Wet weather, high relative humidity, and relatively cool (15–25° C) spring temperatures favor disease development. *P. syringae* pv. *syringae* is spread by wind-driven rain and enters host plants through stomata and wounds. The pathogen also may be dispersed through an association with wheat seed.

Wheat cultivars vary widely in response to *P. syringae* pv. *syringae*. The Canadian cultivar Glenlea is reported to be resistant, whereas Winoka in South Dakota is highly susceptible. Some chemical sprays show promise for leaf blight control.

Selected References

Hayward, A. C., and Waterston, J. M. 1965. *Pseudomonas syringae*. Descriptions of Pathogenic Fungi and Bacteria, No. 46. Commonwealth Mycological Institute, Association of Applied Biologists, Kew, Surrey, England.

Otta, J. D. 1974. *Pseudomonas syringae* incites a leaf necrosis on spring and winter wheats in South Dakota. Plant Dis. Rep. 58:1061–1064.

Otta, J. D. 1977. Occurrence and characteristics of isolates of *Pseudomonas syringae* on winter wheat. Phytopathology 67:22–26.

Sands, D. C., and Rovira, A. D. 1970. Isolation of fluorescent pseudomonads with a selective medium. Appl. Microbiol. 20:513–514.

Sellam, M. A., and Wilcoxson, R. D. 1976. Bacterial leaf blight of wheat in Minnesota. Plant Dis. Rep. 60:242–245.

Pink Seed

In the 1950s, pink seeds appeared in Canadian wheat samples and subsequently in wheat from Europe. The disorder, first presumed to be related to scab, was later attributed to a bacterium. Pink seed occurs infrequently and is considered cosmetic and inconsequential.

Symptoms

In some seed lots, a low percentage of kernels appear pink. The diseased seeds are normal morphologically but have soft, pink endosperms. They often resemble seed given chemical dye (fungicide) treatments but consistently yield a pink bacterium in culture.

Causal Organism

The cause of pink seed is *Erwinia rhapontici* (Millard) Burkh. (syn. *E. carotovora* var. *rhapontici* (Mill.) Dye), a motile, gram-negative rod. The bacterium, $0.5–0.8 \times 1.2–1.5\ \mu m$, has three to seven peritrichous flagella and produces a diffusible pink pigment in culture. The pathogen is catalase-positive and oxidase-negative and ferments glucose.

Etiology and Control

E. rhapontici is an opportunistic pathogen that invades only injured kernels. Its only other known host is rhubarb (*Rheum* spp.), in which it causes crown rot. Pink seed sometimes is associated with gall midge (Cedidomyidae) injury in the field and also with grain harvested prematurely. Currently, no control measures are prescribed.

Selected References

Campbell, W. P. 1958. A cause of pink seeds in wheat. Plant Dis. Rep. 42:1272.

Luisetti, J., and Rapilly, F. 1967. Sur une alteration d'origine bacterienne des graines de blé. Ann. Epiphyt. Phytogenet. 18:483–486.

Roberts, P. 1974. *Erwinia rhapontici* (Millard) Burkholder associated with pink grain of wheat. J. Appl. Bacteriol. 37:353–358.

Spike Blight

Spike blight was first described on wheat as "tundu" in India in 1917. Subsequent descriptions came from Egypt, Iran, Ethiopia, Cyprus, Australia, Canada, and China. Several grasses also contract the disease in Europe and North America, but wheat is the only economic host. *Triticum aestivum, T. durum, T. dicoccum*, and *T. pyramidale* are all susceptible.

Spike blight (also called yellow rot, yellow slime, yellow ear, and tundu disease) is of negligible importance except in portions of India and possibly Ethiopia. Yield loss can be total on a plant basis, but no epidemics of spike blight have been reported. The disease occurs in association with the seed-gall nematode, *Anguina tritici*, usually in low-lying fields.

Symptoms

Initial field symptoms of spike blight include parallel yellow or white streaks along leaf veins. Later a conspicuous yellow exudate on wheat heads is diagnostic. Heads and necks frequently emerge as a distorted, sticky mass. Earlier, leaves may wrinkle or twist as they emerge from the bacteria-laden whorl. When dry, the exudate appears as white flecks on heads and upper leaf surfaces.

The bacterial mass is fluid during wet weather but hard and dry when relative humidity is low and dew is sparse. The dry, hardened exudates mechanically distort leaves, heads, and necks and inhibit or prevent their elongation. Symptoms of seed-gall nematode are a normal part of the spike blight syndrome.

Causal Organism

Clavibacter tritici (Carlson & Vidaver) Davis et al (syns. *Corynebacterium tritici* (Hutch.) Burk., *Phytomonas tritici* (Hutch.) Burk., *Corynebacterium michiganense* pv. *tritici* (Hutch.) Dye & Kemp) is a gram-positive rod. Typically $0.5–0.75 \times 0.95–1.3\ \mu m$, the rods vary from coccoid to club- or wedge-shaped and can appear branched because of the incomplete separation of dividing cells. Cell morphology is influenced somewhat by the culture medium. Pure cultures of the bacterium can be isolated from diseased wheat by dilution plating on yeast-glucose-chalk or other agar.

C. tritici is nonmotile. On most media, it produces a yellow-orange pigment and its colonies are convex, moist, and glistening, with entire margins. The bacterium is aerobic and reduces nitrate but does not hydrolyze starch. It produces acid from mannose, liquefies gelatin, and utilizes acetate. It grows optimally at 23–25° C, with a thermal death point near 50° C. It does not grow below 5 or above 38° C.

Disease Cycle

C. tritici persists in moist soils in association with organic matter. It parasitizes wheat when it enters the protective enclosures of the whorl at the plant apex. Typically this is accomplished via the seed-gall nematode, which acts as a vector. Juveniles of *A. tritici* become contaminated with bacterial cells in soil. Seed galls and nematodes from spike-blighted heads are unavoidably contaminated with the bacterium. Thus, *C. tritici* is disseminated on seed, in soil, and through its association with seed galls. *A. tritici* juveniles and *C. tritici* cells in seed galls can remain viable for more than five years.

Control

Wheat grown on well-drained soils is rarely damaged by spike blight. Controls described for *A. tritici* are applicable for spike blight.

Selected References

Bradbury, J. F. 1973. *Corynebacterium tritici*. Descriptions of Pathogenic Fungi and Bacteria, No. 377. Commonwealth Mycological Institute, Association of Applied Biologists, Kew, Surrey, England.

Commonwealth Mycological Institute. 1978. Distribution Maps of Plant Diseases, No. 156. Commonwealth Agricultural Bureaux, London.

Davis, M. J., Gillaspie, A. G., Jr., Vidaver, A. K., and Harris, R. W. 1984. *Clavibacter*: A new genus containing some phytopathogenic coryneform bacteria, including *Clavibacter xyli* subsp. *xyli* sp. nov. subsp. nov. and *Clavibacter xyli* subsp. *cynodontis* subsp. nov., pathogens that cause ratoon stunting disease of sugarcane and bermuda grass stunting disease. Int. J. Syst. Bacteriol. 34:107–117.

Gupta, P., and Swarup, G. 1968. On the ear-cockle and yellow ear-rot diseases of wheat. I. Symptoms and histopathology. Indian Phytopathol. 21:318–323.

Gupta, P., and Swarup, G. 1972. Ear-cockle and yellow ear-rot diseases of wheat. II. Nematode bacteria association. Nematologica 18:320–324.

Midha, S. K., and Swarup, G. 1974. Factors affecting development of ear-cockle and tundu diseases of wheat. Indian J. Nematol. 2:97–104.

Swarup, G., and Gupta, P. 1971. On the ear-cockle and tundu diseases of wheat. II. Studies on *Anguina tritici* and *Corynebacterium tritici*. Indian Phytopathol. 24:359–365.

Swarup, G., and Singh, N. J. 1962. A note on the nematode-bacterium complex in tundu disease of wheat. Indian Phytopathol. 15:294–295.

White Blotch

White blotch is reported to occur in North Dakota on hard red winter, hard red spring, and durum wheat cultivars. The disease has been observed since 1969 on maturing wheat leaves, often in association with other leaf spot diseases such as bacterial leaf blight, tan spot, and Septoria leaf blotch.

The importance of white blotch is largely unknown. Although there have been no additional reports of the disease, symptoms resembling those induced by white blotch have been observed in Canada (see Physiologic Leaf Spots), Oklahoma, and Brazil.

Symptoms

White to very light tan, irregularly shaped streaks on leaves are typical symptoms of white blotch. The elongated lesions tend to be wider and lighter in color than the streaks of bacterial leaf blight and are not preceded by water-soaking. Such symptoms are most visible on plants beyond the boot stage.

Causal Organism

Bacillus megaterium de Bary pv. *cerealis* Hosford is isolated from washed leaf blotches placed onto potato-dextrose agar medium. This procedure may also recover *Pseudomonas syringae* pv. *syringae*. On agar media, *B. megaterium* pv. *cerealis* produces smooth, white colonies of chained, nonmotile cells that average $3.5\ \mu m$ in length (range $1.5–7.0\ \mu m$) and $1.8\ \mu m$ in width (range $1.1–2.9\ \mu m$). The cells are gram-positive and readily form ellipsoidal spores.

B. megaterium pv. *cerealis* is found in soil and in association with plants, insects, and some fungi. It has been recovered from leaf spots on oats and barley and from wheat glumes and seed. Its frequent association with *P. syringae* pv. *syringae* suggests that it may contribute to a disease complex in addition to acting as an independent wheat pathogen.

Disease Cycle and Control

White blotch develops in the way described for bacterial leaf blight. The pathogen appears unavoidably associated with wheat plants and seed. Specific controls other than resistant cultivars have not been described.

Selected References

Gordon, R. E. 1977. The genus *Bacillus*. Pages 319–336 in: CRC Handbook of Microbiology. 2nd ed. A. I. Laskin and H. A. Lechevalier, eds. CRC Press, Cleveland, OH.

Hosford, R. M., Jr. 1982. White blotch incited in wheat by *Bacillus megaterium* pv. *cerealis*. Phytopathology 72:1453–1459.

Diseases Caused by Fungi

Fungi are lower plants that lack chlorophyll. Most exist as filamentous, branched chains of cells (hyphae) 0.5–100 μm wide with chitinous cell walls and well-differentiated organelles. Nearly all reproduce by spores, produced asexually (imperfectly) through mitosis or sexually (perfectly) through meiosis. Among the large and diverse group of fungi, spore morphology and mechanisms of spore production are most significant taxonomically.

Fungi are broadly adapted for survival in air, soil, and water. More than 10,000 species are pathogenic in plants and animals. They can be highly specialized, invading only specific plant parts or a very narrow range of hosts. Others indiscriminately parasitize many plant parts and species. Most fungal plant pathogens are facultative saprophytes capable of growth on culture media or nonliving plant tissues. Others are obligate parasites that grow only in intimate association with living plants.

Fungi infect wheat through wounds and natural openings or by direct penetration. Infections are manifested as rots, blights, deformations, lesions, spots, rusts, smuts, and mildews. These symptoms and signs of disease result from the depletion of host nutrients, the action of toxic or growth-stimulating by-products, and the mechanical displacement and enzymic breakdown of host tissues. Fungal diseases of wheat such as rusts, smuts, and mildews can be diagnosed as readily from signs of the pathogen as from symptoms of the disease.

Fungi are the largest, oldest, and most investigated group of wheat pathogens. The rusts, though not understood, were cited in biblical records. Diseases like ergot, stem rust, and bunt spawned investigations in the 18th century that became the cornerstones of mycology and plant pathology.

Selected References

Ainsworth, G. C., and Sussman, A. S. 1965–1973. The Fungi—An Advanced Treatise. 4 vols. Academic Press, New York.

Alexopoulos, C. J., and Mims, C. W. 1979. Introductory Mycology. 3rd ed. John Wiley & Sons, New York. 632 pp.

Barnett, H. L., and Hunter, B. B. 1972. Illustrated Genera of Imperfect Fungi. 3rd ed. Burgess Publishing Co., Minneapolis, MN. 241 pp.

Hawksworth, D. L. 1974. Mycologist's Handbook. Commonwealth Mycological Institute, Kew, Surrey, England. 231 pp.

Webster, J. 1980. Introduction to Fungi. 2nd ed. Cambridge University Press, Cambridge. 669 pp.

Fungal Diseases Principally Observed on Seed and Heads

Storage Molds

Storage fungi were known only as downy colored mats or "moulds" on cereal grains until the mid-1800s. Today, their association with detrimental changes in grain appearance and quality in storage is well known, as is their production of by-products (mycotoxins) that are toxic to humans and animals. Although storage fungi may infect wheat seed before harvest, they rarely cause damage in the field. Sometimes seeds unable to germinate in dry soil (see Fig. 113) are subject to attack.

The causal fungi, primarily *Aspergillus* and *Penicillium* spp., are ubiquitous. They persist on innumerable substrates and grow without free water. Their association with grain in the field and in storage is unavoidable. Cereal seeds in equilibrium with relative humidities (RH) above 70% (with equilibrium moisture content above 13%) can support fungal growth and are susceptible to damage.

Symptoms

Casual associations of storage fungi and seed are usually not detrimental. However, when the molds are able to grow, seed infections and decreased seed quality can result. Because the storage molds primarily invade embryos, many kernels that appear sound may be damaged and nongerminable. It is often necessary to examine seed microscopically or incubate seed on agar media (Plate 7) to demonstrate and diagnose infections by storage molds.

When infections are well established, embryos may be darkened and killed. Extensive fungal growth gives seed a musty odor and induces heating, "caking," and seed decay. Sometimes large volumes of grain are darkened or charred ("bin-burned") (Fig. 11C) by heat generated in storage.

Causal Organisms

The major pathogens of seed in storage are species of *Aspergillus* and *Penicillium*. Among *Aspergillus* spp., *A. restrictus* and members of the *A. glaucus* group colonize grain with over 13.5% moisture. Above 15% moisture, species such as *A. candidus* and *A. ochraceus* can grow. *Penicillium* spp. and *A. flavus* begin to develop when grain moisture exceeds 16%.

Aspergillus Link fungi have upright, simple conidiophores that arise from prominent foot cells and terminate in a globose or clavate swelling. The swellings bear clusters of phialides that produce globose, one-celled conidia (phialospores), 2.5–5 μm in diameter, in dry, basipetal chains (Fig. 12). In mass, conidia often are green or black but can be variously colored.

Penicillium Link species have conidiophores without foot cells that occur singly or clustered in synnemata. They are branched and brushlike near their apex, and each branch terminates in a phialide. Conidia (phialospores) are similar morphologically to those of *Aspergillus* and are variously colored (frequently blue or blue-green) in mass.

Members of both genera produce mycotoxins that are toxic to humans and animals and that reduce the palatability of grain. *A. flavus*, the source of aflatoxin, is a noted example. These fungi also produce large numbers of air-disseminated spores that can cause respiratory diseases in humans and animals.

Storage Mold Activity

Wheat seed infections by storage fungi are promoted by warm temperatures, seed injury, debris, and especially moisture. Seed stored at 70% RH has an equilibrium moisture content of 13.5%; at 85% RH, moisture increases to 18%. Within this moisture range, most storage molds grow optimally between 25 and 33° C. Many can grow between 5 and 40° C, and a few species of *Penicillium* grow below 0° C. Seasonal or diurnal temperature fluctuations cause moisture migration and

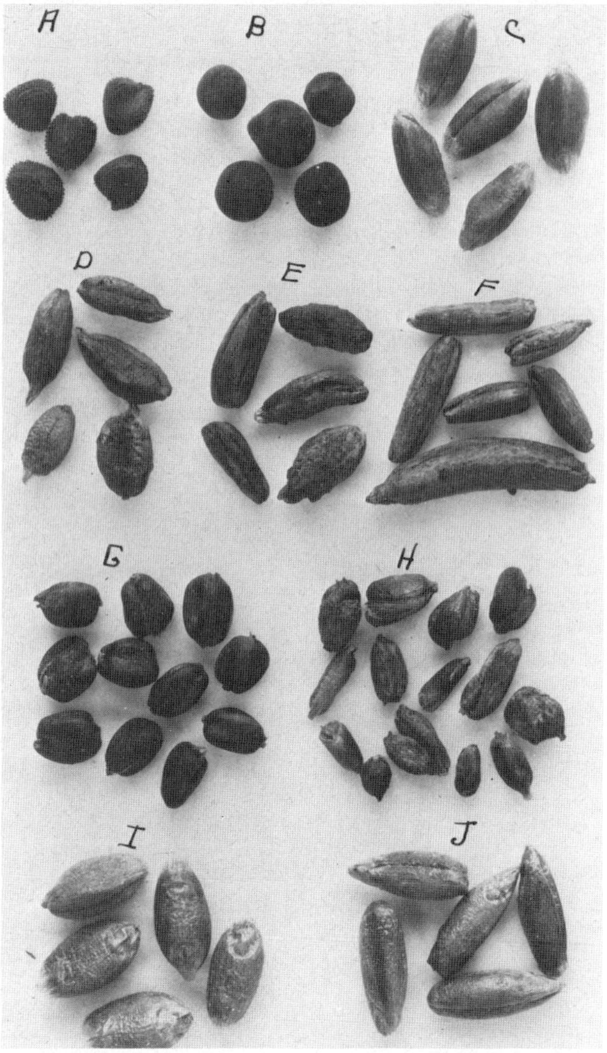

Fig. 11. Common impurities and transformed kernels found in stored wheat grain: **A,** corn cockle seed (*Agrostemma githago*); **B,** vetch seed (*Vicia angustifolia*); **C,** "bin-burned" wheat kernels; **D,** bunt balls; **E,** ergot sclerotia from wheat; **F,** ergot sclerotia from rye; **G,** nematode seed galls from wheat; **H,** nematode seed galls from rye; **I,** healthy wheat kernels; **J,** healthy rye kernels. (Courtesy USDA)

condensation in poorly aerated storage bins. Such moisture movement may permit fungal growth on grain that is otherwise suitably dry.

Control

Storage temperature and seed moisture should be maintained at levels insufficient for fungal growth. Normally, grain with less than 14% moisture (wet-weight basis) stored below 20° C will not be damaged. The suitability of any storage environment can be monitored by periodically sampling seed for fungi, temperature, moisture content, and germinability. In most environments, mechanical aeration is an inexpensive and effective way to cool grain and limit moisture migration.

Harvest or handling operations that mechanically injure seed should be avoided. These operations also should remove as much debris as possible. Treating seed with fungicides can curb storage molds, but treated grain normally is restricted for use only as seed. Some fungicides are inactive when not in solution, and others applied to the seed surface do not contact fungi within the embryo. Vapor-active chemicals such as organic acids that penetrate the seed coat to eradicate internal fungi may also reduce seed germinability. High-moisture grain stored for animal feed can be treated with propionic and acetic acid to inhibit molds.

Fig. 12. Apex of conidiophore of *Aspergillus flavus*. (Courtesy M. F. Brown and H. G. Brotzman)

Selected References

Christensen, C. M. 1973. Loss of viability in storage: Microflora. Seed Sci. Technol. 1:547–562.

Christensen, C. M., and Sauer, D. B. 1982. Microflora. Pages 219–240 in: Storage of Cereal Grains and Their Products. 3rd ed. C. M. Christensen, ed. American Association of Cereal Chemists, St. Paul, MN. 544 pp.

Harman, G. E., and Pfleger, F. L. 1974. Pathogenicity and infection sites of *Aspergillus* species in stored seeds. Phytopathology 64:1339–1344.

Kulik, M. M. 1973. Retention of germinability and invasion by storage fungi of hand-threshed and machine-threshed seeds of wheat in storage. Seed Sci. Technol. 1:805–810.

Sauer, D. B., Storey, C. L., Ecker, O., and Fulk, D. W. 1982. Fungi in U.S. export wheat and corn. Phytopathology 72:1449–1452.

Sauer, D. B., Storey, C. L., and Walker, D. E. 1984. Fungal populations in U.S. farm-stored grain and their relationship to moisture, storage time, regions, and insect infestation. Phytopathology 74:1050–1053.

Wallace, H. A. H., and Sinha, R. N. 1975. Microflora of stored grain in international trade. Mycopathologia 57:171–176.

Black Point (Kernel Smudge)

More than 100 species of fungi, including *Alternaria, Fusarium*, and *Helminthosporium* spp., can be isolated from newly harvested wheat grain. In contrast to storage molds, these fungi are most important in humid field environments, where they infect seed when relative humidity (RH) exceeds 90% and seed moisture content exceeds 20%. Some also are pathogenic in seedlings that develop from infected or infested seed.

Rainfall during seed maturation favors black point. Expanding green kernels are most susceptible. Premature seed senescence also promotes black point because many of the fungi are saprophytic.

Symptoms

Diseased kernels are discolored and appear weathered, black-pointed, or smudged (Plate 8). Black point describes the darkened pericarp and sometimes shriveled embryo end of the seed (see Basal Glume Rot). When embryos are invaded, germinability decreases.

At the market, discolored grain is discounted in value because products made from it have undesirable color or odor characteristics. Flour milled from black-pointed kernels may contain dark specks. Pigments or other compounds of fungal origin may cause illness if infested grain is consumed as feed or food. In the United States, blackened kernels are considered damaged, and only 2% and 4% are permitted in wheat graded as U.S. No. 1 and No. 2, respectively.

Causal Organisms

The pathogens most frequently associated with discolored kernels in the field are *Alternaria* (Fig. 13), *Fusarium*, and *Helminthosporium* spp. *Aspergillus, Chaetomium, Cladosporium, Curvularia, Gloeosporium, Myrothecium, Nigrospora, Penicillium, Plenodomus, Rhizopus*, and *Stemphylium* spp. may also be present. These fungi can coparasitize seed but differ widely in aggressiveness. Without competition from other organisms, each tends to dominate the substrate.

Control

It is impossible to exclude fungi from maturing wheat seed in the field, so control measures attempt only to restrict their activity. Wheat cultivars less prone to damage are a practical deterrent, especially when their resistance is anatomic (mechanical) and physiologic. Eradicant and protectant fungicides applied to maturing heads are effective but rarely economical. Chemical treatment of harvested seed can improve germinability and decrease infection of seedlings grown from diseased kernels.

Although seed damage from black point begins before harvest, it increases if grain is stored under moist or wet conditions (greater than 90% RH and 20% moisture). These conditions also promote the development of storage molds and should be strictly avoided.

Selected References

Adlakha, K. L., and Joshi, L. M. 1974. Black point of wheat. Indian Phytopathol. 27:41–44.

Bhowmik, T. P. 1969. *Alternaria* seed infection of wheat. Plant Dis. Rep. 53:77–80.

Huguelet, J. E., and Kiesling, R. L. 1973. Influence of inoculum composition on the black point disease of durum wheat. Phytopathology 63:1220–1225.

Hyde, M. B. 1950. The sub-epidermal fungi of cereal grains. I. A survey of the world distribution of fungal mycelium in wheat. Ann. Appl. Biol. 37:179–187.

Kilpatrick, R. A. 1968. Factors affecting black point of wheat in Texas, 1964–67. Tex. A&M Univ. Agric. Exp. Stn. Bull. MP-884. 11 pp.

Statler, G. D., Kiesling, R. L., and Busch, R. H. 1975. Inheritance of black point resistance in durum wheat. Phytopathology 65:627–629.

Uoti, J., and Ylimaki, A. 1974. The occurrence of *Fusarium* species in cereal grain in Finland. Ann. Agric. Fenn. 13:5–17.

Black (Sooty) Head Molds

A succession of saprophytic and weakly parasitic fungi, including *Alternaria, Cladosporium*, and *Sporobolomyces* spp., are associated with wheat plants throughout their life span in the field. When wet weather accompanies wheat maturation and especially when harvest is delayed, dark green, black, and sometimes pink or white superficial molds may develop on wheat heads (Plate 9). In humid climates, epiphytic molds may also appear on maturing leaves. Sometimes, in addition to colonizing senescing or damaged tissues, the fungi may cause mild infections in seed and cause black point or smudge. Heads that are shaded, weakened, undersized, or prematurely ripe (see Plate 34) are prone to colonization by these fungi. Plants that are nutritionally deficient, lodged, or damaged by diseases tend to support epiphytic fungi and appear sooty. Thus, these molds often indicate that damage from another cause has already occurred.

The mass and composition of epiphytic fungi on wheat plants are related to available nutrients and weather conditions. Nutrients on the plant surface or from pollen or insect honeydew support *Alternaria* and *Cladosporium* spp. as initial colonizers. In addition, *Sporobolomyces* and other fungi use nutrients on the plant surface and may interfere with the establishment of pathogens such as *Leptosphaeria nodorum* and *Cochliobolus sativus*.

Causal Organisms

The principal "sooty" molds are saprophytic and mildly parasitic species of *Cladosporium* and *Alternaria*. However, species of *Stemphylium, Epicoccum, Aureobasidium, Sporobolomyces*, and *Cryptosporium* can contribute to the syndrome.

Cladosporium fungi, principally *C. herbarum* (Link) Fr., produce olive brown mycelium and spores. Conidiophores are upright, clustered, or single and are variously branched near their apex. Conidia (blastospores) are dark and one- or two-celled. They vary in shape and measure approximately 4–7 × 10–15 μm. They develop terminally and laterally in acropetal, simple or branched chains.

Members of the genus *Alternaria* Nees produce dark brown, mostly simple conidiophores and simple or branched chains of conidia. The conidia (porospores) develop acropetally, are brown, and vary in shape from clavate to elliptical. They are multiseptate, and many have a simple or branched apical beak (Fig. 13).

Sporobolomyces Klur. and van Niel spp. are yeastlike, appear red, pink, or white in culture, and reproduce by budding (blastospores) directly from cells or from sterigmata. The spores are asymmetric, frequently kidney-shaped, and 2–6 μm in diameter and are often forcibly discharged (ballistospores). *Cryptococcus* and *Aureobasidium* spp. share many of the characteristics of *Sporobolomyces* but tend to be white.

Stemphylium spp., typically *S. botryosum* Wall., produce dark conidiophores 3–6 μm in diameter and swollen at their apex. Conidia (porospores) develop singly as swellings through the apex and are olive brown with transverse, longitudinal, and oblique septa. They are globose, oblong, ovoid, and 15–20 × 18–35 μm and are frequently constricted at a median septum.

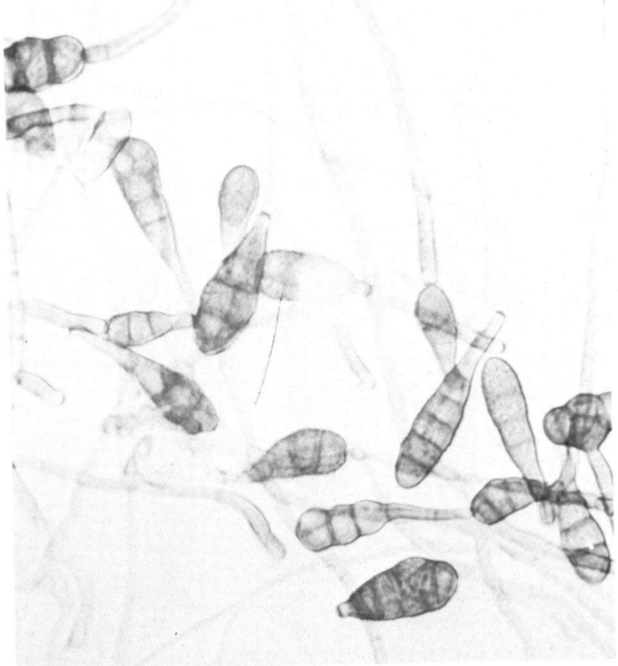

Fig. 13. Hyphae and conidia of *Alternaria alternata*. (Courtesy W. G. Fields)

One species described in 1910 as *S. tritici* Patt caused floret sterility in wheat in the southwestern United States.

Epicoccum Link spp. are dark and are distinguished by short, stout conidiophores arising from shallow sporodochial pads. Their conidia (dictyospores) are globose, multiseptate, 6–17 × 9–25 μm, and one- to several-celled.

Control

Several fungicides control sooty molds, but their use is rarely economical. Resistant cultivars are not known, but differences in tolerance are recognized.

Selected References

Ajayi, O., and Dewar, A. M. 1983. The effects of barley yellow dwarf virus, aphids and honeydew on *Cladosporium* infection of winter wheat and barley. Ann. Appl. Biol. 102:57–65.

Arya, H. C., and Panwar, K. S. 1955. Some studies on a virulent strain of *Cladosporium herbarum* (Link) Fr. on wheat. Indian Phytopathol. 8:176–183.

Blakeman, J. P., ed. 1981. Microbial Ecology of the Phylloplane. Academic Press, New York. 502 pp.

Blakeman, J. P., and Fokkema, N. J. 1982. Potential for biological control of plant diseases on the phylloplane. Annu. Rev. Phytopathol. 20:167–192.

Jenkyn, J. F., and Prew, R. D. 1973. The effect of fungicides on incidence of *Sporobolomyces* spp. and *Cladosporium* spp. on flag leaves of winter wheat. Ann. Appl. Biol. 75:253–256.

Schol-Schwarz, M. B. 1959. The genus *Epicoccum* Link. Mycologia 41:311–319.

Simmons, E. G. 1967. Typification of *Alternaria, Stemphylium* and *Ulocladium*. Mycologia 59:67–92.

Wiltshire, S. P. 1938. The origin and modern conceptions of *Stemphylium*. Trans. Br. Mycol. Soc. 21:211–239.

Ergot

Ergot was described in 16th century European literature, and its disease cycle on small grains has been known for over 200 years. In 400 B.C., Hippocrates apparently prescribed potions of ergoty grain to "further childbirth." As recently as 1978, ingestion of grain products contaminated with ergot alkaloids resulted in ergotism (illness and death) in humans and animals. Ergot draws attention because of its conspicuous signs and symptoms in the field. Also, the causal fungus progresses through diverse morphological forms and ecological associations during its life cycle. Its sclerotial stage has important implications in medicine and toxicology.

Ergot occurs the world over on wheat, triticale, barley, oats, and many cultivated and wild grasses. Rye is the principal economic host. Normally, the incidence of ergot on wheat is low, but the disease is a constant threat. Recent outbreaks have occurred on triticale and on open-floreted, male-sterile wheat lines used for hybrid seed production.

Fig. 14. Ergots (sclerotia) of *Claviceps purpurea* developed from infected florets and protruding from spikelets. (Courtesy T. G. Atkinson)

Symptoms

The most characteristic and noticeable signs of ergot are the purple-black, hornlike sclerotia (ergots) that replace one or more seeds in the head (Fig. 14). They protrude from the glumes on maturing heads and are up to 10 times larger than normal seed. This sclerotial stage of the fungus is preceded by a "honeydew" stage visible at flowering. Infected florets exude a sugary slime that accumulates in sticky, yellowish droplets. Insects, attracted to and feeding on the exudate, conspicuously congregate about infected heads and serve to disseminate the pathogen.

Before sclerotia develop, infected ovaries swell and become stromalike. Their surface is convoluted with a superficial layer of dense conidiophores. Not all infections progress through the honeydew or sclerotial stages, so sterility may be the only indication of floret infection.

Intact or broken sclerotia are easily seen among harvested seed (Fig. 11E). In minute quantities, however, they are best detected by chemical assays for ergot alkaloids. Grain that exceeds market tolerances for ergot is discounted in value and may be toxic if eaten by humans or animals. Not all sclerotia in grain originate from wheat. Infected grasses in the crop frequently contribute sclerotia to grain. Normally smaller and more slender than wheat ergots, grass ergots sometimes are solely responsible for ergoty grain samples.

Causal Organism

Claviceps purpurea (Fr.) Tul. produces three dissimilar morphological stages, which led early investigators to surmise that three different fungi were involved. A conidial and a sclerotial stage occur on wheat, and an ascogenous stage develops from sclerotia in soil.

Sclerotia are blue-black and 2–20 mm long, with white-gray contents. They germinate after cold-temperature treatment in

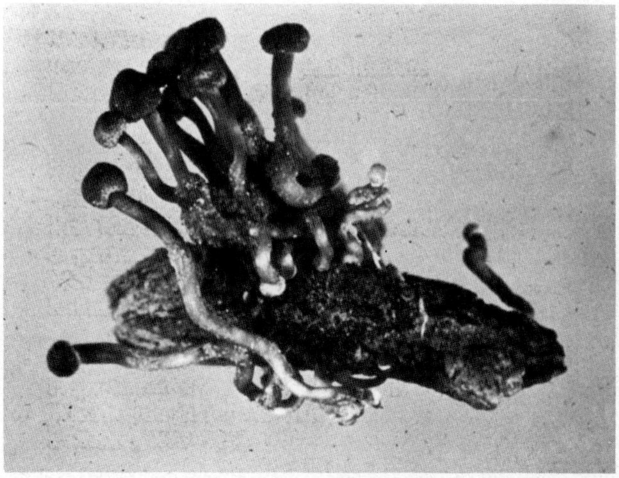

Fig. 15. Germinated sclerotium of *Claviceps purpurea*. The knoblike stromata that form atop tall stalks produce perithecia and ascospores. (Courtesy Plant Pathology Department, North Dakota State University)

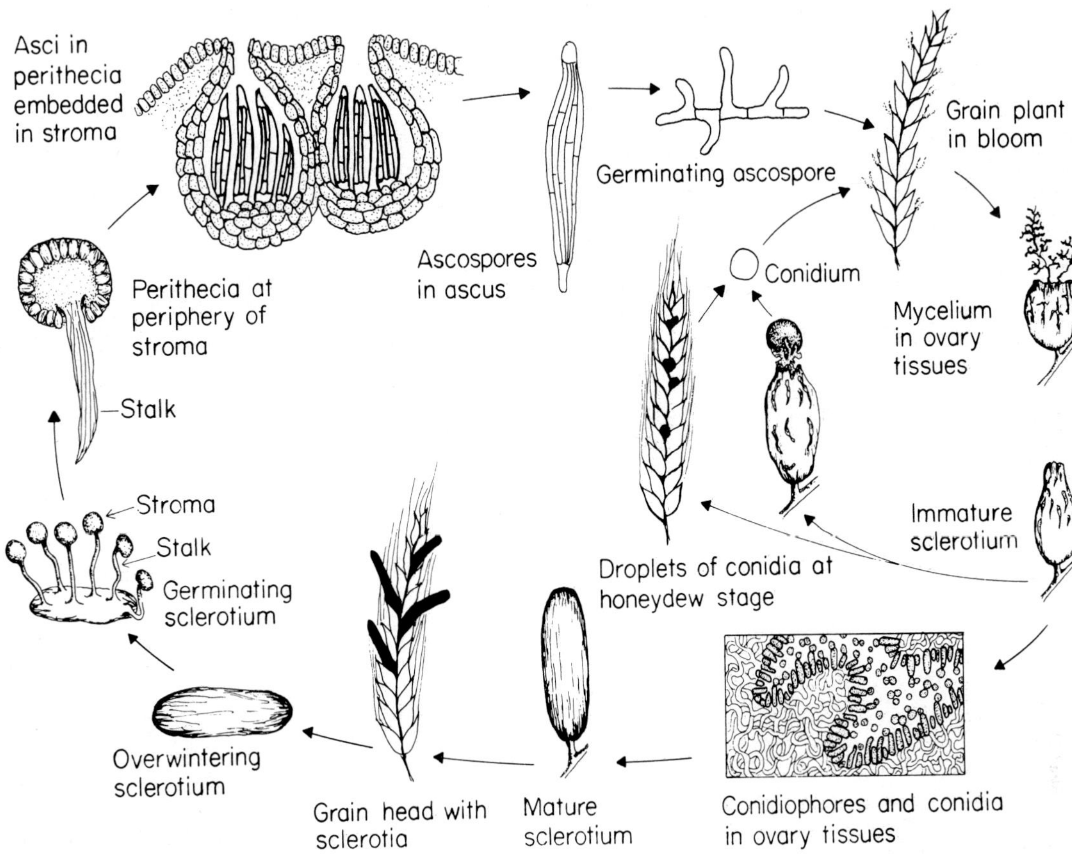

Fig. 16. Disease cycle of ergot of wheat caused by *Claviceps purpurea*. (Reprinted, by permission, from G. N. Agrios, Plant Pathology, Fig. 97C, 2nd ed., copyright 1978, Academic Press, New York)

response to moisture and form one or more stalked stromata (Fig. 15). Initially white, the stromata darken with age and reach 5–20 mm in length, and flask-shaped perithecia are embedded in their knoblike apex. The perithecia measure approximately 150 × 200 μm, protrude slightly from the stroma, and contain numerous hyaline, club-shaped asci. Ascospores (eight per ascus) are filiform, finely septate, and approximately 0.6 × 60 μm. Conidia of the honeydew stage are 2–3 × 4–6 μm, one-celled, hyaline, and elliptical.

C. purpurea isolates produce alkaloid chemicals in varying amounts. Sclerotia in particular contain a number of alkaloids that cause convulsive or gangrenous ergotism and death in humans and animals. The toxins (some of which are related to LSD) accumulate in the body, so their detrimental effects sometimes occur after long periods of low-level ingestion. Many of the alkaloids induce contraction of terminal arteries and smooth (nonstriped) muscle fibers. In controlled dosages, they are of medicinal value in stopping blood flow, in postpartum contraction of the uterus, and in treatment of migraine.

Disease Cycle

Ergot is indigenous to grasses in temperate areas. These grasses furnish most of the primary inoculum for infection of wheat and other cultivated hosts. Sclerotia remain viable for approximately one year in soil and for longer periods with grain in storage. Sclerotia in soil or sown with seed germinate during spring and early summer to produce stromata and ascospores (Fig. 16). Ascospores are the primary inoculum for floret infection. They are dispersed by wind and splashing rain, and those that contact a stigma germinate and penetrate the ovary within 24 hr. Within five days, conidia form on the ovary surface (honeydew stage) and serve as secondary inoculum. They are disseminated to other florets by contact, rain splash, and insects. With time, conidial production declines and the swollen, convoluted ovaries enlarge further and are converted progressively, base-to-tip, into sclerotia (Fig. 16).

Ergot is favored by wet, cool weather that accompanies and prolongs the flowering period. These conditions also favor "honeydew" formation, which attracts insect vectors. Florets are most susceptible to infection just before anthesis.

Control

Ergot can be avoided by the use of seed free of sclerotia, crop rotation, deep soil tillage, and clean cultivation. Modern grain-cleaning equipment can separate most sclerotia from seed. On a smaller scale, sclerotia can be removed by brine flotation techniques. Tillage operations should bury sclerotia deeper than 4 cm so that stromata will not form or will not reach the soil surface to liberate ascospores. Similarly, a one-year break between susceptible grain or grass crops reduces soilborne sclerotia to negligible numbers. Mowing headlands or roadways before grasses mature eliminates potential ergot reservoirs, as does spraying grasses with maleic hydrazide to prevent them from heading.

As with loose smut and Karnal bunt, certain wheat cultivars mechanically escape infection by having florets that remain closed or that open only for short periods during anthesis.

Some fungi, including *Fusarium roseum* 'Sambucinum', hyperparasitize *C. purpurea* and offer potential for biological control.

Selected References

Bove, F. J. 1970. The Story of Ergot. S. Karger AG, Basel, Switzerland. 297 pp.

Campbell, W. P., and Freisen, H. A. 1959. The control of ergot in cereal crops. Plant Dis. Rep. 43:1266–1267.

Darlington, L. C., and Mathre, D. E. 1976. Resistance of male sterile wheat to ergot as related to pollination and host genotype. Crop Sci. 16:728–730.

Mower, R. L., Snyder, W. C., and Hancock, J. G. 1975. Biological control of ergot by *Fusarium*. Phytopathology 65:5–10.

Peach, J. M., and Loveless, A. R. 1975. A comparison of two methods

of inoculating *Triticum aestivum* with spore suspensions of *Claviceps purpurea*. Trans. Br. Mycol. Soc. 64:328–331.
Platford, R. G., and Bernier, C. C. 1970. Resistance to *Claviceps purpurea* in spring durum wheat. Nature 222:770.
Puranik, S. B., and Mathre, D. E. 1971. Biology and control of ergot on male sterile wheat and barley. Phytopathology 61:1075–1080.
Roylance, H. B., and Hall, R. F. 1974. Ergot, a loser for grain growers and livestock owners. Univ. Idaho Coop. Ext. Serv. Curr. Inf. Ser. No. 145. 2 pp.
Young, J. C. 1981. Variability in the content and composition of alkaloids found in Canadian ergot. J. Environ. Sci. Health B16:381–393.

Scab (Head Blight)

Scab (also called head blight, pink mold, whiteheads, and tombstone scab) is manifested by the premature death or blighting of spikelets. The disease occurs on all small-grain cereal crops and is especially important in humid regions. Scab is frequently severe in wheat sown in the residue from a previous host crop such as corn. In such crops, significant yield losses may result from floret sterility and poor seed filling.

Fig. 17. Healthy spike (left) and spikes variously blighted and bleached by scab (caused by *Fusarium* spp.). (Courtesy Illinois Agricultural Experiment Station)

The causal fungi, *Fusarium* spp. and *Microdochium nivale* (*F. nivale*), are facultative, nonspecific parasites that may also infect other plant parts. Thus, damage to heads may be coincident with root or leaf infections or associated with subsequent seedling blights (see Common Root and Foot Rot). The pathogens have simple nutritional requirements and are omnipresent as saprophytes. Moisture usually is the limiting factor for parasitism.

Symptoms

Scab is best recognized on emerged immature heads where one or more spikelets or the entire head appears prematurely bleached (Fig. 17). If the rachis is infected, all tissues above that point are faded. Small, dark fruiting bodies (perithecia) (Fig. 18) and superficial (frequently pink or orange) mycelium and spore masses may be seen on, and especially at the base of, diseased spikelets (Plate 10). Bleached spikelets usually are sterile or contain shriveled and/or discolored seed (Fig. 19).

Grain from head-blighted fields may be less palatable to livestock than normal grain and may contain mycotoxins that induce muscle spasms and vomiting in humans and other nonruminant animals. The toxins apparently remain stable for years in stored grain. Bread made from scabby wheat has been described as "intoxicating."

Fig. 18. Dark perithecia of *Gibberella zeae* on blighted wheat head. (Courtesy Illinois Agricultural Experiment Station)

Causal Organisms

In nearly all cases, fungi in the genus *Fusarium* cause head blight or scab. The principal pathogens are *F. graminearum* Schwabe (syn. *F. roseum* Lk. emend. Snyd. & Hans. f. sp. *cerealis* (Cke.) Snyd & Hans. 'Graminearum') and its teleomorph, *Gibberella zeae* (Schw.) Petch (syns. *G. roseum* f. sp. *cerealis* 'Graminearum,' *G. saubinettii* (Mont.) Sacc.); *F. avenaceum* (Corda ex Fr.) Sacc. (syn. *F. roseum* Lk. emend. Snyd. & Hans. f. sp. *cerealis* (Cke.) Snyd. & Hans. 'Avenaceum') and its teleomorph, *G. avenacea* Cook; *F. culmorum* (Smith) Sacc. (syn. *F. roseum* Lk. emend. Snyd. & Hans. f. sp. *cerealis* (Cke.) Snyd. & Hans. 'Culmorum'); and *Microdochium nivale* (Ces. ex Berl. & Vogl.) Sammuels & Hallett (syns. *Gerlachia nivalis* (Ces. ex Berl. & Vogl.) Gams & Müller, *F. nivale* (Fr.) Ces. ex Berl. & Vogl.) and its teleomorph *Monographella nivalis* (Schaff.) Müller (syn. *Calonectria nivalis* Schaff.) (see Pink Snow Mold). *Bipolaris sorokiniana* (Sacc. in Sorok.) Shoem. (syn. *Helminthosporium sativum* P.K. & B.) may also induce head blighting, but it is better known as a cause of seed decay, root rot, and leaf spots (see Common Root and Foot Rot).

Fusarium spp. produce extensive mycelium in culture, often in shades of pink or purple-yellow. Conidia (phialospores) are hyaline and of two types. Microconidia, not always present, are oval and typically one-celled. Macroconidia are distinctive and canoe-shaped, have several cells, and are often held in moist heads (Fig. 20).

F. graminearum is variably gray, pink, brown, or burgundy in culture and grows optimally on most agar media between 24 and 26° C. Its macroconidia are longer and proportionally narrower than those of *F. culmorum*. Phialides are lateral and short (3.5–4.5 × 10–14 μm). Conidia are sickle-shaped, 2.5–5 × 35–62 μm, and commonly three- to seven-septate, with a well-marked foot cell. Sporodochia are uncommon. Globose chlamydospores 10–12 μm in diameter are formed singly or in chains by some strains.

Perithecia of *G. zeae* on wheat glumes are superficial, gregarious, and dark purple or black. They arise from an inconspicuous stroma and are ovoid, papillate, and 150–350 μm in diameter. Asci are clavate, measure 8–11 × 60–65 μm, and contain eight hyaline spores. Ascospores are zero- to three-septate but normally three-celled and measure 3–5 × 17–25 μm.

F. avenaceum (teleomorph *G. avenacea* Cook) is usually present in cool, moist climates. Colonies in culture often appear rose red, fringed with white. Slender diagnostic conidia develop on aerial mycelium or in sporodochia. Microconidia (when present) are narrow and curved, one- to three-septate, and 3.0–4.4 × 8–50 μm. Macroconidia are uniform, 3.5–4 × 40–80 μm, fusoid, curved, and four- to seven-septate, with an elongated apical cell and a conspicuous foot cell. Chlamydospores are not formed.

F. culmorum tends to be yellow-red in culture and causes most agar media to turn red-brown. Microconidia are absent but macroconidia are abundant, thick, and bluntly pointed at their apex. They develop singly from phialides (5 × 15–20 μm) that are loose at first and are later aligned in sporodochia. The macroconidia have a prominent foot, average three to five septations, and range from 4 to 7 μm wide and from 25 to 50 μm long. Thick-walled, globose chlamydospores 9–14 μm in diameter are common. *F. culmorum* is one of the most stable and uniform members of the genus.

M. nivale (*F. nivale*), more than *F. avenaceum*, is limited to areas with cool, moist climates. In the United States, scab caused by this fungus has been reported in northwestern Washington under conditions of heavy fog and mist in June and early July. The fungus produces narrow hyphae (1.5–5.0 μm in diameter) and white to peach-colored colonies. Macroconidia are small (2.8–4 × 16–25 μm) and typically one- to three-septate, without an evident heel. Phialides (2–3 × 7–9 μm) are borne on hyphal branches or in loose sporodochia. Chlamydospores do not occur.

Disease Cycle

Fungi that cause scab overwinter on host residues (Fig. 21). Grass residue, cornstalks, and wheat stubble are typical sources of primary inoculum. There is evidence also for pathogen survival on wheat seed. Conidia or ascospores from these sources are carried by air currents to wheat heads. During moist, warm weather, the spores germinate and invade flower parts, glumes, or other portions of the spike. Infections are most frequent and serious at anthesis. At this time, anthers and pollen may serve the pathogen as a food base. Blight symptoms develop within three days after infection when temperatures range between 25 and 30° C and moisture is continuous. In areas that otherwise are too dry for scab development, sprinkler irrigation may predispose wheat plants to the disease.

Secondary infections may result from airborne conidia. However, ascospores normally are produced too late to function as secondary inoculum. Ascospores appear better adapted than conidia to persist in the diseased host and contaminate seed.

Control

No highly resistant cultivars are available, but breeders in Yugoslavia have enhanced resistance in their wheats by applying recurrent selection. Some cultivars are infected less frequently, apparently because of physical barriers to floret and spikelet infection. Not all differences in the incidence of head blight among cultivars growing in adjacent fields are genetically based. The time of anthesis and prevailing weather conditions at anthesis can also influence scab development.

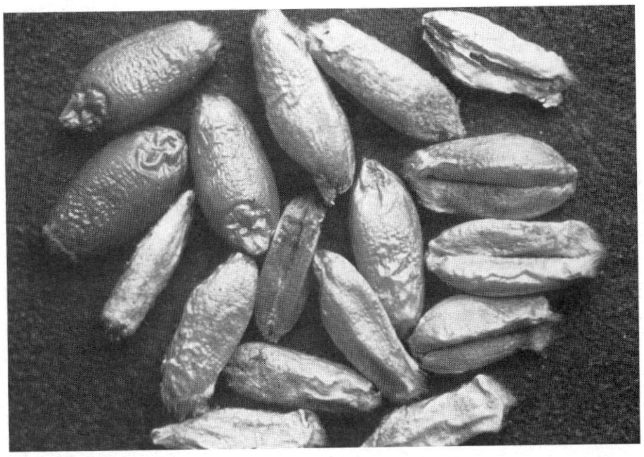

Fig. 19. Shrunken seed from a head affected by scab (caused by *Fusarium* spp.). (Courtesy Plant Pathology Department, University of Nebraska, Lincoln)

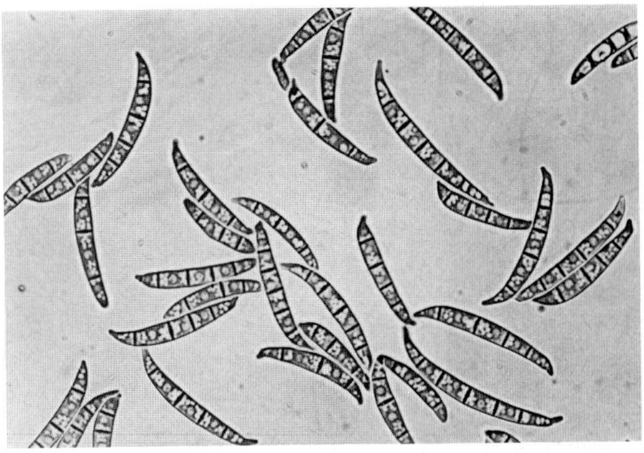

Fig. 20. Macroconidia of *Fusarium graminearum*. (Courtesy T. A. Toussoun, Fusarium Research Center, The Pennsylvania State University)

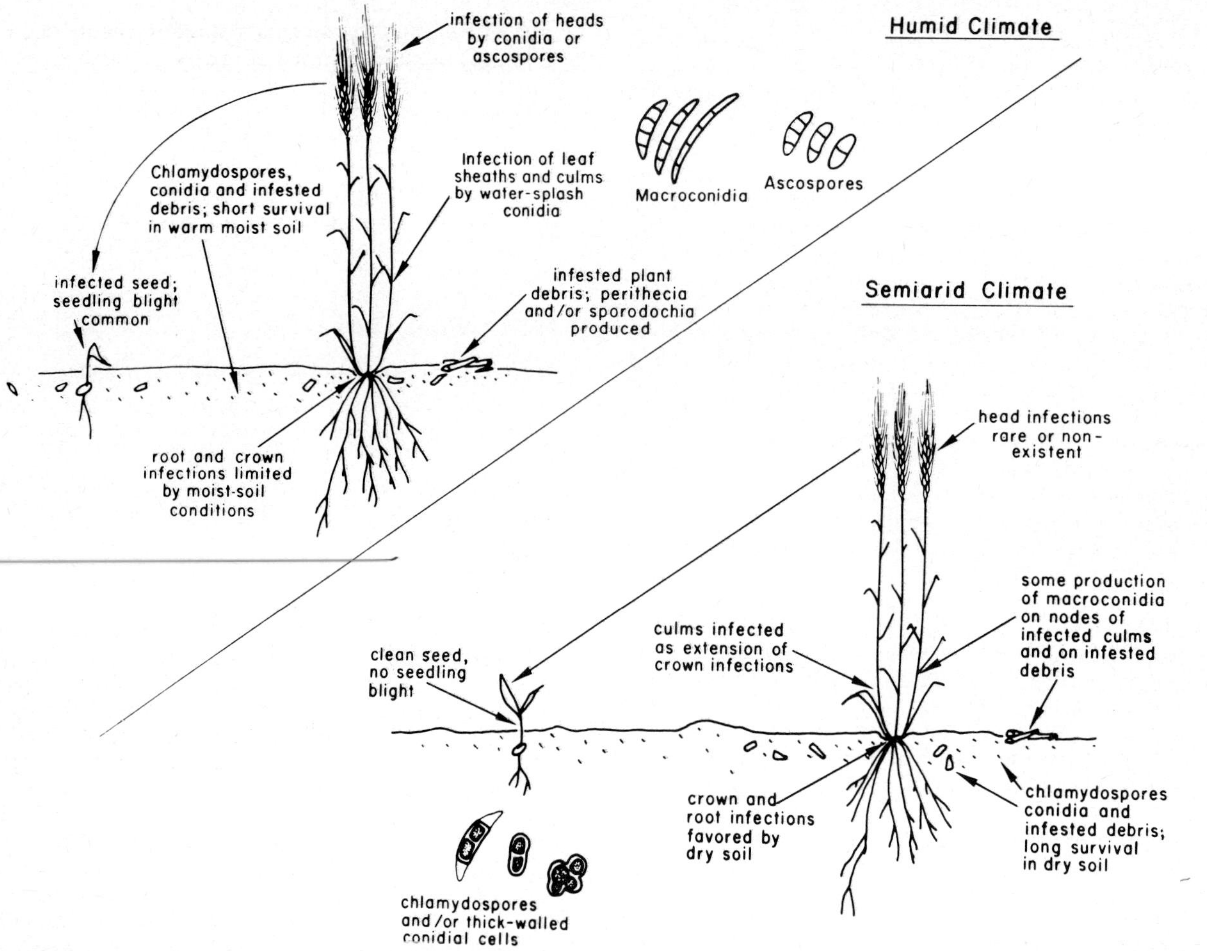

Fig. 21. Disease cycle of *Fusarium* species (*F. graminearum, F. culmorum, F. avenaceum*) on wheat. (Reprinted, by permission, from P. E. Nelson, T. A. Toussoun, and R. J. Cook, eds., *Fusarium*: Diseases, Biology and Taxonomy, Fig. 4-1, 1981, The Pennsylvania State University Press, University Park, PA)

Chemical seed treatments such as TCMTB (2-[thiocyanomethylthio] benzothiazole) and thiram can deter seed and seedling infections but will not control head blight. Many fungicides are ineffective against internal inoculum, but most eradicate superficial inoculum and protect seedlings in infested soil. Fungicides applied to newly emerged heads are effective controls but may not be economical. MBC (methyl 2-benzimidazolecarbamate) is used to control scab in some countries.

Even though the pathogens are ubiquitous, crop rotation with at least a one-year break from cereal and grass cultivation is advised. Plowing to bury crop residues is also desirable because the fungi survive best on surface debris. Applying lime to soil reduces inoculum levels in some instances.

Selected References

Andersen, A. L. 1948. The development of *Gibberella zeae* headblight of wheat. Phytopathology 38:595–611.

Cook, R. J. 1981. Fusarium diseases of wheat and other small grains in North America. Pages 39–55 in: *Fusarium*: Diseases, Biology and Taxonomy. P. E. Nelson, T. A. Toussoun, and R. J. Cook, eds. The Pennsylvania State University Press, University Park, PA. 457 pp.

Inglis, D. A., and Cook, R. J. 1981. *Calonectria nivalis* causes scab in the Pacific Northwest. Plant Dis. 65:923–924.

Martin, R. A., and Johnston, H. W. 1982. Effects and control of Fusarium diseases of cereal grains in the Atlantic Provinces. Can. J. Plant Pathol. 4:210–216.

Nelson, P. E., Toussoun, T. A., and Marasas, W. F. O. 1983. *Fusarium* Species, An Illustrated Manual for Identification. The Pennsylvania State University Press, University Park, PA. 193 pp.

Schroeder, H. W., and Christensen, J. J. 1963. Factors affecting resistance of wheat to scab caused by *Gibberella zeae*. Phytopathology 53:831–838.

Seaman, W. L. 1982. Epidemiology and control of mycotoxigenic fusaria on cereal grains. Can. J. Plant Pathol. 4:187–190.

Stack, R. W., and McMullen, M. P. 1985. Head blighting potential of *Fusarium* species associated with spring wheat heads. Can. J. Plant Pathol. 7:79–82.

Sutton, J. C. 1982. Epidemiology of wheat head blight and maize ear rot caused by *Fusarium graminearum*. Can. J. Plant Pathol. 4:195–209.

Smuts

Smut diseases of cereal crops have been known and studied for more than two centuries. Today wheat smuts are less damaging than they were 50 years ago because resistant cultivars and chemical and cultural controls are widely employed. However, wheat smuts still occur worldwide and cause significant yield and quality losses.

Wheat smuts are caused by six highly specialized fungi: *Tilletia tritici* (syn. *T. caries*) and *T. laevis* (syn. *T. foetida*) cause common bunt (stinking smut); *T. controversa* causes dwarf bunt (also known as TCK); *T. indica* (syn. *Neovossia indica*) causes Karnal bunt (partial bunt); *Urocystis agropyri* (syn. *U. tritici*) causes flag smut; and *Ustilago tritici* causes loose smut. Most species of smut fungi are composed of races differentiated by virulence. The variability within each species and generation of smut is a product of genetic recombination during sexual reproduction. The progeny from a single generation may differ from each other and from the parent in virulence, host specificity, and other characteristics.

Smut fungi penetrate wheat directly and progress intracellularly and intercellularly without haustoria. Smut fungi make extensive mycelial growth within compatible hosts, and more than one smut disease may be present in the same wheat plant. Host plants are rarely killed, except possibly by flag smut, and tissue necrosis is uncommon. Chlorosis also is infrequent, but leaf flecking may occur in some cultivars with dwarf and/or common bunt. Plants may be severely stunted in the case of dwarf bunt. Usually heading and maturity are not altered (except in the case of flag smut) and are necessary for the completion of pathogen life cycles.

Except for *Urocystis agropyri*, the haploid stages of smut fungi grow on simple salt-sugar media. However, most smut fungi cannot complete all stages of their life cycle in vitro, and some require special growth factors to do so.

Selected Reference

Fischer, G. W., and Holton, C. S. 1957. Biology and Control of the Smut Fungi. Ronald Press, New York. 622 pp.

Common Bunt (Stinking Smut)

Common bunt (also called stinking smut and covered smut) was recognized as an infectious disease in the 18th century, and by 1807, a microorganism was associated with its cause. Common bunt occurs worldwide and, because of its economic importance, has been extensively investigated. Because it requires cool, moist soil conditions, the disease is less frequent and usually less damaging on spring-sown wheat than on fall-sown wheat.

Common bunt is caused by two fungi—*Tilletia tritici* (syn. *T. caries*) and *T. laevis* (*T. foetida*)—that are distinguished by spore morphology and to some extent by geographic distribution. In addition to wheat, both pathogens infect rye, triticale, and grasses, including *Aegilops, Lolium, Elymus, Agropyron*, and *Hordeum* spp.

Common bunt reduces wheat yields and grain quality. Wheat seed contaminated with bunt spores has a pungent, fishy odor and a darkened appearance. Smutted grain is discounted in value and sometimes not accepted at the market. Furthermore, bunt spores released during threshing (Plate 11) are combustible, and explosions and fires have resulted from their ignition by sparks from mechanical harvesters.

Fig. 22. Mixture of fragile bunt balls (produced by *Tilletia tritici, T. laevis*, or *T. controversa*) and healthy wheat kernels. (Courtesy Illinois Agricultural Experiment Station)

Symptoms

Plants with common bunt may be moderately stunted but are not readily distinguished until heads emerge. Bunted heads are slender and maintain their green color longer than healthy heads. The glumes of some or all spikelets become conspicuously spread apart, exposing the plump bunt balls they contain (Plate 12; Figs. 22 and 26). This symptom is particularly apparent in awned cultivars. Sometimes, healthy kernels and bunt balls are found on the same spike.

Bunt balls approximate the shape of normal kernels and are dull gray-brown (Fig. 11, D and I). Their fragile covers (pericarps) remain intact initially (covered smut) but rupture at harvest, releasing black, powdery spores (Plate 11) that have a fishy odor. Dark spore clouds sometimes are produced when severely affected fields are mechanically harvested (this is possible also for sooty molds).

Causal Organisms

Two closely related fungi, *T. tritici* (Bjerk.) Wint. (syn. *T. caries* (DC.) Tul.) and *T. laevis* Kühn (syns. *T. levis, T. foetida* (Wallr.) Liro, *T. foetens* (Berk. & Curt.) Schroet.), cause common bunt. While both species are widely distributed, *T. tritici* is more common. The two fungi have similar life cycles and may occur together in the same infected plant. Their hyaline, binucleate mycelium matures and fragments into dark masses of thick-walled, globose teliospores 15–23 μm in diameter. Teliospores of *T. laevis* are smooth-walled, whereas those of *T. tritici* are reticulate. The spores contain trimethylamine, a volatile, malodorous chemical that is the basis for the term "stinking smut." Not all strains of common bunt fungi produce trimethylamine in consistent amounts. Teliospores contaminating wheat seed in quantities not visible to the unaided eye are detected microscopically in concentrated seed washes.

Teliospores of common bunt fungi germinate to form a basidium (promycelium), on which eight to 16 terminal, hyaline basidiospores (primary sporidia) develop. The uninucleate, filiform basidiospores fuse near their middle in compatible pairs to form H-shaped structures and establish a dikaryon (Fig. 23). The dikaryon then yields infectious hyphae or hyaline, sickle-shaped, secondary sporidia. The sporidia are forcibly discharged and, in turn, produce infectious hyphae or more secondary sporidia.

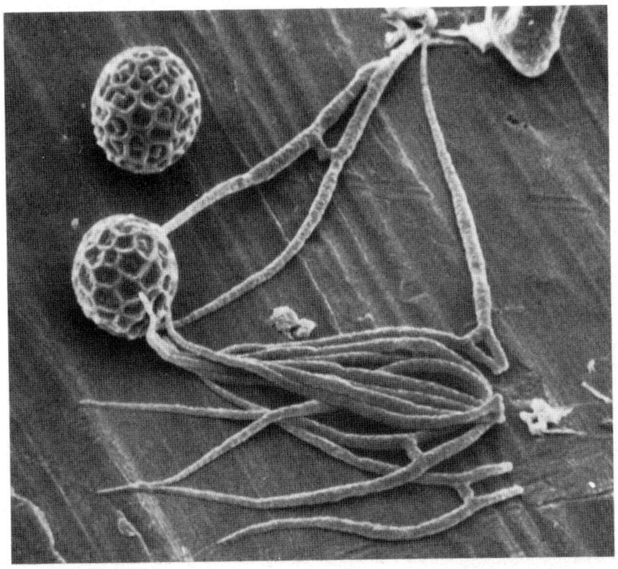

Fig. 23. Teliospores of *Tilletia tritici*. Note reticulate walls and fused primary sporidia. (Courtesy M. F. Brown and H. G. Brotzman)

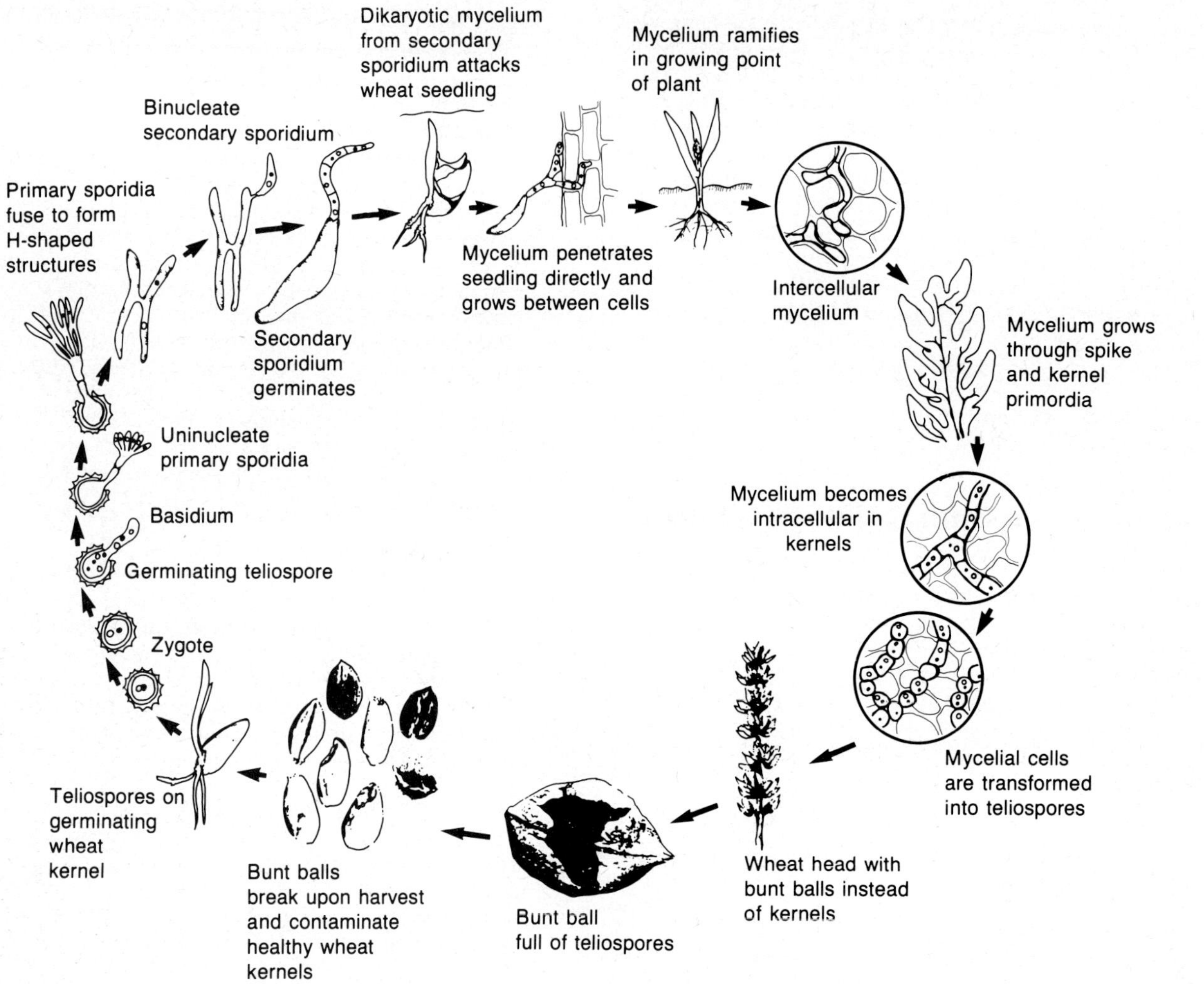

Fig. 24. Disease cycle of common bunt (stinking smut) of wheat caused by *Tilletia tritici* or *T. laevis*. (Modified by J. J. Nielsen from G. N. Agrios, Plant Pathology, Fig. 136, 2nd ed., copyright 1978, Academic Press, New York)

Disease Cycle

Common bunt fungi persist as teliospores on seed and in soil (Fig. 24). Soilborne teliospores are significant epidemiologically in winter-wheat regions with arid summers. Teliospores, originally in soil or placed there on contaminated seed, germinate in response to moisture. Cool temperatures (5–15° C) favor spore germination and production of infectious hyphae, which penetrate coleoptiles before seedlings emerge. In growing host plants, the pathogen progresses to and inhabits terminal meristematic tissues, especially the flower primordia of the spike.

In susceptible cultivars, mycelium inhabits the developing ovary and displaces all tissues within the pericarp. At harvest, mature bunt balls are broken and teliospores are released to contaminate soil and seed. Teliospores are dispersed by wind and especially by their association with seed (Fig. 22). Dispersal on seed is especially important in distributing new strains of bunt fungi.

Control

Common bunt is controlled by the use of resistant cultivars, clean seed, and chemical seed treatment. However, most resistant cultivars are short-lived because of the development of new virulent races of bunt fungi. Historically, seed treatment with organomercury fungicides reduced common bunt to manageable levels where soilborne spores were unimportant. More recently, chemicals such as hexachlorobenzene, pentachloronitrobenzene, and carboxin have proved effective against both seedborne and soilborne spores. Where fungicides are not used or where regulations have restricted their use, bunt often recurs. In Australia and Greece, strains of *T. laevis* have developed resistance to polychlorobenzene fungicides.

Wheat sown early in autumn when soil temperatures exceed 15° C may escape infection.

Selected References

Hoffmann, J. A. 1982. Bunt of wheat. Plant Dis. 66:979–986.

Hoffmann, J. A., and Metzger, R. J. 1976. Current status of virulence genes and pathogenic races of the wheat bunt fungi in the northwestern USA. Phytopathology 66:657–660.

Hoffmann, J. A., and Waldher, J. T. 1981. Chemical seed treatments for controlling seedborne and soilborne common bunt of wheat. Plant Dis. 65:256–259.

Holton, C. S., and Heald, F. D. 1941. Bunt or Stinking Smut of Wheat. Burgess Publishing Co., Minneapolis, MN. 211 pp.

Kendrick, E. L. 1965. The reaction of varieties and hybrid selections of winter wheat to pathogenic races of *Tilletia caries* and *T. foetida*. Plant Dis. Rep. 49:843–846.

Purdy, L. H., and Kendrick, E. L. 1963. Influence of environmental factors on the development of wheat bunt in the Pacific Northwest. IV. Effect of soil temperature and soil moisture on infection by soil-borne spores. Phytopathology 53:416–418.

Singh, J., and Trione, E. J. 1969. In vitro culture of the dikaryons of the wheat bunt pathogens, *Tilletia caries* and *T. controversa*. Phytopathology 59:648–652.

Trione, E. J. 1974. Morphology and cytology of *Tilletia caries* and *T. controversa* in axenic culture. Am. J. Bot. 61:914–919.

Dwarf Bunt

Early investigations of wheat common bunt uncovered a marked inconsistency in the degree of stunting of diseased plants. The pathogen causing the severe stunting was given species status in 1935. Originally described as a form of *Tilletia tritici* and then as *T. brevifaciens* Fischer, the fungus eventually was assigned to *T. controversa* Kühn in agreement with original 1874 literature that described the fungus on *Agropyron repens* in Europe. In addition to wheat, *T. controversa* infects rye, winter-sown barley, and a variety of wild and cultivated grasses. However, natural infection of hosts other than wheat and rye is not common.

Dwarf bunt occurs in the United States, Canada, Argentina, Uruguay, Europe, and central Asia. It appears to be limited to areas where winter wheat is subject to prolonged snow cover. Infection of spring-sown wheat has not been observed.

Symptoms

Dwarf bunt symptoms resemble those of common bunt except that plants infected with *T. controversa* typically achieve only one-fourth to one-half of normal size and have an increased number of tillers (Fig. 25). Also, spikes with dwarf bunt tend to be broader, with glumes open wider (Fig. 26), than uninfected spikes or spikes with common bunt.

Causal Organism

Teliospores of *T. controversa* are morphologically similar to those of *T. tritici* and *T. laevis*. They have reticulate walls like those of *T. tritici*, but the reticulations are typically broader and deeper and are shrouded by a gelatinous sheath. Under fluorescent microscopy, teliospores of *T. controversa* tend to fluoresce and appear lighter (yellow-brown) in color than those of common bunt fungi. When dry, teliospores of *T. controversa* are less prone to collapse than those of *T. tritici* and *T. laevis*. Serologic techniques (monoclonal antibodies) are showing promise for rapid, specific, and sensitive detection of common and dwarf bunt teliospores in grain samples.

Disease Cycle

Soilborne teliospores are the most important source of primary inoculum. Teliospores (Plate 11) of *T. controversa* persist in soil up to 10 years and have a lower temperature optimum (3–8° C) and longer incubation period (three to 10 weeks) for germination than those of common bunt fungi. After seedlings emerge, infection originates from teliospores germinating under snow at or near the soil surface. Plants in the two- to three-leaf stage are most susceptible.

Control

Dwarf bunt is controlled by using resistant cultivars. The disease also may be controlled by applying certain fungicides (e.g., polychlorobenzenes) to the soil surface after seeding. Because *T. controversa* infects plants after emergence, fungicides that protect seed are not effective. Systemic fungicides, such as thiabendazole, may be effective as seed treatments if seeding is delayed.

Some strict international quarantines limit the dissemination of *T. controversa* on wheat seed.

Selected References

Conners, I. L. 1954. The organism causing dwarf bunt of wheat. Can. J. Bot. 32:426–431

Grey, W. E., Mathre, D. E., Hoffmann, J. A., Powelson, R. L., and Fernández, J. A. 1986. Importance of seedborne *Tilletia controversa* for infection of winter wheat and its relationship to international commerce. Plant Dis. 70:122–125.

Hardison, J. R. 1963. Incidence and control of *Tilletia controversa* in perennial grasses. Phytopathology 53:579–585.

Hoffmann, J. A. 1982. Bunt of wheat. Plant Dis. 66:979–986.

Hoffmann, J. A., and Purdy, L. H. 1967. Effect of stage of development of winter wheat on infection by *Tilletia controversa*. Phytopathology 57:410–413.

Hoffmann, J. A., Kendrick, E. L., and Metzger, R. J. 1967. A revised classification of pathogenic races of *Tilletia controversa*. Phytopathology 57:279–281.

Metzger, R. J., and Hoffmann, J. A. 1978. New races of common bunt useful to determine resistance of wheat to dwarf bunt. Crop Sci. 18:49–51.

Purdy, L. H., Kendrick, E. L., Hoffmann, J. A., and Holton, C. S. 1963. Dwarf bunt of wheat. Annu. Rev. Microbiol. 71:199–222.

Trione, E. J. 1982. Dwarf bunt of wheat and its importance in international wheat trade. Plant Dis. 66:1083–1088.

Fig. 25. Stunting and increased tillering (left) induced by dwarf bunt (caused by *Tilletia controversa*) and healthy plant (right). (Courtesy J. A. Hoffmann)

Fig. 26. Mature, healthy head (left) and head with glumes spread by sori of dwarf bunt (*Tilletia controversa*) (right). (Courtesy J. A. Hoffmann)

Karnal Bunt

Karnal bunt (also called partial bunt) was first described in Karnal, Punjab, India in 1931. Caused by *Tilletia indica*, the disease is widespread in northwest India and in adjacent areas of Pakistan and Afghanistan. More recently it has been introduced into Mexico. Karnal bunt affects common wheat and to a lesser degree durum wheat, triticale, and related species. The disease significantly reduces seed quality, but only minimal effects on yield have been reported.

Symptoms

Developing wheat kernels are randomly attacked and incompletely converted to smut sori. Normally, only a few seeds are attacked per head. Diseased heads are not conspicuous because the glumes are not noticeably distorted by infected kernels (see Common Bunt). Infected kernels are usually only partially eroded at their embryo end (Plate 13). Larger sori may extend along the crease and occasionally envelop the whole kernel. The delicate pericarp surrounding each sorus of teliospores is easily broken during harvest. The brown-black teliospores of *T. indica* have the same fishy odor as those of common bunt fungi.

Causal Organism

T. indica Mitra (syn. *Neovossia indica* (Mitra) Mundk.) produces large, globose, dark-brown teliospores 22–49 μm in diameter (Plate 13). The teliospores bear truncate projections surrounded by a delicate membranous sheath. Upon germination, stout promycelia emerge and numerous (frequently over 100) haploid sporidia form at their apex. These primary sporidia do not fuse as with other bunt fungi but germinate directly to form hyphae and/or uninucleate secondary sporidia, which are forcibly discharged.

Disease Cycle

Teliospores of *T. indica* persist in soil and on seed. Soilborne teliospores are the principal source of primary inoculum. Teliospores germinate at or near the soil surface in response to free moisture and produce primary and secondary sporidia. The sporidia are wind-dispersed and appear to infect spikelets by directly penetrating the glumes and ovary wall. Promoted by cool, humid or wet weather, infections lead to partial and, occasionally, complete conversion of developing kernels into darkened masses of teliospores. At harvest, the pericarp of bunted kernels is easily broken, liberating teliospores that contaminate soil and seed. Partially bunted seeds often retain their capacity to germinate and produce healthy plants.

Control

Where the required conditions of temperature and moisture occur regularly, Karnal bunt is difficult to control. Chemical seed treatments inhibit the germination of seedborne teliospores, and some fungicides applied at heading protect against infection. Wheat cultivars differ in susceptibility to Karnal bunt, but none are known to be immune. Durum wheats and triticales are less susceptible than bread wheats. Many countries, including the United States, have imposed quarantines to prevent the introduction or spread of Karnal bunt.

Selected References

Anonymous. 1983. U.S. quarantine of Mexican wheat: Dilemma for researchers and exporters. Diversity 5:13–14.

Aujla, S. S., Grewal, A. S., and Sharma, I. 1983. Relative efficiency of Karnal bunt inoculation techniques. Indian J. Mycol. Plant Pathol. 13:99–100.

Chatrath, M. S., and Adlakha, K. L. 1963. Karnal bunt of wheat (*Neovossia indica*). Agric. Res. 3:260.

Dhaliwal, H. S., Randhawa, A. S., Chand, K., and Singh, D. 1983. Primary infection and further development of Karnal bunt of wheat. Indian J. Agric. Sci. 53:239–244.

Dhiman, J. S., Bedi, P. S., and Mavi, H. S. 1984. Relationship among temperature, humidity and incidence of Karnal bunt of wheat. Indian J. Ecol. 11:134–138.

Gardner, J. S., Allen, J. V., and Hess, W. M. 1983. Sheath structure of *Tilletia indica* teliospores. Mycologia 75:333–336.

Joshi, L. M., Singh, D. V., Srivastava, K. D., and Wilcoxson, R. D. 1983. Karnal bunt: A minor disease that is now a threat to wheat. Bot. Rev. 4:309–330.

Mitra, M. 1985. Stinking smut (bunt) of wheat with special reference to *Tilletia indica* Mitra. Indian J. Agric. Sci. 5:1–24.

Munjal, R. L. 1974. Technique for keeping the cultures of *Neovossia indica* in sporulating condition. Indian Phytopathol. 27:248–249.

Royer, M. H., and Rytter, J. 1985. Artificial inoculation of wheat with *Tilletia indica* from Mexico and India. Plant Dis. 69:317–319.

Smilanick, J. L., Hoffmann, J. A., and Royer, M. H. 1985. Effect of temperature, pH, light, and desiccation on teliospore germination of *Tilletia indica*. Phytopathology 75:1428–1431.

Zhang, Z., Lange, L., and Mathur, S. B. 1984. Teliospore survival and plant quarantine significance of *Tilletia indica* (causal agent of Karnal bunt) particularly in relation to China. EPPO Bull. 14:119–128.

Loose Smut

Loose smut of wheat is easily recognized in the field and has been known for centuries. The causal fungus, *Ustilago tritici*, is unique in that it is first incorporated into developing kernels and persists within seed embryos. Plants that develop from such seeds produce conspicuously black, loose-smutted spikes.

Loose smut occurs throughout the world and reduces yields in proportion to the incidence of smutted heads. Yield losses usually are less than 1% but are as high as 27% in some fields. In contrast to the bunts, loose smut has little or no effect on grain quality for food or feed. However, grain from fields in which smutted heads occurred should not be used as seed without fungicide treatment to eradicate the pathogen.

Although wheat is the principal host, triticale and rye can be infected. Some grasses in the genera *Aegilops, Agropyron, Elymus, Haynaldia*, and *Hordeum* also serve as hosts.

Fig. 27. Healthy head (left) and heads with early (center) and late (right) symptoms of loose smut caused by *Ustilago tritici*. (Courtesy E. D. Hansing)

Symptoms

Loose smut symptoms are most obvious just after heading. Initially, diseased heads are blackened and clearly visible among newly emerged green, healthy heads. The spikelets of infected heads are transformed into a dry, olive black teliospore mass. As heads emerge, teliospores are dispersed by wind and washed off by rain. Within a few days, only the bare rachis remains (Fig. 27)—thus the term "loose" as opposed to "covered" smut. When spikelet tissues are not completely destroyed, the rachis may bear remnants of glumes or awns.

Before heading, infected plants may have dark green, erect leaves, sometimes with chlorotic streaks. Microscopic examination is necessary to reveal mycelium of the pathogen harbored at nodes and culm apices. Infected seed is fully germinable and not visibly altered. Mycelium within the embryo occupies the cotyledon or scutellum (Fig. 28).

Causal Organism

U. tritici (Pers.) Rostr. (syn. *U. nuda* var. *tritici* Schaf.) produces hyaline, dikaryotic mycelium in culture and in host plants. At maturity its hyphae thicken and fragment into brown, spherical, echinulate teliospores (chlamydospores) 5–9 μm in diameter. Upon germination the diploid spores produce a basidium (promycelium) and four hyphae of uninucleate cells, but no basidiospores (sporidia) appear. The fusion of such compatible hyphae yields infectious, dikaryotic hyphae.

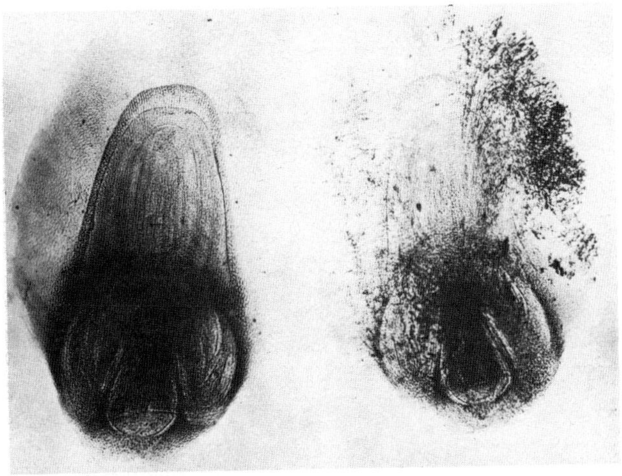

Fig. 28. Healthy embryo (left) and embryo infected with *Ustilago tritici* (cause of loose smut) (right).

Disease Cycle

U. tritici survives as dormant mycelium within the embryo of wheat seed (Fig. 29). When infected seed germinates, the pathogen is activated, progresses toward the shoot apex, and spreads through culm nodes and seed primordia. Usually, all head tissues except the rachis are invaded intracellularly and converted to sori. The mycelium in such sori differentiates and fragments into dry, dark brown teliospores that are dispersed by wind and rain when diseased heads emerge (Fig. 27). Open flowers on neighboring plants become infected when germinating teliospores penetrate the ovary wall and possibly the stigma. Thus, the fungus is reestablished within developing kernels (Fig. 28) that are otherwise normal, visibly unaltered, and fully germinable.

Infections occur only during flowering and are favored by humid weather and cool to moderate temperatures (16–22° C). Within one week after flowering, the ovary becomes resistant to infection.

Control

U. tritici is unaffected by surface-active fungicides used as seed treatments. Its control has hinged on cultivar resistance, environmental obstacles, cultural practices, hot-water or heat treatment of seed, and seed certification. More recently, systemic chemicals, such as carboxin, have been developed that eradicate the fungus within germlings. Both effective and low in cost, such chemicals have revolutionized loose smut control. Valuable seed lots are sometimes examined microscopically (embryo staining) to determine the percentage of infected seed and whether chemical treatment is necessary.

Despite the availability of systemic fungicides, selecting resistant cultivars and using pathogen-free seed are of value. Because most cultivars are susceptible, loose smut is held in check through field inspections and seed certification programs

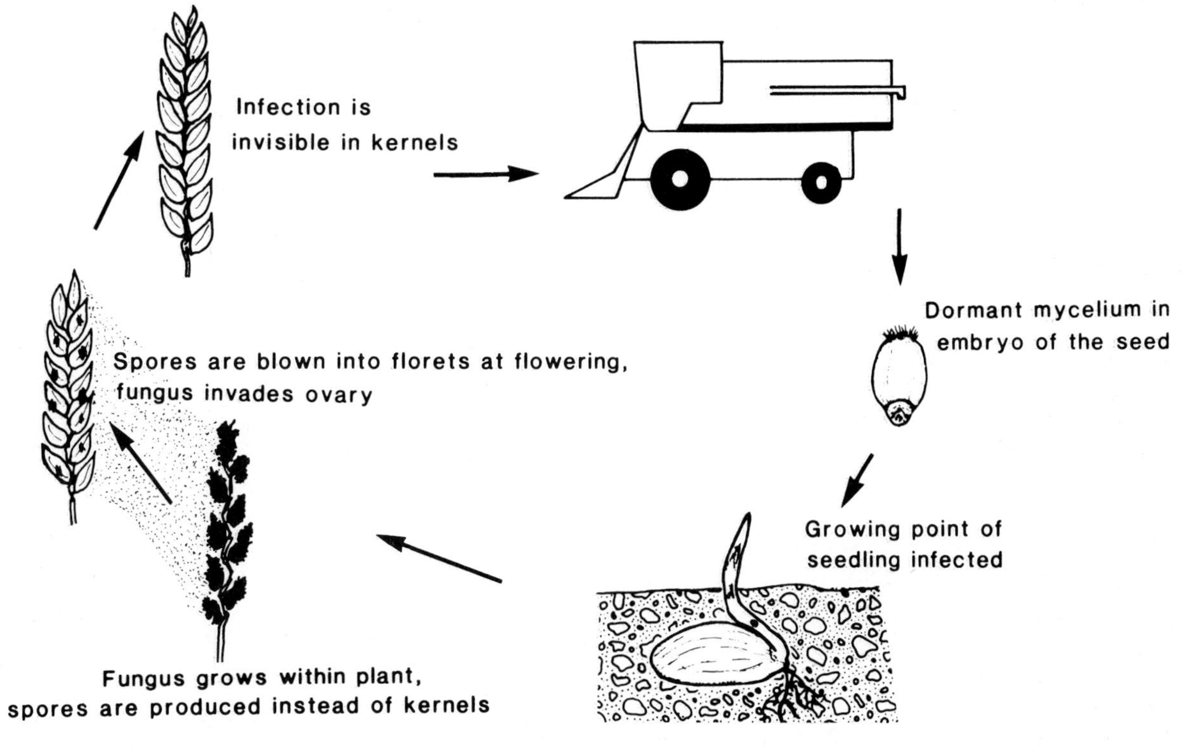

Fig. 29. Disease cycle of loose smut of wheat caused by *Ustilago tritici*. (Courtesy J. J. Nielsen)

that identify sources of pathogen-free seed. Some cultivars avoid infection by having unexposed florets or florets exposed only for short periods. Others physiologically impede the progress of the pathogen toward seed primordia and, although infected, set normal heads and seed.

Where systemic fungicides are not available, seedborne inoculum can be destroyed by hot-water treatments. The process is laborious, and numerous time-temperature schemes have been proposed. Solar- and dry-heat treatments of seed also can be beneficial. All heat treatments aim to kill the fungus without reducing seed germinability. All heat-treated seed should be checked for germinability, and seeding rates should be adjusted to compensate for any decrease in viability.

Selected References

Atkins, I. M., Merkle, O. G., Porter, K. B., Lahr, K. A., and Weibel, D. E. 1963. The influence of environment on loose smut percentages, reinfection, and grain yields of winter wheat at four locations in Texas. Plant Dis. Rep. 47:192–196.

Gaskin, T. A., and Schafer, J. F. 1962. Some histological and genetic relationships of resistance of wheat to loose smut. Phytopathology 52:602–607.

Jones, J. P., and Collins, F. C. 1971. Control of loose smut of wheat with carboxin and benomyl. Plant Dis. Rep. 55:1053–1055.

Kavanagh, T. 1961. Temperature in relation to loose smut in barley and wheat. Phytopathology 51:189–193.

Khanzada, A. K., Rennie, W. J., Mathur, S. B., and Neergard, P. 1980. Evolution of two routine embryo test procedures for assessing the incidence of loose smut infection in seed samples of wheat, *Triticum aestivum*. Seed Sci. Technol. 8:363–370.

Loria, R., Wiese, M. V., and Jones, A. L. 1982. Effects of free moisture, head development, and embryo accessibility on infection of wheat by *Ustilago tritici*. Phytopathology 72:1270–1272.

Nielsen, J. 1969. A race of *Ustilago triciti* virulent on Thatcher wheat and its derivatives. Plant Dis. Rep. 53:393–395.

Nielsen, J. 1972. Isolation and culture of monokaryotic haploids of *Ustilago tritici*, observations on their physiology, and the taxonomic relationship between *U. tritici* and *U. nuda*. Can. J. Bot. 50:1775–1781.

Nielsen, J. 1978. Host range of the smut species *Ustilago nuda* and *Ustilago tritici* in the tribe Triticeae. Can. J. Bot. 56:901–915.

Nielsen, J. 1983. Spring wheats immune or highly resistant to *Ustilago tritici*. Plant Dis. 67:860–863.

Niemann, E. 1962. Evaluation of the thermal and anaerobic methods for the control of loose smut in barley and wheat. Angew. Bot. 36:1–15. (In German, English summary)

Punithalingam, E., and Waterston, J. M. 1970. *Ustilago nuda*. Descriptions of Pathogenic Fungi and Bacteria, No. 280. Commonwealth Mycological Institute, Association of Applied Biologists, Kew, Surrey, England.

Shinohara, M. 1976. Behavior of *Ustilago nuda* (Jens.) Rostrup and *U. tritici* (Pers.) Rostrup in their host tissues. Rev. Plant Prot. Res. 9:124–142.

Tapke, V. F. 1948. Environment and the cereal smuts. Bot. Rev. 14:359–412.

Tyler, L. J. 1965. Failure of loose smut to build up in winter wheats exposed to abundant inoculum naturally disseminated. Plant Dis. Rep. 49:239–241.

Fungal Diseases Principally Observed on Foliage

Flag Smut

Since its initial description in Australia in 1868 and in the United States in 1918, flag smut has been observed on all wheat-growing continents. The disease preferentially occurs on autumn-sown club and white wheats. In portions of Australia and the western United States, it once reduced winter wheat yields an average of 1% annually and by over 50% in some fields. Currently the disease is less significant because of resistant cultivars, chemical seed treatments, and quarantines.

The flag smut fungus, *Urocystis agropyri*, consists of strains that differ in host specificity. Most isolates that infect wheat do so exclusively. Many grasses (especially *Poa* spp.) are susceptible to other strains of the pathogen.

Symptoms

Flag smut is conspicuous on leaves in spring and summer. Between jointing and heading, long, gray-black streaks develop on leaf blades and sheaths (Fig. 30). The streaks are linear, subepidermal smut sori that develop between leaf veins (see Stripe Rust). With time, the sori enlarge, break through the epidermis, and liberate gray-black spore masses. The process mechanically weakens leaves, often causing them to split or fray longitudinally.

Sorus development begins in tissues within the whorl and mechanically alters the emergence of leaves and heads. Leaves often are twisted laterally, and heading may be prevented. Heads that emerge may have striped glumes and necks. Diseased plants usually are stunted and tiller excessively. Symptoms may not be evident on all tillers of infected plants.

Causal Organism

Urocystis agropyri (Preuss) Schroet. (syn. *U. tritici* Koern.) produces linear, black, erumpent sori in host leaves (Fig. 30).

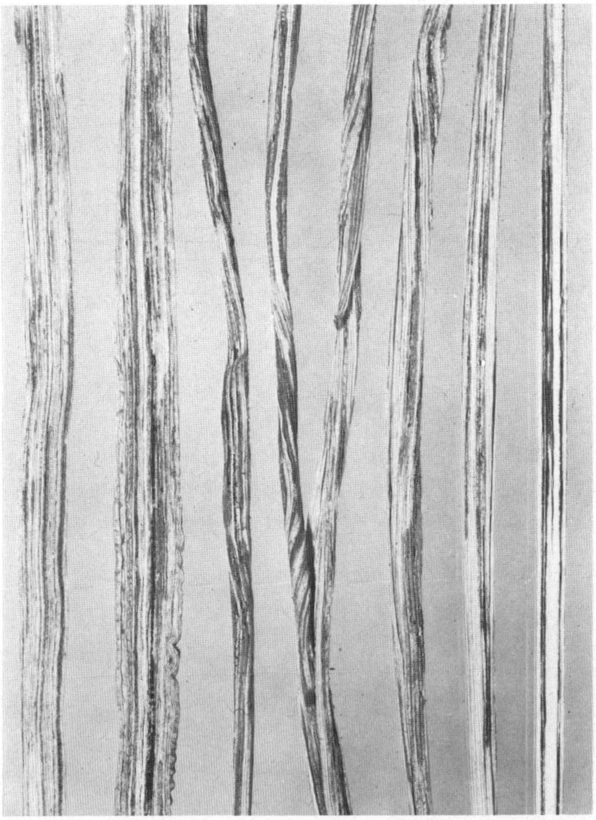

Fig. 30. Leaves (left) and culms (right) with linear, gray-black, erumpent sori of the flag smut fungus, *Urocystis agropyri*. (Courtesy Illinois Natural History Survey)

Within sori, generative hyphal tips branch, intertwine, and differentiate into unique, multicellular spore balls. Spore balls are nearly spherical, are 18–50 μm in diameter, and have one to four teliospores within a sheath of smaller, hyaline, flattened, sterile cells. Teliospores are red-brown, smooth, globose, and 10–20 μm in diameter. They germinate in place to form a short promycelium that bears three or four hyaline, cylindrical sporidia at its tip. Germination is highly variable but is optimal between 18 and 24°C and does not occur below 5 or above 30°C.

Four races of *U. agropyri* have been characterized on wheat based on geographic distribution.

Disease Cycle

Spore balls liberated from tattered leaves and other plant parts contaminate soil and seed. Teliospores in soil or introduced on seed can remain viable for three years before they germinate and produce sporidia that infect wheat coleoptiles. Infection occurs before seedling emergence and is favored by soil temperatures between 10 and 20°C. Temperature limits for infection are near 5 and 28°C. *U. agropyri* overwinters as mycelium within seedlings, then systemically invades and sporulates within the upper portion of host plants in spring.

Control

Resistant cultivars and chemical seed treatments provide the most practical control of flag smut. Seeding in cool soil is beneficial, as is a one- to two-year break in wheat cropping. However, some soilborne teliospores remain viable for three years or more. Shallow seeding (less than 2.5 cm deep) results in fewer infections than deep seeding. Seed treatment with systemic fungicides such as carboxin limits infections from both seedborne and soilborne inoculum. Quarantines also effectively limit the dissemination of contaminated grain.

Selected References

Allan, R. E. 1976. Flag smut reaction in wheat—Its genetic control and association with other traits. Crop Sci. 16:685–687.

Greenhalgh, F. C., and Brown, J. S. 1984. A method for determining the reactions of wheat breeding lines to flag smut. Australas. Plant Pathol. 13:36–38.

Johnson, A. G. 1959. Further studies of physiologic races in *Urocystis tritici*. Phytopathology 49:299–302.

Line, R. F. 1972. Chemical control of flag smut of wheat. Plant Dis. Rep. 56:636–640.

Nelson, B. D., Jr., and Durán, R. 1984. Cytology and morphological development of basidia, dikaryons, and infective structures of *Urocystis agropyri* from wheat. Phytopathology 74:299–304.

Purdy, L. H. 1965. Flag smut of wheat. Bot. Rev. 32:565–606.

Purdy, L. H., and Allan, R. E. 1967. Heritability of flag smut resistance in three wheat crosses. Phytopathology 57:324–325.

Alternaria Leaf Blight

Alternaria leaf blight occurs mainly in India and has recently been reported in southern Italy and the Yaqui Valley of Mexico. The disease was first reported in 1924 but remained incompletely characterized for several years. Early studies associated *Alternaria* spp. with the disease, but the causal organism, *A. triticina*, was not identified until 1962. From 1960 to 1964, leaf blight damaged all commercial cultivars on the Indian subcontinent. It developed on plants approaching maturity, caused premature death of the uppermost leaves and heads, and reduced yield significantly. Today durum wheats, their derivatives, and introduced Mexican wheats are considered most susceptible. In addition to wheat, the disease affects triticale in India and other graminaceous hosts in the Middle East and Nigeria.

Symptoms

Alternaria leaf blight is characterized by small, oval, chlorotic lesions scattered on lower leaves (Fig. 31). The lesions darken to brown-gray, enlarge, become sunken, assume irregular shapes, and may have a yellow margin. The lesions progressively develop from lower to upper leaves, and blighting may extend to heads and leaf sheaths. Under humid conditions, the lesions support visible clusters of dark, powdery conidia.

Alternaria leaf blight is likely to develop near irrigation ditches, in low areas, or wherever humidity and soil moisture are high. It develops rapidly once wheat plants are six to eight weeks old and especially as the crop approaches maturity. From a distance, blighted fields appear dull and bronzed.

Causal Organism

Isolations from leaf lesions routinely yield *Alternaria* spp., many of which are saprophytic and mask the pathogen. *A. triticina* Pras. & Prab. is distinguished by its wheat-specific virulence and somewhat by cultural characters. The hyphae, conidiophores, and conidia of *A. triticina* are hyaline initially and later deep olive buff. Conidiophores measure 3–6 × 17–28 μm and are single (clustered when protruding through stomata). Conidia are acrogenous and single or in short chains of two to four spores. They vary from 7 to 30 μm in width and from 15 to 90 μm in length, are ellipsoid to conical, and taper to a beak (see *A. alternata*, Fig. 13). Each has one to 10 transverse septa and up to five longitudinal septa. Upon germination up to four germ tubes emerge.

A. triticina grows on a variety of simple media. Growth is optimal between 20 and 24°C, with limits near 5 and 35°C. Six races of the pathogen have been reported. Nonspecific phytotoxins produced by the pathogen apparently play a role in wheat pathogenesis.

Disease Cycle

A. triticina is poorly adapted to oversummer on host residue in soil but survives well as conidia on, and as mycelium within, seed harvested from blighted fields. At approximately four weeks of age, seedlings begin to lose their apparent resistance,

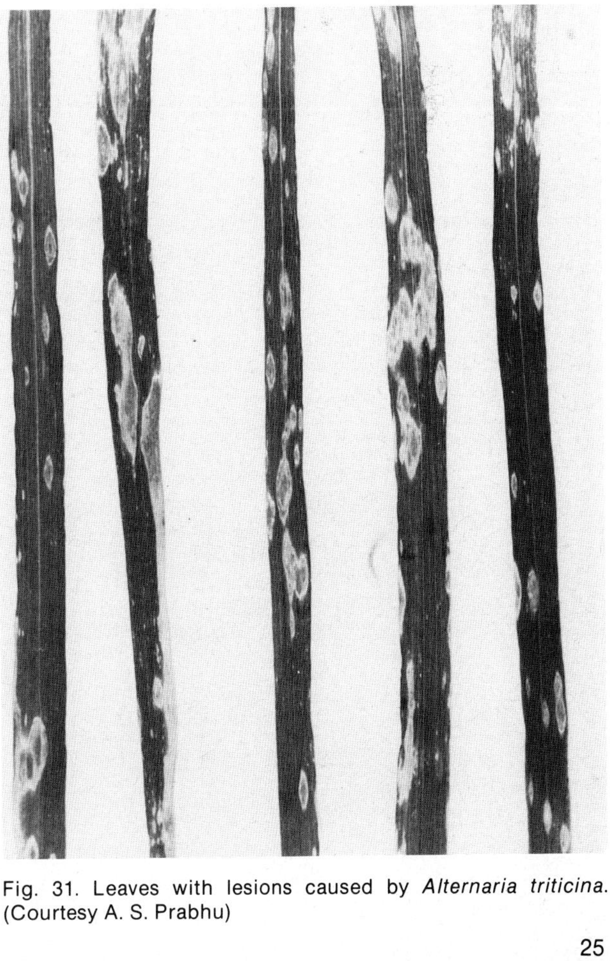

Fig. 31. Leaves with lesions caused by *Alternaria triticina*. (Courtesy A. S. Prabhu)

and leaves, especially those in contact with soil, become infected. Germ tubes with appressoria directly penetrate upper and lower leaf surfaces. Mycelial invasion is intercellular and intracellular. Sporulation within lesions and on mature foliage contributes secondary inoculum, which is disseminated by wind. Late-season head infections account for most seedborne inoculum.

Temperatures between 20 and 25° C are optimal for infection and disease development. Infection requires 10 hr of continuous leaf wetness, after which symptoms appear within four to six days. Leaf blight also is favored by high doses of nitrogen that promote lush vegetative growth.

Control

Resistant cultivars are in use on a limited basis. Low-cost fungicides, applied to foliage before symptoms appear, also provide economical blight control. Because the risk of new infections extends over a two-month period, sprays at 10-day intervals ensure protection but may be uneconomical. When blight can be forecast using weather data, fewer and more timely sprays control the disease.

Chemical and hot-water seed treatments reduce external but not internal inoculum. The ability of *A. triticina* to multiply rapidly from low inoculum densities often negates the value of seed treatments.

Selected References

Anahosur, K. H. 1978. *Alternaria triticina*. Descriptions of Pathogenic Fungi and Bacteria, No. 583. Commonwealth Mycological Institute, Association of Applied Biologists, Kew, Surrey, England.

Bhowmik, T. P. 1974. Fungicidal control of Alternaria leaf blight of wheat. Indian Phytopathol. 27:162–167.

Chaudhuri, S., Maiti, S. S., and Saha, P. 1976. Leaf blight of triticale caused by *Alternaria triticina*. Plant Dis. Rep. 60:133–134.

Frisullo, S. 1982. Parassiti fungini delle piante nell'Italia meridionale. 1. *Alternaria triticina* Pras. et Prab. sufrumento duro. Phytopathol. Mediterr. 21:113–115.

Jain, K. L., and Prabhu, A. S. 1976. Occurrence of a chromogenic variant in *Alternaria triticina*. Indian Phytopathol. 29:22–27.

Kumar, C. S. K. V., and Rao, A. S. 1979. Production of phytotoxic substances by *Alternaria triticina*. Can. J. Bot. 57:1255–1258.

Kumar, V. R., and Arya, H. C. 1973. Certain aspects of perpetuation and recurrence of leaf blight of wheat in Rajasthan. Indian J. Mycol. Plant Pathol. 3:93–94.

Nema, K. G., and Joshi, L. M. 1971. Symptoms and diagnosis of the 'spot blotch' and 'leaf blight' diseases of wheat. Indian Phytopathol. 24:418–419.

Prabhu, A. S., and Prakash, V. 1973. The relation of temperature and leaf wetness to the development of leaf blight of wheat. Plant Dis. Rep. 57:1000–1004.

Prasada, R., and Prabhu, A. S. 1962. Leaf blight of wheat caused by a new species of *Alternaria*. Indian Phytopathol. 15:292–293.

Rao, A. S., and Subrahmanyam, P. 1974. A modified medium for preferential isolation of *Alternaria triticina* on wheat. Indian Phytopathol. 27:133–134.

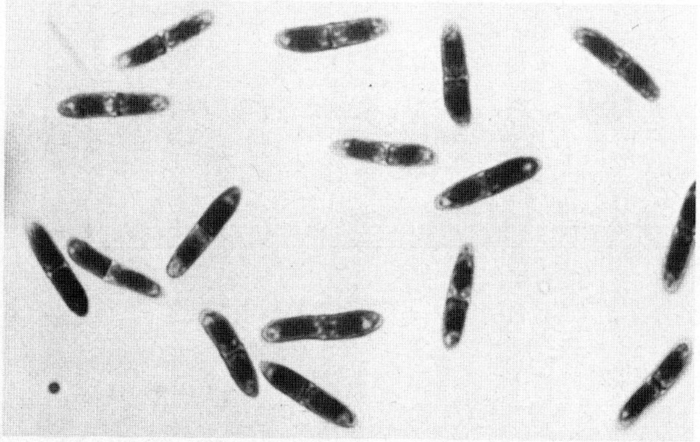

Fig. 32. Conidia (pycnidiospores) of *Ascochyta tritici*. (Courtesy A. L. Scharen)

Sokhi, S. S. 1974. Alternaria blight on wheat in India. PANS 20:55–57.

Waller, J. M. 1981. The recent spread of some tropical plant diseases. Trop. Pest Manage. 27:360–362.

Ascochyta Leaf Spot

Ascochyta leaf spot, caused by *Ascochyta tritici*, is inconspicuous and often overlooked in association with other leaf spot diseases. It is reported to be of minor economic importance on wheat in Japan, Europe, and North America. However, its distribution and frequency may be greater than realized, because most cereals and grasses throughout the world are susceptible to *Ascochyta* spp. Also, *A. tritici* isolates from wheat appear to have a broad host range among the Gramineae.

Symptoms

Ascochyta leaf spot appears primarily on lower leaves (Plate 14). There the first visible symptoms are chlorotic flecks that develop into distinct, chlorotic, ellipsoidal or round lesions 1–5 mm wide. Later the lesions become diffuse and gray-brown internally and resemble those caused by *Septoria nodorum*. The pathogen sometimes produces pycnidia that appear as black dots within necrotic lesions. Pycnidia are submerged in host tissues except for a papillate projection.

Causal Organism

A. tritici Hori & Enj. is generally accepted as the cause of Ascochyta leaf spot, but *A. graminicola* Sacc. and *A. sorghi* Sacc. are cited as wheat pathogens in some literature. *A. tritici* produces separate or clustered, dark, globose, ostiolate pycnidia immersed in host tissue or on various artificial media. In culture, pycnidia measure 103–320 × 124–172 μm. Conidia (pycnidiospores) are straight, hyaline, and oblong; they measure 3–6 × 14–27 μm and typically have one median septum (Fig. 32).

Disease Cycle and Control

A. tritici apparently persists as mycelium and pycnidia in host debris. Primary inoculum has not been identified, but pycnidiospores liberated during wet weather are suspect. Leaf spotting is often associated with high humidity, dense foliage, and leaves in contact with soil. Wheat tissues are not invaded beyond lesion areas, and secondary spread, if it occurs, is by pycnidiospores.

Currently, no specific controls are prescribed for Ascochyta leaf blight. However, controls listed for Septoria diseases should be effective (see Septoria Leaf and Glume Blotches).

Selected References

Cook, R. J. 1970. *Ascochyta graminicola* on wheat and barley. Plant Pathol. 19:48–49.

Roane, C. W., Roane, M. K., and Starling, T. M. 1974. *Ascochyta* species on barley and wheat in Virginia. Plant Dis. Rep. 58:455–456.

Scharen, A. L., and Krupinsky, J. M. 1971. *Ascochyta tritici* on wheat. Phytopathology 61:675–680.

Sprague, R. 1950. Ascochyta leaf spot of cereals and grasses in the United States. Mycologia 42:523–553.

Cephalosporium Stripe

Initial descriptions of Cephalosporium stripe came from Japan in 1930. By 1955 the disease also was known in the United Kingdom and across the northern winter wheat belt in North America. In the United States, Cephalosporium stripe is frequent in the Northwest, the Great Lakes states, and Kansas.

The incidence of diseased plants may range to 100% in some fields, and yields may be reduced as much as 80%. The disease is especially severe where wheat follows susceptible cereal or grass crops. Most winter cereals (oats, barley, rye, and triticale) and several grasses (*Bromus, Dactylis,* and *Poa* spp.) are susceptible, but wheat is the major economic host. Spring grains and annual grasses are susceptible but either escape infection or do not permit infections to build to damaging proportions.

The causal fungus, *Cephalosporium gramineum*, is soilborne and is the only true vascular fungal pathogen of wheat. It inhabits and occludes xylem vessels and impedes the transport of water and nutrients through nodes, leaf veins, and interveinal tissues. Yield losses result from reduced seed set and weight and premature death of tillers.

Symptoms

Cephalosporium stripe tends to be more prevalent in lower, wetter field areas and on acid soils. During jointing and heading, conspicuous, chlorotic, longitudinal stripes appear on leaves (Plate 15). The stripes may be diffuse or mottled initially but eventually become prominently yellow and necrotic on maturing green leaves. The continuity of stripes, and one or more darkened veins within them, through the culm, leaf sheath, and blade is diagnostic. Usually, only one or two stripes and discolored veins are present per leaf. Stripes may not develop on all leaves or on all tillers of diseased plants.

Infected seedling leaves sometimes show a mosaiclike yellowing in late winter or early spring. Such leaves may be short-lived and may die before showy stripes develop. Near harvest, the culm of infected plants may darken at and below nodes. Longitudinal sections cut through upper nodes often show that the entire nodal plate is darkened. Such plants typically are or will be stunted, prematurely ripe, and white-headed (see Plate 34).

Causal Organism

C. gramineum Nis. & Ika. (syn. *Hymenula cerealis* Ell. & Ev.) is slow-growing relative to other soil fungi. It is easily recovered from striped leaves and, with selective media, can be recovered directly from soil. In host plants and on culture media, it sporulates profusely (phialospores and blastospores) and makes limited mycelial growth. Colonies in culture are wet, mostly submerged, restricted, and white, gray, or pale yellow. Conidiophores (phialides) are short, 1–2 × 4–10 μm, hyaline, simple, and inconspicuous. Conidia are unicellular, 2–3 × 3–7 μm, and clumped in slime (Fig. 33).

The fungus produces superficial sporodochia up to 1 mm in diameter on wheat straw (Fig. 34). This sporodochial stage is saprophytic and is identified as *H. cerealis* in some literature. The sporodochia have a hymenial surface of branched, hyaline phialides, 1.2–1.5 × 10–30 μm, densely clustered on a thin stroma (Fig. 35). Dry sporodochia on wheat straw are flat and gray-black and are easily dislodged. When moist and actively sporulating, they are raised, yellow-brown, and glistening with mounds of slime-bound conidia (Fig. 34). Sporodochial conidia are ovoid, unicellular, hyaline, 0.8–1.5 × 2–4.5 μm, and smaller than conidia produced in vitro.

Disease Cycle

Cephalosporium stripe is favored by wet soils, fluctuating winter temperatures, and repeated cropping of susceptible cereals or grasses in the same field. The fungus persists in association with host residues on or within 8 cm of the soil surface. Soil pH in the range 3.9–5.5 prolongs pathogen survival and may increase disease.

Soilborne conidia serve as primary inoculum and enter roots during winter and early spring. Root infection apparently is incidental through wounds caused by soil heaving, frost injury, and perhaps other mechanical stresses. In the United Kingdom, the disease sometimes is associated with wireworm (Elateridae) damage.

C. gramineum must enter vascular tissues to colonize a living host plant. Conidia in xylem vessels are carried upward in transpirational streams and lodge and multiply at nodes and in leaves. The fungus emits metabolites and occludes vessels in the course of leaf stripe formation. It does not persist, and appears to be of no consequence, in roots.

Harvest and tillage operations return host debris and inoculum to soil. *C. gramineum* grows saprophytically and indiscriminately in moist straw and produces antibiotics that discourage other microbial growth in the straw. It eventually emerges through stomata or other openings to sporulate on the straw surface. Saprophytic conidial production is highly efficient via phialides even when sporodochia do not form. In cool, wet climates, conidial populations can exceed 100,000/g (dry weight) of soil.

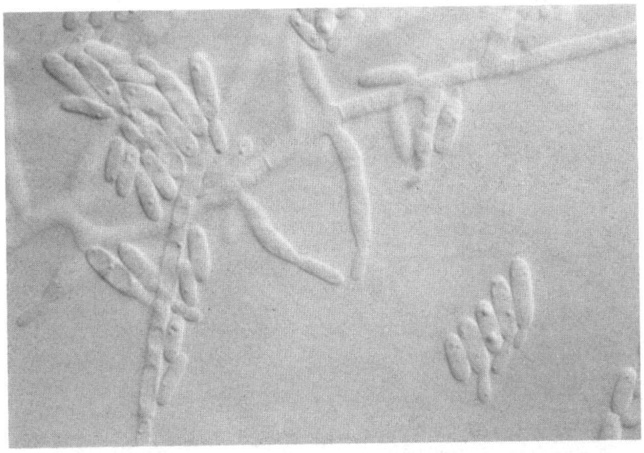

Fig. 33. Conidia (phialospores and blastospores) of *Cephalosporium gramineum* in agar culture. (Courtesy T. D. Murray)

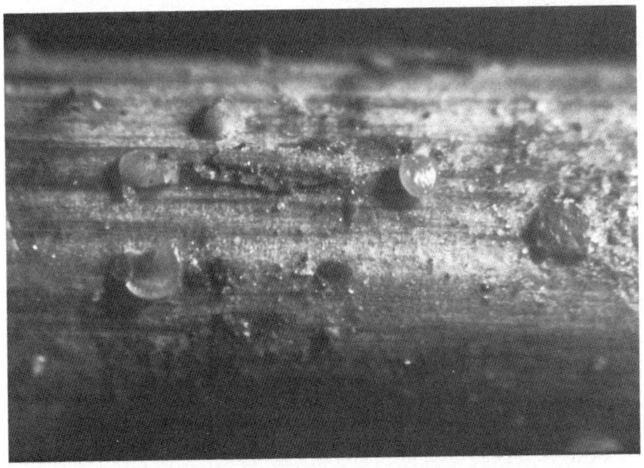

Fig. 34. Sporodochia of *Cephalosporium gramineum* on moist straw. (Courtesy M. V. Wiese and A. V. Ravenscroft)

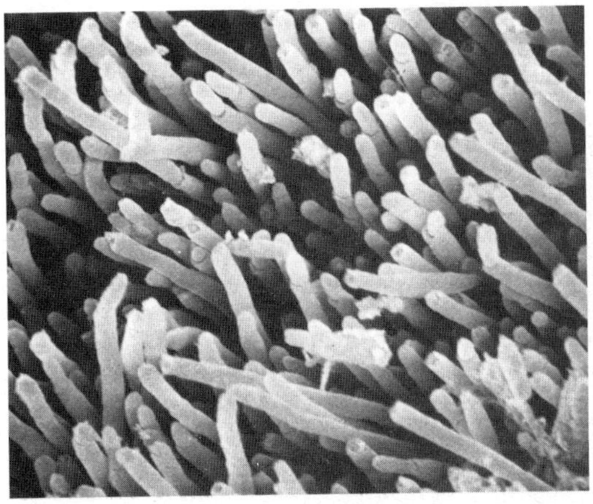

Fig. 35. Clustered sporulating phialides within a sporodochium of *Cephalosporium gramineum*. (Courtesy M. V. Wiese)

Control

Reducing soilborne inoculum by crop rotation and residue management controls leaf stripe. Winter wheat is damaged less when it is grown in rotations with spring cereals or nonhosts such as corn or legumes. Avoiding host crops for at least two years may be necessary to reduce the disease to inconsequential levels. Clean cultivation of all crops is important to prevent multiplication of the pathogen on grassy weeds.

Where rotations and nonhost crops are impractical, infested residues should be removed or plowed deeper than 8 cm. The pathogen is sustained on host debris that remains on or near the soil surface. Late-autumn seeding reduces leaf striping, apparently by limiting autumn root growth and minimizing sites for injury and infection during the winter. Some wheat cultivars, like Crest, Nugaines, Winridge, Luke, Lewjain, and PI 178383, are tolerant in the United States, but none are highly resistant.

Selected References

Bailey, J. E., Lockwood, J. L., and Wiese, M. V. 1982. Infection of wheat by *Cephalosporium gramineum* as influenced by freezing of roots. Phytopathology 72:1324–1328.

Bockus, W. W., O'Connor, J. P., and Raymond, P. J. 1983. Effect of residue management method on incidence of Cephalosporium stripe under continuous winter wheat production. Plant Dis. 67:1323–1324.

Bockus, W. W., and Sim, T., IV. 1982. Quantifying Cephalosporium stripe disease severity on winter wheat. Phytopathology 72:493–495.

Bruehl, G. W. 1957. Cephalosporium stripe disease of wheat. Phytopathology 47:641–649.

Bruehl, G. W. 1963. *Hymenula cerealis*, the sporodochial stage of *Cephalosporium gramineum*. Phytopathology 53:205–208.

Bruehl, G. W., and Lai, P. 1968. Influence of soil pH on survival of *Cephalosporium gramineum* in infested wheat straw. Can. J. Plant Sci. 48:245–262.

Howell, M. J., and Burgess, P. A. 1969. *Cephalosporium gramineum* causing leaf stripe of grasses and its sporodochial stage, *Hymenula cerealis*, on cereals and grasses. Plant Pathol. 18:67–70.

Lai, P., and Bruehl, G. W. 1968. Antagonism among *Cephalosporium gramineum, Trichoderma* spp., and *Fusarium culmorum*. Phytopathology 58:562–566.

Latin, R. X., Harder, R. W., and Wiese, M. V. 1982. Incidence of Cephalosporium stripe as influenced by winter wheat management practices. Plant Dis. 66:229–230.

Love, C. S. 1985. Effect of soil pH on infection of wheat by *Cephalosporium gramineum*. (Abstr.) Phytopathology 75:1296.

Mathre, D. E., and Johnston, R. H. 1975a. Cephalosporium stripe of winter wheat: Infection processes and host response. Phytopathology 65:1244–1249.

Mathre, D. E., and Johnston, R. H. 1975b. Cephalosporium stripe of winter wheat: Procedures for determining host response. Crop. Sci. 15:591–594.

Mathre, D. E., and Johnston, R. H. 1979. Decomposition of wheat straw infected by *Cephalosporium gramineum*. Soil Biol. Biochem. 11:577–580.

Pool, R. A. F., and Sharp, E. L. 1969a. Some environmental and cultural factors affecting Cephalosporium stripe of winter wheat. Plant Dis. Rep. 53:898–902.

Pool, R. A. F., and Sharp, E. L. 1969b. Possible association of a polysaccharide and an antibiotic with the disease cycle of Cephalosporium stripe. Phytopathology 59:1763–1764.

Slope, D. B., and Bardner, R. 1965. Cephalosporium stripe of wheat and root damage by insects. Plant Pathol. 14:184–187.

Wiese, M. V. 1972. Colonization of wheat seedlings by *Cephalosporium gramineum* in relation to symptom development. Phytopathology 62:1013–1018.

Wiese, M. V., and Ravenscroft, A. V. 1975. *Cephalosporium gramineum* populations in soil under winter wheat cultivation. Phytopathology 65:1129–1133.

Wiese, M. V., and Ravenscroft, A. V. 1978. Sporodochium development and conidium production in *Cephalosporium gramineum*. Phytopathology 68:395–401.

Anthracnose

Anthracnose diseases are caused by fungi that produce spores in dark acervuli and principally cause dark local lesions on host plants. Several *Colletotrichum* species cause anthracnose among the Gramineae with some degree of host specificity. Sorghum, Sudan grass, and rye are the most important hosts, but wheat, barley, oats, corn, and about 20 genera of temperate-climate grasses are susceptible.

Anthracnose on wheat attracted attention in the early 1900s. From 1910 to 1940, it frequently caused severe plant infections and reduced yields as much as 25%. Anthracnose currently occurs worldwide but is damaging only in isolated areas, especially on crops that are nutritionally stressed or grown under reduced tillage. In current literature, anthracnose is more often reported on corn than on small-grain cereals.

Symptoms

Anthracnose infections occur below and above ground as dark, roughly elliptical lesions 1–2 cm long. The lesions are first water-soaked, then bleached and necrotic. They are obscure and may resemble eyespot or sharp eyespot until dark acervuli appear within them. The acervuli contain diagnostic dark spines visible under low magnification (Fig. 36). Lesions on wheat normally are confined to the lower half of stems, and acervuli usually do not develop until the host matures (Fig. 37). Severe infections on culms or crowns reduce plant vigor and promote head blighting, lodging, and shriveled grain.

Causal Organism

C. graminicola (Ces.) Wilson (syns. *Dicladium graminicolum* Ces., *C. cereale* Manns), the cause of anthracnose in wheat, produces dark, elongated acervuli within lesions on grass hosts. Stromata are 70–300 μm in diameter and bear prominent, dark, septate spines (setae) up to 100 μm long. At the base of the spines, a dense hymenium produces hyaline, unicellular, sickle-shaped conidia measuring 3–5 × 19–29 μm (Fig. 36). Phialides are 4–8 × 8–20 μm, unicellular, hyaline, and cylindrical.

Colonies on potato-dextrose agar are gray and feltlike. Conidia and appressoria are numerous when cultures are well aerated, and sclerotia sometimes occur. Appressoria are diagnostic, tawny brown, prominent, and terminal on

Fig. 36. Mature anthracnose lesion on wheat culm. Acervulus of *Colletotrichum graminicola* at the center of the lesion produces sickle-shaped conidia and dark spines. (Courtesy M. F. Brown and H. G. Brotzman)

thickened hyphae. They assume irregular shapes, average 11.2 × 15.6 μm, and have a single germ pore.

C. graminicola strains vary in host specificity and in virulence on wheat. Pathogenic races of the fungus were characterized as early as 1914. *Glomerella graminicola* Politis, a rare ascomycete, was recently described as its perfect stage.

Disease Cycle

C. graminicola survives in soil as mycelium and conidia on numerous cereal and grass host residues. Conidia are dispersed by wind or rain and serve as primary inoculum. They germinate in water films, develop prominent appressoria, and penetrate wheat directly. Because inoculum is primarily soilborne, root, crown, and basal stem infections are most frequent. Upper plant parts may be infected secondarily. Conidia may become seedborne during harvest or as a result of head infections.

Alternative grassy hosts, alkaline soil, and continuous wheat cultivation promote anthracnose. Wet weather favors infection and subsequent sporulation. Under optimal conditions (25° C, free moisture, and susceptible cultivars), acervuli develop within 10 days of infection. The fungus does not invade tissues beyond lesion areas.

Control

Anthracnose is controlled by crop rotation and residue management. More specific controls have not been devised.

Fig. 37. Mature head and culms darkened by anthracnose (caused by *Colletotrichum graminicola*). (Courtesy Illinois Natural History Survey)

Rotations with legumes or other poor hosts or nonhosts limit soilborne inoculum. Clean cultivation also eliminates grasses on which the fungus can survive.

Cleaning or chemically treating seed before planting is advantageous, and crops should be fertilized adequately. Some wheat cultivars are highly resistant to *C. graminicola*, but they are not in widespread use.

Selected References

Ali, M. M. 1963a. Factors influencing formation of acervuli and conidia by *Colletotrichum graminicola*. Mycopathol. Mycol. Appl. 19:94–98.

Ali, M. M. 1963b. Factors influencing pathogenicity of three isolates of *Colletotrichum graminicola* on wheat. Mycopathol. Mycol. Appl. 19:161–166.

Bruehl, G. W., and Dickson, J. G. 1950. Anthracnose of cereals and grasses. U.S. Dep. Agric. Tech. Bull. 1005.

Luke, H. H., and Sechler, D. T. 1963. Rye anthracnose. Plant Dis. Rep. 47:936–937.

Mordue, J. E. M. 1967. *Colletotrichum graminicola*. Descriptions of Pathogenic Fungi and Bacteria, No. 132. Commonwealth Mycological Institute, Association of Applied Biologists, Kew, Surrey, England.

Politis, D. J. 1975. The identity and perfect state of *Colletotrichum graminicola*. Mycologia 67:56–62.

Dilophospora Leaf Spot (Twist)

Like the seed-gall nematode (*Anguina tritici*) commonly associated with it, Dilophospora leaf spot is primarily of historical interest. During the first half of this century, it damaged wheat in the United States and Canada and caused minor yield losses in parts of Europe. Its most recent occurrence on wheat was reported in India in 1961. The causal fungus, *Dilophospora alopecuri*, sometimes occurs on meadow and range grasses but is rare on cultivated cereals.

Symptoms

Spotted and distorted leaves are diagnostic. Leaf spots initially are small, yellow, elongate or spindle-shaped flecks that increase to 1–1.5 × 4–8 mm and become tan-brown with black, crusty centers. The spots tend to be confined to blades and sheaths but may occur on peduncles and heads. Leaves are killed when lesions are numerous.

Deformations occur when seedling infections reach the whorl. This phase of the disease is common where plants are parasitized by the seed-gall nematode. *D. alopecuri* colonizes tissue and surfaces and prevents leaves from emerging or causes them to emerge distorted and covered with downy, gray mycelium. When exposed to light and drying, the fungus darkens into stromata, which bear pycnidia. Infected organs resemble those damaged by flag smut, especially when stromata occur in streaks.

Causal Organism

D. alopecuri (Fr.) Fr. (syns. *D. graminis* Desm., *Dilophia graminis* Fuckel) produces conspicuous, erumpent, black stromata and globose, ostiolate pycnidia within host lesions. Mature pycnidia, 60–300 μm in diameter, have a broad (25–40 μm), torn ostiole and simple, hyaline, short conidiophores. Conidia (pycnidiospores) are zero- to three-septate, hyaline, cylindrical to ellipsoid, and 1.5–2.5 × 8–15 μm, with distinctive claw-shaped appendages (0.5 × 5–7 μm) at each end.

Conidial germination requires water and nutrients and is optimal between 20 and 25° C, with limits at 15 and 35° C. Germ tubes have the same diameter as the parent conidium. *D. alopecuri* produces pycnidia and its distinctive pycnidiospores on potato-dextrose agar fortified with yeast extract. It also can be cultured on moistened autoclaved wheat or barley kernels.

Disease Cycle and Control

D. alopecuri survives as mycelium in host debris and as conidia on seed. Conidia act as primary inoculum, and those produced secondarily appear to account for later infections on upper plant parts. Conidia are dispersed by wind and splashing rain, and their appendages may function in attachment to seed, host plants, and nematode vectors. The fungus usually reaches the whorl and initiates "twist" symptoms only via larvae of *A. tritici*. However, leaf spotting can occur without seed-gall nematodes.

Controlling *A. tritici* also retards *D. alopecuri*. In the past, the use of clean seed, eradicative chemical seed treatments, and burning or avoiding infested host residues were recommended. Currently, crop rotations, clean cultivation, and the use of clean seed keep the disease at inconsequential levels.

Selected References

Atanasoff, D. 1925. The Dilophospora disease of cereals. Phytopathology 15:11–40.

Munjal, R. L., and Kaul, T. N. 1961. Dilophospora leafspot of wheat in India. Indian Phytopathol. 14:13–15.

Powdery Mildew

Powdery mildews on wheat, other cereals, and numerous grasses produce showy symptoms that were recognized centuries ago. Powdery mildews are caused by obligately parasitic, host-specific fungi within the species *Erysiphe graminis* DC. ex Merat. These fungi are presumed to have developed through prolonged association with their grass hosts. The fungal strains that attack wheat do so exclusively (except in special circumstances) and are included within *E. graminis* f. sp. *tritici*. This form species is widely distributed in humid and semiarid regions of the world.

E. graminis f. sp. *tritici* utilizes the nutrients, reduces the photosynthesis, and increases the respiration and transpiration of its hosts. Infected plants lose vigor, and their growth, heading, and seed filling are impaired. Heavily infected leaves and even entire plants can be killed prematurely. Yield losses are related to the intensity of attack and result from reduced head numbers and kernel weights. Younger tillers especially may fail to produce heads. Yield losses up to 40% are recorded and are greatest when disease development precedes or accompanies flowering.

Symptoms

Powdery mildews on wheat and on other cereals and grasses are remarkably similar. *E. graminis* f. sp. *tritici* may infect all aerial portions of the plant, but it is usually most prevalent on the upper surface of the lower leaves. Symptoms can appear at any time after seedlings emerge.

The pathogen is entirely superficial except for haustoria that penetrate epidermal cells. It is conspicuous, therefore, as effuse patches (colonies) of cottony mycelium and conidia on the host surface (Fig. 38). Colonies, initially white and sporulating, later turn dull gray-brown. Chlorotic patches normally appear on the leaf surface directly opposite mildew colonies.

Sexual fruiting bodies (cleistothecia) are visible without magnification as distinct brown-black dots within aging colonies on maturing plants (Plate 16). Immature cleistothecia are inconspicuous, light-colored, and spherical.

Causal Organism

E. graminis DC. f. sp. *tritici* E. Marchal is heterothallic, and hybridization between genetically dissimilar strains occurs. This process, within cleistothecia, generates new virulent types (physiologic races) that can be distinguished by differential reactions on host series.

E. graminis f. sp. *tritici* produces superficial (epiphytic) colonies of persistent, interlaced hyphae 5–10 μm wide. Conidiophores are short, simple, and 8–10 × 25–30 μm and arise from slightly swollen basal cells. Conidiophores have a terminal generative cell from which hyaline, unicellular, ellipsoid to ovate conidia (8–10 × 20–35 μm) are borne basipetally and diurnally in long chains (meristem arthrospores) (Fig. 39). This conidial stage is named *Oidium monilioides* Desm. in some literature.

Infected wheat epidermal cells contain elliptical haustoria (5–10 × 10–30 μm) that bear distinctive fingerlike appendages up to 20 μm long on each apex (Fig. 40). The haustorium forms a close physical association with, but does not penetrate, the host cell plasmalemma.

Mature cleistothecia are globose and 135–280 μm in diameter and are composed initially of interlaced, light-colored mycelium. They become dark and sunken when mature and develop diagnostic, simple or sparingly branched appendages on their surface. Asci are numerous (normally 15–20 per cleistothecium). When mature, asci (25–40 × 70–110 μm) are cylindrical to pedicellate and contain eight hyaline ascospores measuring 10–13 × 20–23 μm.

Disease Cycle

E. graminis f. sp. *tritici* overwinters as cleistothecia on straw and, in milder climates, also as mycelium and conidia. Windborne ascospores or conidia are primary inoculum. Ascospores

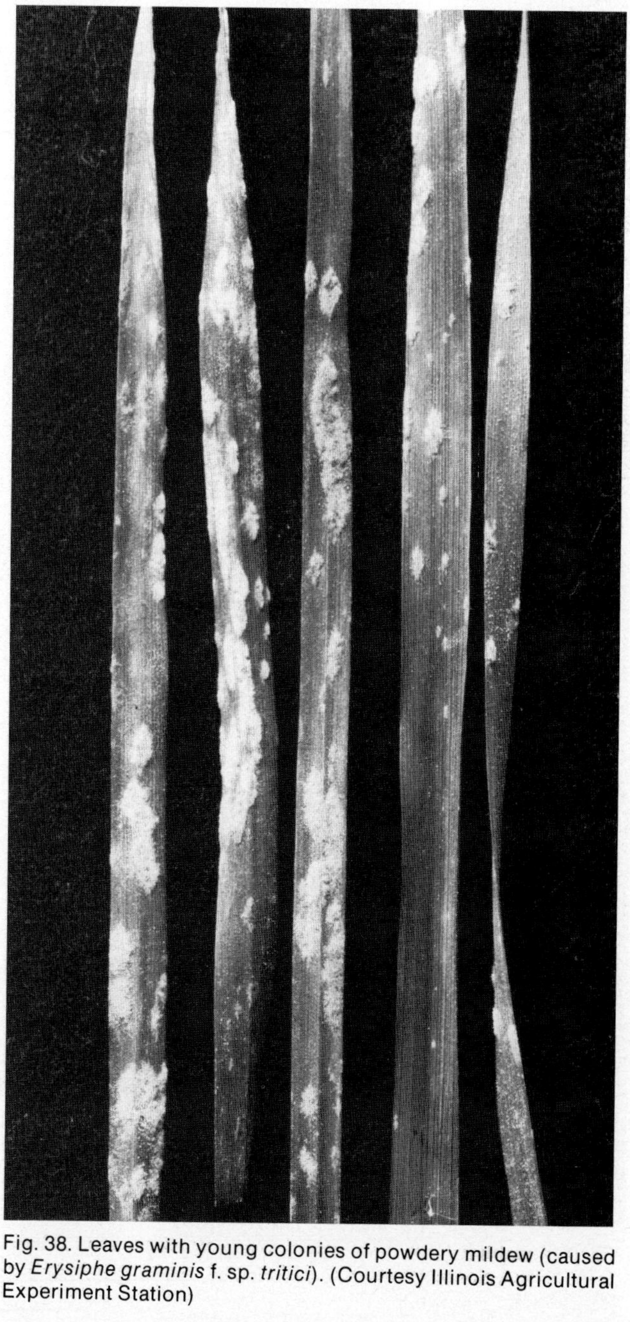

Fig. 38. Leaves with young colonies of powdery mildew (caused by *Erysiphe graminis* f. sp. *tritici*). (Courtesy Illinois Agricultural Experiment Station)

are produced in midsummer, and conidia are most frequent in spring. The germ tubes of both spore types penetrate wheat directly, establish haustoria in penetrated cells, and give rise to superficial, sporulating colonies. Resultant conidia are dispersed by wind and induce secondary infections. Conidial production declines markedly as colonies age.

Cleistothecia develop as temperatures increase or as the fungus and host mature or become moisture-stressed. Ascospores, released after rains, are sparse relative to conidia but sometimes infect wheat in autumn. Volunteer wheat plants are opportune hosts and support the pathogen between summer and winter crops.

Conidia are produced in great numbers and are most important epidemiologically. They survive only a few days but withstand dispersal over several kilometers. They germinate over a wide temperature range (1–30° C) and without free moisture. Some germinate using only endogenous moisture.

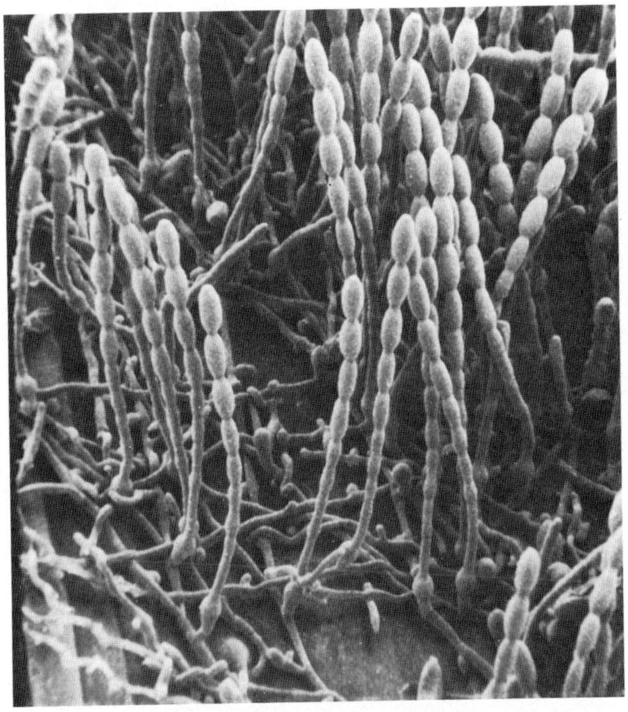

Fig. 39. Chains of conidia within a sporulating colony of *Erysiphe graminis* f. sp. *tritici*. (Courtesy M. F. Brown and H. G. Brotzman)

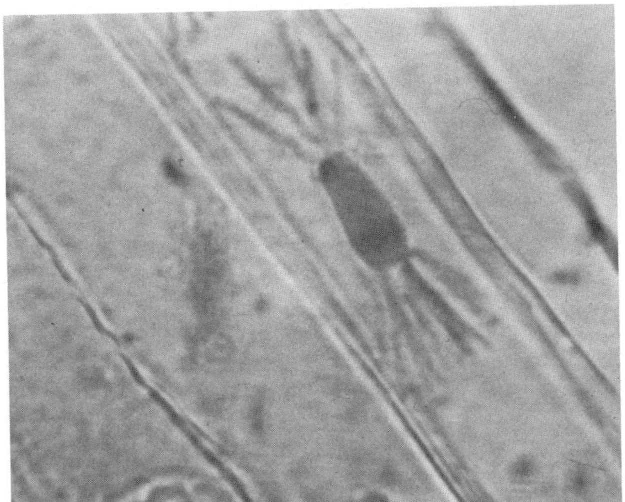

Fig. 40. Haustorium of *Erysiphe graminis* f. sp. *tritici* in a host cell. (Courtesy Plant Pathology Department, Washington State University)

Germination is optimal at 100% relative humidity (RH) but can occur at 85% RH. In favorable field environments, germination, infection, and secondary sporulation are completed within seven to 10 days.

Powdery mildew development is optimal between 15 and 22° C and is markedly retarded above 25° C. Most stages of infection proceed in darkness, except for host penetration and conidial formation, which require light. Wheat is most susceptible during periods of rapid growth. Dense stands of susceptible cultivars, heavy nitrogen fertilization, high humidity, and cool temperatures favor disease development.

Control

Fungicides are used to control powdery mildew where costs are not prohibitive. In parts of Europe, selective, systemic fungicides are often employed.

Resistant cultivars are perhaps the best defense against powdery mildew, and they are useful so long as the prevalent races of the fungus are controlled. The host cuticle may be a mechanical barrier to penetration, or invaded cells may quickly die and be nonsupportive. Some cultivars permit penetration and infection but retard or prevent sporulation. Others become resistant or immune only as adult plants. Most cultivars are susceptible to new races or shifts in pathogen virulence. Monogenic resistance has generally proved to be short-lived. The reduced rate of mildew development on some "slow-mildewing" cultivars appears to be more durable.

Crop rotations, clean cultivation (especially the destruction of volunteer wheat), and destruction of host residues reduce overwintering inoculum. However, the benefits of such practices may be negated by airborne inoculum or environmental conditions that permit rapid disease development. The pathogen thrives on lush, nitrogen-rich plants but is less damaging when nitrogen, potassium, and phosphorus fertilization is correctly proportioned.

Selected References

Ellingboe, A. H. 1972. Genetics and physiology of primary infection by *Erysiphe graminis*. Phytopathology 62:401–406.

Eshed, N., and Wahl, I. 1975. Role of wild grasses in epidemics of powdery mildew on small grains in Israel. Phytopathology 65:57–63.

Gustafson, G. D., and Shaner, G. 1982. The influence of plant age on the expression of slow-mildewing resistance in wheat. Phytopathology 72:746–749.

Hyde, P. M., and Colhoun, J. 1975. Mechanisms of resistance of wheat to *Erysiphe graminis* f. sp. *tritici*. Phytopathol. Z. 82:185–206.

Johnson, H. W. 1970. Control of powdery mildew of wheat by soil-applied benomyl. Plant Dis. Rep. 54:91–93.

Johnson, H. W. 1974. Overwintering of *Erysiphe graminis* f. sp. *tritici* on maritime grown winter wheat. Can. Plant Dis. Surv. 54:71–73.

Kapoor, J. N. 1967. *Erysiphe graminis*. Descriptions of Pathogenic Fungi and Bacteria, No. 153. Commonwealth Mycological Institute, Association of Applied Biologists, Kew, Surrey, England.

Last, F. T. 1962. Effects of nutrition on the incidence of barley powdery mildew. Plant Pathol. 11:133–135.

McKeen, W. E. 1974. The interface between powdery mildew haustorium and the cytoplasm of the susceptible barley epidermal cell. Can. J. Microbiol. 20:1475–1478.

Parmentier, G., and Rixhon, L. 1973. Effects of crop rotation on powdery mildew infection in winter wheat. Parasitica 29:129–133.

Powers, H. R., Jr., and Moseman, J. G. 1957. Pathogenic variability within cleistothecia of *Erysiphe graminis*. Phytopathology 47:136–138.

Shaner, G. 1973. Reduced infectability and inoculum production as factors of slow mildewing in Knox wheat. Phytopathology 63:1307–1311.

Shaner, G., and Finney, R. E. 1977. The effect of nitrogen fertilization on the expression of slow-mildewing resistance in Knox wheat. Phytopathology 67:1051–1056.

Singh, A., and Saxena, S. C. 1973. Chemical control of powdery mildew of wheat. Indian J. Mycol. Plant Pathol. 3:202–203.

Smith, H. C., and Smith, M. 1974. Surveys of powdery mildew in wheat and an estimate of national yield losses. N.Z. J. Exp. Agric. 2:441–445.

Snow Molds

Snow molds are caused by fungi that grow at or near the soil surface and beneath persistent snow cover. Wheat, rye, and barley as winter cereals and numerous grasses are susceptible. Snow protects dormant plants from freezing and desiccation but favors the growth of low-temperature parasitic fungi. Snow deeper than 30 cm maintains the surface temperature of unfrozen soil at $0 \pm 0.5°C$ when air temperatures fluctuate between $+20$ and $-20°C$. Only within the last 50 years have fungi beneath snow been recognized as an important factor in the winter survival of cereals. Previously they were dismissed as saprophytes on winter-injured plants.

Snow molds occur at higher elevations and latitudes where growing seasons are short and marginal for spring wheat cultivation. They occur routinely in portions of the Soviet Union, Japan, Canada, and northern Europe. In Scandinavia and the northern United States, the diseases are especially damaging to winter wheat crops.

Symptoms

Snow mold symptoms are best seen by examining plants beneath snow cover or immediately after snow melt (see Snow Rot). Conspicuous, sometimes slimy mycelial growth on living and dead plant parts is diagnostic. Also, numerous brown-black, irregular or spherical sclerotia may be present (Plates 17 and 18; Fig. 41) if *Typhula* or *Sclerotinia* spp. are present. Dead leaves are common, but damage is highly variable. Leaves may be wholly or partly necrotic, flaccid, or rotted (Plates 19 and 20). If the crown is invaded, the entire plant can be killed. Snow molds usually damage plants in areas corresponding to the pattern of deep snow cover. This distinguishes plants damaged by snow mold from winter-injured plants, which usually occur in areas that are barren of snow or covered with ice.

Despite extensive leaf destruction, wheat plants often recover and produce satisfactory yields if crowns are not damaged. However, resultant stands may be thin and may be vulnerable to competition from weeds. Dry, warm weather after snow melt favors plant recovery, whereas cool, damp weather permits damage to progress.

Causal Organisms

The important snow mold fungi on wheat are *Microdochium nivale* (*Fusarium nivale*) (pink snow mold), *Typhula* spp. (speckled snow mold), and *Sclerotinia borealis* (Sclerotinia snow mold). In North America, *Coprinus psychromorbidus* also is pathogenic. This latter pathogen was identified only from white mycelium and clamp connections as a "low-temperature basidiomycete" before its *Coprinus*-like fructification was discovered. It attacks cereals, alfalfa, cultivated turfgrasses, and several wild grasses, including *Bromus* and *Festuca* spp.

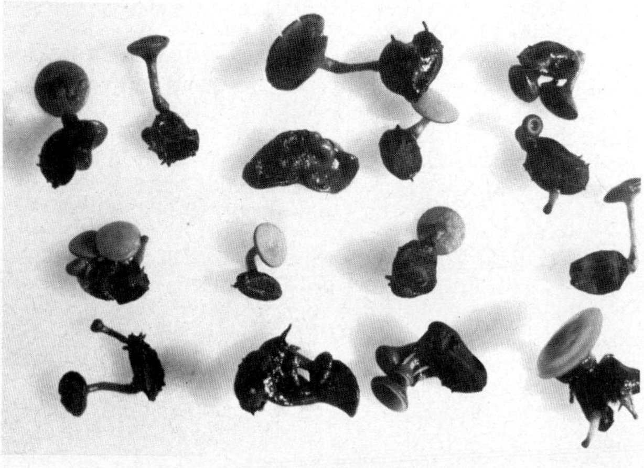

Fig. 41. Apothecia arising from sclerotia of *Sclerotinia borealis*, the cause of Sclerotinia snow mold. (Courtesy J. D. Smith)

Snow mold fungi are nonspecific pathogens that may occur in combinations on a given plant. The major parasites reproduce sexually and have potential for variability, but no pathovars (races) have been reported. Although they grow on most laboratory media, snow mold fungi are weak saprophytes in the field and persist as sclerotia, mycelium, or perithecia in association with host debris. Sclerotia can survive in soil for a year or more.

Beneath deep snow, photosynthesis is impaired and host plants slowly deplete their carbohydrate and then their protein reserves. In this weakened state, they are predisposed to disease caused by *M. nivale* and *Typhula* spp. *S. borealis*, however, is a more aggressive pathogen; it can invade wheat below freezing, whereas *Typhula* spp. are most pathogenic near the freezing point and *M. nivale* is most pathogenic above 0° C.

Control

Growing spring wheat is often the only recourse against snow mold. Winter cultivars differ in tolerance, but none are immune to damage and all can be killed. Resistance is nonspecific and is related to factors such as plant size, carbohydrate reserves, and low metabolic rates.

Fungicides are useful in chronic snow mold areas if they offer protection from the time of application (late autumn) through the winter. In the case of *M. nivale*, seed treatments reduce seedborne inoculum and for a time protect seedlings developing in infested soil.

Snow molds are less successful when wheat is rotated with spring cereals or legumes. Inoculum declines in soils without winter hosts, and legumes tend to decrease the viability of *Typhula* sclerotia. Fertilization and seeding practices should ensure crown development but avoid dense foliar growth in autumn.

Hastening snow melt by aerial application of blackening agents such as ashes or coal dust decreases snow mold injury. The practice is economical only in areas highly prone to snow mold damage.

Selected References

Amano, Y., and Osanai, I. 1983. Winter wheat breeding for resistance to snow mold and cold hardiness. III. Varietal differences of ecological characteristics on cold acclimation and relationships of them to resistance. Bull. Hokkaido Prefect. Agric. Exp. Stn. 50:83–97.

Bruehl, G. W. 1967a. Lack of significant pathogenic specialization within *Fusarium nivale, Typhula idahoensis* and *T. incarnata* and correlation of resistance in winter wheat to these fungi. Plant Dis. Rep. 51:810–814.

Bruehl, G. W. 1967b. Effect of plant size on resistance to snow mold of winter wheat. Plant Dis. Rep. 51:815–819.

Bruehl, G. W., and Cunfer, B. 1971. Physiologic and environmental factors that affect the severity of snow mold of wheat. Phytopathology 61:792–799.

Bruehl, G. W., Kiyomoto, R., Peterson, C., and Nagamitsu, M. 1975. Testing winter wheats for snow mold resistance in Washington. Plant Dis. Rep. 59:566–570.

Cormack, M. W. 1948. Winter crown rot or snow mold of alfalfa, clovers and grasses in Alberta. Can. J. Res. 26:71–85.

Jamalainen, E. A. 1964. Control of low temperature parasitic fungi in winter cereals by fungicidal treatment of stands. Ann. Agric. Fenn. 3:1–54.

Jamalainen, E. A. 1974. Resistance in winter cereals and grasses to low-temperature parasitic fungi. Annu. Rev. Phytopathol. 12:281–302.

Smith, J. D. 1975. Snow molds on winter cereals in northern Saskatchewan in 1974. Can. Plant Dis. Surv. 55:91–96.

Traquair, J. A., and Smith, J. D. 1982. Sclerotial strains of *Coprinus psychromorbidus*, a snow mold basidiomycete. Can. J. Plant Pathol. 4:27–36.

Pink Snow Mold

Pink snow mold occurs on wheat in central Europe, Scandinavia, Canada, and the United States. Its distribution is

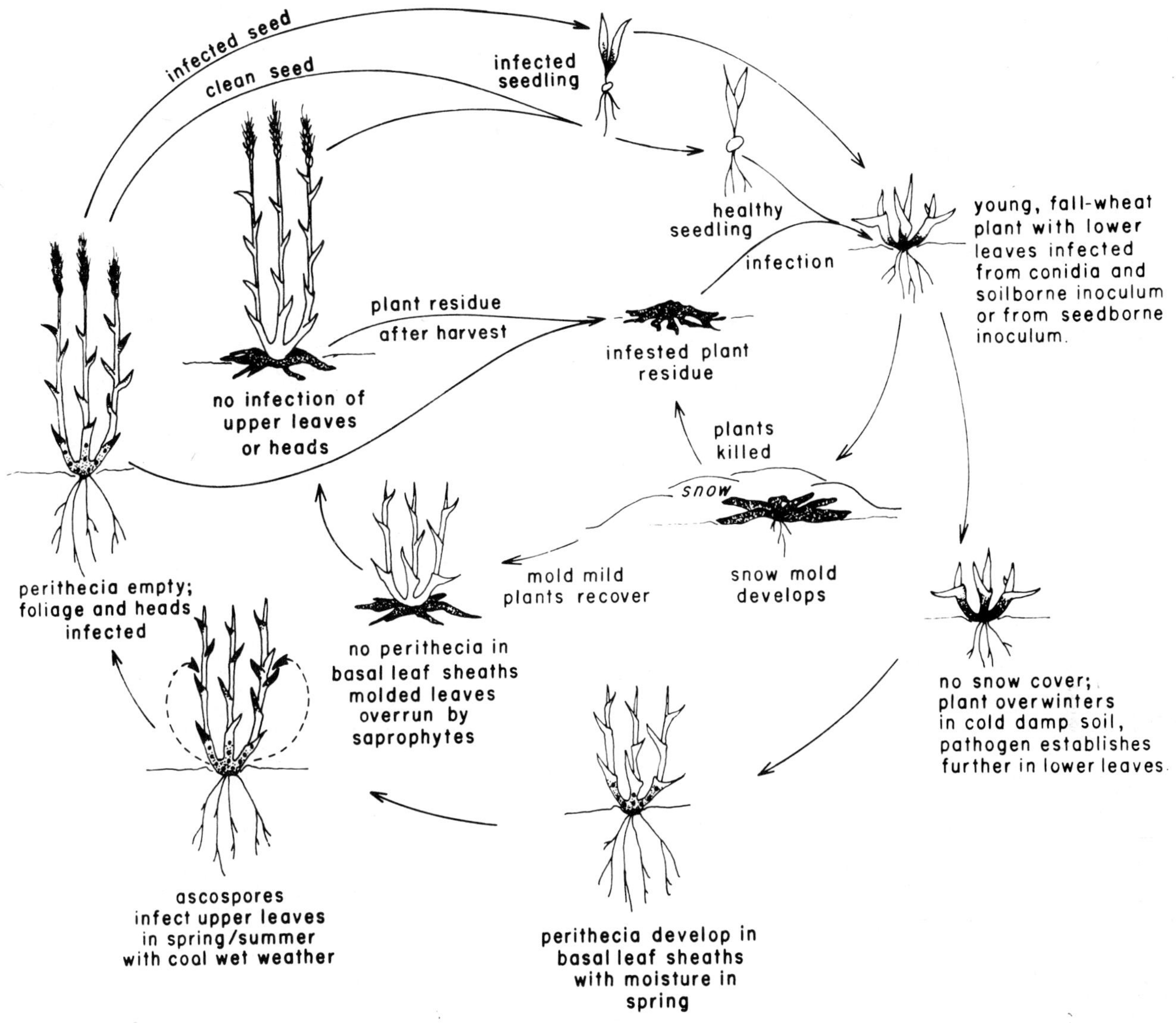

Fig. 42. Disease cycle of *Monographella nivalis* (anamorph *Microdochium nivale* [*Fusarium nivale*]) on wheat. (Reprinted, by permission, from P. E. Nelson, T. A. Toussoun, and R. J. Cook, eds., *Fusarium*: Diseases, Biology and Taxonomy, Fig. 4-2, 1981, The Pennsylvania State University Press, University Park, PA)

wider still on grasses. Infected plants have chlorotic and dry, necrotic leaves that remain intact (Plates 19 and 20), as opposed to crumbling, as with speckled snow mold. The most important diagnostic feature is the visible pink color of mycelium and the presence of sporodochia at snow melt (Plates 19 and 20).

The causal fungus, long known as *Fusarium nivale* (Fr.) Ces., is now considered to belong to the genus *Microdochium* and is designated *M. nivale* (Ces. ex Berl. & Vogl.) Samuels & Hallett (syns. *Gerlachia nivalis* (Ces. ex Berl. & Vogl.) Gams & Müller, *F. nivale* Ces. ex Berl. & Vogl.). Its teleomorph (perfect stage) is *Monographella nivalis* (Schaff.) Müller (syn. *Calonectria nivalis* Schaff.).

In addition to its characteristic salmon pink color, *M. nivale* is less dependent on cool temperatures than other snow mold fungi. Like fusaria, it can attack all plant parts during wet, cool periods in autumn and spring (see Scab). It is less virulent than other snow mold fungi at temperatures near or below freezing.

Perithecia of *Monographella nivalis* develop in late spring and summer in response to cool, humid weather. The perithecia are dark brown, oval to globose, and 100–150 × 120–180 μm and usually appear embedded within lower leaf sheaths. Each bears a hyaline, apical papilla that reaches the host surface, but no stroma is present. *Monographella nivalis* is homothallic and may produce perithecia within two weeks on sterilized moist straw at 18° C in the laboratory.

Asci are hyaline, clavate, thin-walled, and 6–9 × 60–70 μm. They normally contain six to eight ascospores. Mature ascospores are hyaline, ellipsoid, three- or four-celled, and 3.5–4.5 × 10–17 μm.

Wheat is invaded in autumn by hyphae from perithecia and by ascospores or mycelium from host debris (Fig. 42). Primary infections occur on leaf sheaths and blades near soil level. Infections expand via mycelial extension during cool, wet periods and beneath snow. Airborne conidia, and sometimes ascospores, are secondary inoculum and initiate infections elsewhere on plants in spring (see Scab).

Selected References

Booth, C. 1971. *Micronectriella nivalis*. Descriptions of Pathogenic Fungi and Bacteria, No. 309. Commonwealth Mycological Institute, Association of Applied Biologists, Kew, Surrey, England.

Cook, R. J., and Bruehl, G. W. 1968. Ecology and possible significance of perithecia of *Calonectria nivalis* in the Pacific Northwest. Phytopathology 58:702–703.

Galea, V. J., Price, T. V., and Sutton, B. C. 1986. Taxonomy and biology of the lettuce anthracnose fungus. Trans. Br. Mycol. Soc. 86:619–628.

Lebeau, J. B. 1968. Pink snow mold in southern Alberta. Can. Plant Dis. Surv. 48:130–131.

Richardson, M. J., and Zillinsky, F. J. 1972. A leaf blight caused by *Fusarium nivale*. Plant Dis. Rep. 56:803–804.

Speckled Snow Mold (Typhula Blight)

In contrast to leaves killed by pink snow mold, leaves killed by *Typhula* spp. are easily shattered and normally show dense, gray-white mycelium. Speckled snow mold also tends to be more closely associated with snow cover than pink snow mold. Numerous dark sclerotia, 0.5–2 mm in diameter, are a key diagnostic feature (Plate 17). They appear within tissues or scattered among mycelium, giving plants a speckled appearance.

Speckled snow mold is caused by *T. ishikariensis* Imai, *T. idahoensis* Remsb. (syn. *T. borealis* Ekstr.), and *T. incarnata* Lasch ex Fr. (syns. *T. itoana* Imai, *T. graminum* Karst.). The first two species are sometimes regarded as synonymous, but intermating among all three pathogens is rare. All three fungi thrive in the field where snow is deep and persistent. In the laboratory, they can be cultured on host plants or on simple agar media.

Mature sclerotia of *T. ishikariensis* and *T. idahoensis* are erumpent or superficial on aerial plant parts. They are dark brown-black, globose, and 0.5–1.5 mm in diameter. Their basidiocarps are clavate, 3–8 mm tall, and powdery white-gray (*T. ishikariensis*) or tan-brown (*T. idahoensis*). Basidiospores are 4×10 μm and ovoid.

T. incarnata sclerotia are spherical and red-brown and occur on roots and leaves. Within soil they are 1–3 mm in diameter, compared to 0.5–1.5 mm on leaves. Basidiocarps are pink, occasionally branched, and 7–12 mm tall. Basidiospores are ovoid, flattened on one side, and $3.5–7 \times 8–12$ μm.

The host range of *T. ishikariensis* is the widest among snow mold fungi and extends beyond cereals and grasses to legumes, sugar beets, and broadleaf weeds. *T. ishikariensis* is prominent in reclaimed forest soils in the northwest United States.

T. idahoensis primarily inhabits grassland soils. It tends to occur predominantly in areas with high light intensity and in well-hardened wheat. *T. ishikariensis*, on the other hand, is persistent in areas of lower light intensity and in less hardened or unvernalized wheat. Among snow mold fungi, *T. incarnata* and *Microdochium nivale* have the widest geographic distribution. They can attack wheat plants below the soil surface with or without snow cover.

Typhula spp. are poor saprophytes and survive as parasites or sclerotia. Sclerotia germinate in autumn in response to moisture and form upright, club-shaped basidiocarps (Fig. 43). The basidiocarp surface is a hymenium from which basidiospores are forcibly ejected and airborne. Autumn-sown wheat is infected either by basidiospores or by infectious hyphae from soilborne sclerotia. In the United States, basidiospores are most frequent in mid-November and maintain their viability for up to 60 days.

Legume crops tend to decrease the longevity of *Typhula* sclerotia. Their use in crop rotation schemes is encouraged.

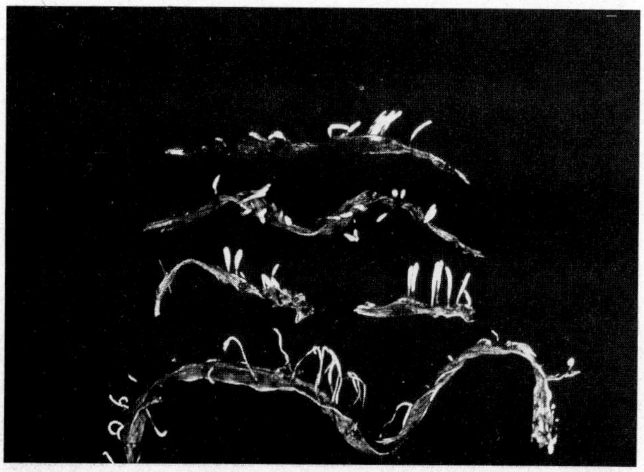

Fig. 43. Erect basidiocarps of a *Typhula* sp. arising from sclerotia within leaves killed by speckled snow mold. (Courtesy R. Sprague)

Selected References

Bruehl, G. W., and Cunfer, B. M. 1975. *Typhula* species pathogenic to wheat in the Pacific Northwest. Phytopathology 65:755–760.

Bruehl, G. W., Machtmes, R., and Kiyomoto, R. 1975. Taxonomic relationships among *Typhula* species as revealed by mating experiments. Phytopathology 65:1108–1114.

Cunfer, B. M., and Bruehl, G. W. 1973. Role of basidiospores as propagules and observations on sporophores of *Typhula idahoensis*. Phytopathology 63:115–120.

Dejardin, R. A., and Ward, E. W. B. 1971. Growth and respiration of psychrophilic species of the genus *Typhula*. Can. J. Bot. 49:339–347.

Huber, D. M., and McKay, H. C. 1968. Effect of temperature, crop, and depth of burial on the survival of *Typhula idahoensis* sclerotia. Phytopathology 58:961–962.

Matsumoto, N., and Sato, T. 1983. Niche separation in the pathogenic species of *Typhula*. Ann. Phytopathol. Soc. Jpn. 49:293–298.

Sunderman, D. W. 1964. Modification of the Cormack and Lebeau technique for inoculating winter wheat with snow mold-causing *Typhula* species. Plant Dis. Rep. 48:394–395.

Sclerotinia Snow Mold (Snow Scald)

Sclerotinia borealis Bub. & Vleug. (syn. *S. graminearum* Elen.) is an ascomycete sometimes placed in the genus *Whetzelinia*. It causes snow mold of wheat and grasses, especially when damp, cold autumns are followed by deep snow cover that persists five months or more on slightly frozen ground. Such conditions exist primarily in regions of Japan, the Soviet Union, Scandinavia, Europe, and Canada.

Sclerotinia snow mold occurs in isolated patches. In the spring, dead leaves and plants are gray and thready. Numerous black, irregular sclerotia, 2–4 mm long, usually are present in and on diseased tissues.

Sclerotia of *S. borealis* germinate in autumn and produce cup-shaped apothecia lined with a fertile layer of asci (Fig. 41). Ascospores are forcibly ejected and windborne. Wheat infection occurs as described for *Typhula* spp. (see Speckled Snow Mold).

S. borealis is best cultured on media with low water potentials (-10 to -30 bars) and grows at temperatures down to $-5°$C.

Selected References

Groves, J. W., and Bowerman, C. A. 1955. *Sclerotinia borealis* in Canada. Can. J. Bot. 33:591–594.

Nissinen, O., and Salonen, A. 1972. The effect of weather conditions on the incidence of *Sclerotinia borealis* and of the species and variety of the grass on the wintering of ley. J. Sci. Agric. Soc. Finl. 44:98–114.

Tomiyama, K. 1955. Studies on the snow blight disease of winter cereals. Hokkaido Natl. Agric. Exp. Stn. Rep. 47. 234 pp.

Ward, E. W. B. 1966. Preliminary studies of physiology of *Sclerotinia borealis*, a highly psychrophilic fungus. Can. J. Bot. 44:237–246.

Snow Rot

Snow rot of winter wheat was first reported in Japan in 1929. In 1978, snow rot was also confirmed in the United States (Washington State). Snow rot, like snow mold, occurs in areas that have extended periods of snow cover. However, snow rot affects plants only in areas where cold water from snow melt runs or accumulates (Plate 21). Yield losses to snow rot are minimal and are confined to plants in drainageways or low areas. Barley is sometimes attacked as well as wheat.

Symptoms

As wheat plants emerge from beneath snow cover, the leaves bear dark green, water-soaked blotches. Leaves submerged in water may be entirely water-soaked, dark green, and flaccid. Leaves in contact with soil are more extensively rotted than upright leaves, and entire plants may be killed if the growing point is invaded. The soft-rotted tissues eventually dry to a light brown color, and spherical oospores and sporangia (Fig. 44) can be observed within these tissues with a microscope. Roots are rarely invaded and only when washed free of soil (see Pythium Root Rot).

Causal Organisms

Three *Pythium* spp.—*P. iwayamai* Ito, *P. okanoganense* Lipps, and *P. aristosporum* Vanterpool—are the principal snow rot pathogens. A fourth species, *P. paddicum*, is an additional snow rot pathogen in Japan.

Pythium spp. can best be identified on leaf blades in water culture. *P. iwayamai* has one or two (rarely three or four) antheridia (male cells) per oogonium (female cell). The antheridial stalks are monoclinous or diclinous (originate from the oogonial stalk or from another hypha). Oospores are thick-walled and aplerotic (only partially fill the oogonium). Sporangia of *P. iwayamai* are variously shaped (spherical, ellipsoidal, citriform, or ovoid) and usually papillate. The number and origin of the antheridial stalks plus the multiple shapes of the sporangia separate this species from other *Pythium* spp.

P. aristosporum has smooth, terminal, subspherical oogonia with aplerotic oospores. This species has one to five (usually two to three) crook-necked, clavate antheridia per oogonium. Antheridial stalks are monoclinous or diclinous, sometimes branching to two or three antheridial cells. Sporangia are lobate or digitate or have toruloid enlargements.

The oogonial walls of *P. okanoganense* have a single protrusion and an antheridial stalk that originates a short distance from the oogonium. The oospores of *P. okanoganense* are conspicuously aplerotic compared to those of *P. iwayamai* and *P. paddicum*.

All four *Pythium* spp. are capable of growth at near-freezing temperatures. Under snow cover, water from snow melt is essentially 0.5° C.

The pathogens are best isolated from diseased plants collected from beneath snow cover. Diseased tissues should be washed, blotted dry, and plated on 2% water agar at 5–10° C. The pathogens cannot be isolated from dried plants collected later in the season, even when oospores are detected in the tissues.

Disease Cycle

The pathogens presumably survive the summer as dormant oospores in soil or plant debris. They normally cannot be recovered from plants until snow melt occurs. Wheat plants must be predisposed for an extended period of time under snow cover before they become susceptible. Zoospores, produced and distributed in water from snow melt, are the primary inoculum. They invade leaf tissues via stomata and encyst in stomatal openings (Fig. 44). New sporangia are produced and secondary zoospores are released within two to three days. Oospores develop as host tissues begin to dry (see Downy Mildew and Pythium Root Rot).

Control

Snow rot controls other than water management to avoid snow melt and runoff are not available. No resistant cultivars are available. Resistance to snow rot is not related to snow mold resistance or to winterhardiness. Late-seeded wheat, which overwinters as small plants, may escape infection. Spring wheats are susceptible but unaffected.

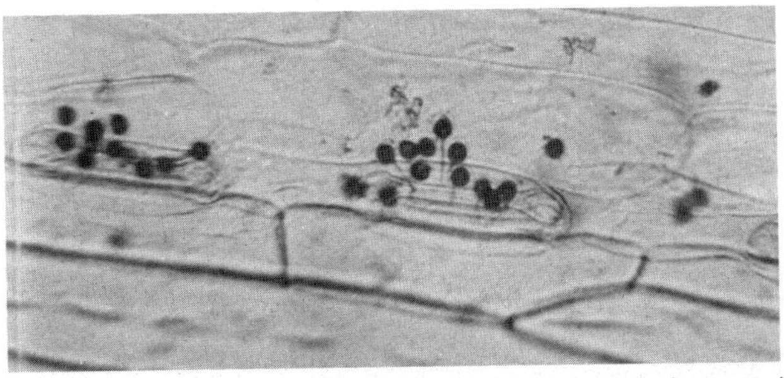

Fig. 44. Zoospores of *Pythium iwayamai* encysted above and penetrating the stomata on a wheat leaf. (Courtesy P. E. Lipps)

Selected References

Hirane, S. 1960. Studies on Pythium snow blight of wheat and barley, with special reference to the taxonomy of the pathogens. Trans. Mycol. Soc. Jpn. 2:82–87.

Lipps, P. E. 1980. The influence of temperature and water potential on asexual reproduction by *Pythium* spp. zoospores in snow rot of wheat. Phytopathology 70:794–797.

Lipps, P. E., and Bruehl, G. W. 1978. Snow rot of winter wheat in Washington. Phytopathology 68:1120–1127.

Lipps, P. E., and Bruehl, G. W. 1980a. Infectivity of *Pythium* spp. zoospores in snow rot of wheat. Phytopathology 70:723–726.

Lipps, P. E., and Bruehl, G. W. 1980b. Reaction of winter wheat to Pythium snow rot. Plant Dis. 64:555–558.

Takamatsu, S., and Ichitani, T. 1986. Occurrence of Pythium snow rot caused by *Pythium okanoganense* in wheat and barley in Japan. Ann. Phytopathol. Soc. Jpn. 52:82–85.

Downy Mildew (Crazy Top)

Downy mildews are caused by fungi called "water molds" because they depend on free moisture for at least part of their life cycle. On wheat, however, the sign of downy fungal growth (downy mildew) is rarely visible. Instead, the pathogen, *Sclerophthora macrospora*, causes a proliferation of florets and a distortion of spikes and upper leaves (crazy top).

Downy mildew was first reported on wheat in Italy in 1900 as "grano incipollito" (sodden grain). The disease today is not economically important, except perhaps in localized field areas subject to flooding.

In addition to wheat, downy mildew occurs on barley, rice, corn, oats, sorghum, and about 140 species of perennial and annual grasses. Among perennial grasses, species of *Agropyron, Bromus, Poa, Phalaris*, and *Stenotaphrum* are likely reservoirs for the pathogen.

Symptoms

Diseased plants typically are scattered in or along standing water, and many are "flagged" with prominently yellowed leaves. Before heading, diseased plants show stunting, yellowing, excessive tillering, and thickened, leathery, or warty leaves. Such plants normally do not form heads or seed, and many die before jointing. After heading, diseased plants exhibit variously twisted heads and leaves (crazy top) (Fig. 45). Many have proliferated flower parts. A virus carried by the fungus may influence symptom expression in host plants.

Numerous spherical, pale yellow oospores within diseased tissues are diagnostic (Fig. 46) (see also Pythium Root Rot), as are substomatal sporangiophoric pads. These fungal structures are best seen in decolorized leaf tissues stained with acid fuchsin. Diagnostic signs of the pathogen are sometimes visible on leaves as patches of milky white, asexual zoosporangia (downy mildew) (Fig. 47).

Causal Organism

S. macrospora (Sacc.) T. S. & N. (syns. *Sclerospora macrospora* Sacc., *Phytophthora macrospora* (Sacc.) Ito et

Fig. 45. Plants with "crazy top" symptoms of downy mildew, caused by *Sclerophthora macrospora*. (Courtesy W. G. Willis)

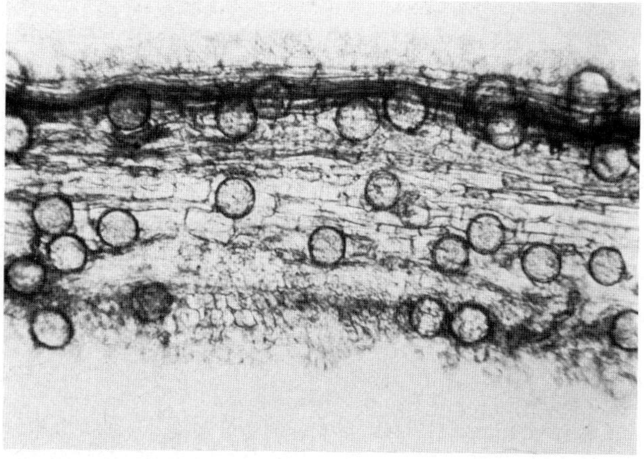

Fig. 46. Oospores of *Sclerophthora macrospora* in tissues affected by downy mildew. (Courtesy L. K. Edmunds)

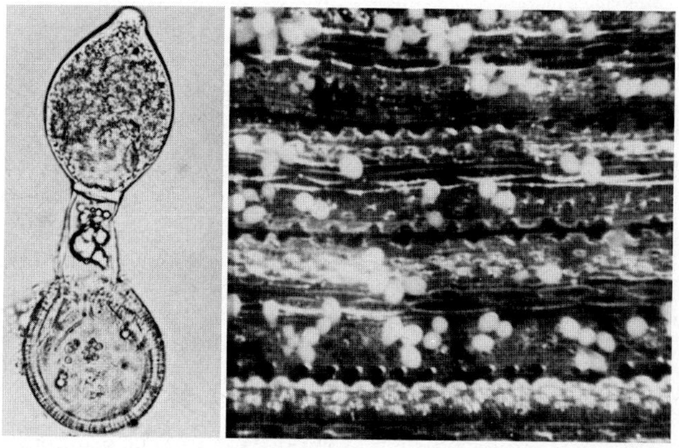

Fig. 47. Reproductive structures of *Sclerophthora macrospora*: sporangium from germinated oospore (left) and asexual sporangia on a leaf segment (right). (Reprinted, by permission, from Semeniuk and Mankin, 1964)

Tanaka) causes wheat downy mildew. It is a biotrophic (obligate) parasite. Although its sporangia resemble those of *Phytophthora* spp., it is currently included in the Pythiaceae.

Oospores in host tissues are smooth-walled, spherical, and 40–70 μm in diameter (Fig. 46). They mature in senescing host tissues, especially leaves and glumes. Oospores germinate in water or wet soil and produce lemon-shaped sporangia, 43–64 × 60–100 μm, on determinate sporangiophores (Fig. 47). Within 1 or 2 hr of formation, the sporangia liberate 30–90 zoospores that measure 9–12 μm in diameter and are motile by two flagella. In some instances zoospore formation is aborted and secondary sporangia are produced. Sporangia develop optimally in water between 10 and 25° C, with limits near 7 and 31° C. At optimal temperatures, zoospores are released and germinate within 2 hr to form slender germ tubes (2–3 μm wide) and coenocytic, infectious hyphae. In host tissues such hyphae vary in width from 2 to 50 μm.

Asexual sporangia, although rare on wheat, appear in small clusters immediately above stomata (Fig. 47). They are ovoid to pear-shaped, measure approximately 47 × 75 μm, and liberate zoospores in the same manner as oosporic sporangia.

Disease Cycle and Control

Plant parts, primarily leaf sheaths, in standing water are subject to infection. Seedlings are more susceptible than adult plants, and infection requires 4-hr exposure to zoospores, which serve as primary inoculum. Zoospores derived sexually or asexually (from fertilized or unfertilized oospores) are disseminated in moving water. Their germ tubes penetrate wheat tissues directly and proliferate intercellular and intracellular hyphae.

Wheat infections are most frequent in flooded areas near ditches and in lowlands where the pathogen may occur on other hosts. Infection apart from water is unlikely, although oospores (but not sporangia or zoospores) can persist for months in well-drained soils.

Control measures other than water management are not warranted. In problem areas, improved drainage is beneficial. Avoiding soils infested with host debris adds an additional margin of control through sanitation.

Selected References

Kenneth, R. G. 1981. Downy mildew of graminaceous crops. Pages 367–394 in: The Downy Mildews. D. M. Spencer, ed. Academic Press, London.

Payak, M. N., Renfro, B. L., and Lal, S. 1970. Downy mildew diseases incited by *Sclerophthora*. Indian Phytopathol. 23:183–193.

Semeniuk, G. 1976. *Sclerophthora macrospora* infection of three annual grasses by oospore and asexual inocula. Plant Dis. Rep. 60:745–748.

Semeniuk, G., and Mankin, C. J. 1964. Occurrence and development of *Sclerophthora macrospora* on cereals and grasses in South Dakota. Phytopathology 54:409–416.

Troutman, J. L., and Matejka, J. C. 1972. Downy mildew of small grains and sorghum in Arizona. Plant Dis. Rep. 56:773–774.

Whitehead, M. D. 1958. Pathology and pathological histology of downy mildew, *Sclerophthora macrospora*, on six graminicolous hosts. Phytopathology 48:485–493.

Rusts

Three different rust diseases—stem rust, leaf rust, and stripe rust—occur on wheat. They are named for the dry, dusty, yellow-red or black spots and stripes (sori or pustules) that erupt through the plant epidermis. The size and surrounding coloration of rust pustules determine the specific infection types, which can vary with different wheat cultivars, temperatures, and rust races (Plates 22–24). Some infections are visible only as chlorotic flecks or brown, necrotic spots. Others may result in sporulating pustules of various sizes. Rust symptoms and signs are most obvious in spring and summer but may occur at any time after seedlings emerge. All aerial parts of wheat plants are susceptible to infection, and all may bear symptoms of more than one rust.

Rusts have been of great importance historically and are mentioned in the earliest records of wheat cultivation. Wheat rusts changed the course of early civilizations by destroying the major food source.

The capacity of wheat rusts to develop into widespread epidemics is well documented. In past decades in North America, rusts were conservatively estimated to decrease wheat yields by over 1 million metric tons annually. Similar statistics could be quoted for most wheat-growing regions of the world. Rusts, by a wide margin, account for the bulk of wheat disease literature.

Rust epidemics that occur before or during flowering are most detrimental. Head infections are especially damaging, even if infections do not occur elsewhere on the plant. Besides reducing seed yield, rusts lower the crop's forage value and winterhardiness and predispose plants to certain other diseases. Like powdery mildew, the rusts modify the host epidermis and increase transpiration and respiration and decrease photosynthesis. Overall, they reduce plant vigor, seed filling, and root growth. Rusted wheat plants are less palatable and are sometimes mildly toxic to livestock.

Causal Organisms

Wheat rusts are caused by three highly specialized fungi. Stem rust is caused by *Puccinia graminis* f. sp. *tritici*, leaf rust by *P. recondita* f. sp. *tritici*, and stripe rust by *P. striiformis*. Each of these species is composed of numerous physiologic races separable by patterns of pathogenicity on differential hosts. The interaction of specific genes within each race and host combination determines the resultant infection type (Plates 22–24) and distinguishes races of rust. Many rust races that were prevalent in the past are insignificant today because of breeding efforts that incorporated specific resistances into modern wheats. However, new rust races continually surface to threaten wheat production because of the pathogens' capacity for mutation and sexual reproduction.

Life Cycles

Most wheat rust fungi are considered to be obligate parasites, although some strains have a limited capacity for vegetative growth on agar media (Fig. 48). A few unique strains can sporulate and complete their life cycle apart from host plants.

Wheat rusts have complex life cycles that involve alternate hosts and as many as five spore stages (Figs. 49 and 50). Urediospores, produced in great numbers in the spring and summer, are important epidemiologically. They are dispersed by wind to other plants, where they generate new infections and secondary urediospores in intervals as short as seven days.

Urediospores are one-celled, spiny-walled, and dikaryotic (Figs. 48, 50, and 51). They are nutrient-independent and germinate in contact with water films. Germ tubes penetrate stomata directly (stripe rust) or via an appressorial peg. The formation of substomatal vesicles and intercellular hyphae with globose or lobed haustoria that establish physiologic contact with host cell membranes (see Powdery Mildew) completes the infection process.

Urediospore production on host plants may be followed by teliospore development within uredia or within separate telial sori. Teliospores are brown-black, binucleate, and two-celled and have thick, smooth walls (Fig. 50). They remain within the sorus and persist over the winter. Their germination in spring involves nuclear fusion, reductive division (meiosis), and production of a promycelium (basidium) with four haploid sporidia (basidiospores). Basidiospores cannot reinfect wheat but are windborne to alternate hosts in the cases of leaf and stem rust fungi (Fig. 49); basidiospores of the stripe rust fungus are presumed functionless.

Infections on alternate hosts produce yellow-orange sori (pycnia) on upper leaf surfaces. Within the pycnia, uninucleate, hyaline spores (pycniospores) and receptive hyphae fuse in compatible pairs when brought into contact by rain or insects. Their fusion restores the dikaryon, which proliferates and erupts through the opposite leaf surface as an aecial cup (Figs. 49 and 53). Dikaryotic, dry, yellow spores (aeciospores) are

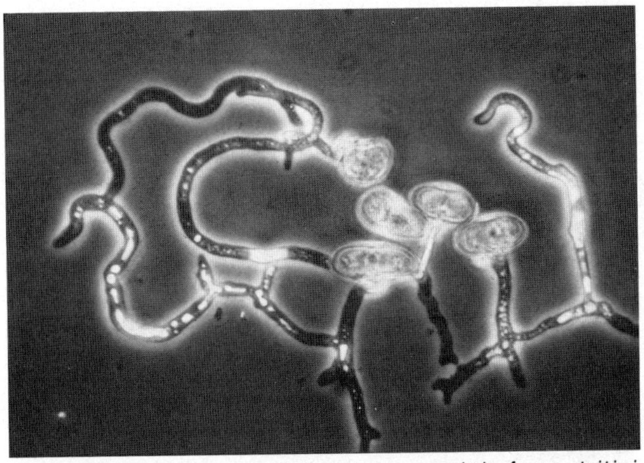

Fig. 48. Urediospores of *Puccinia graminis* f. sp. *tritici* germinating in vitro. (Courtesy G. W. Buchenau and G. A. Myers)

liberated from aecia and windborne to wheat. They penetrate wheat through stomata, and infection results in urediospore production.

Epidemiology

Rust epidemics develop when compatible wheat plants and rust fungi occur together over large areas. When this circumstance is coupled with free moisture and temperatures between 15 and 25°C, infection is completed in 6–8 hr and secondary urediospores are produced in seven to 10 days. Because urediospores are numerous and are readily disseminated vast distances by wind, an infrequent race can become prominent within a few weeks (urediospores increase more than 10,000-fold per generation). This combination of large numbers of viable spores and efficient wind dissemination makes the rusts remarkably successful parasites.

Uredia have limited survival ability compared to telia but can survive year-round on hosts in milder climates. Urediospores of the stripe rust fungus maintain their infectivity at cooler temperatures than those of leaf rust and stem rust fungi. Urediospores often serve as sources of primary inoculum in spring and summer by virtue of long-distance wind dispersal. The annual progression of urediospores across continents is well documented. Thus, teliospores and alternate hosts are required for the completion of pathogen life cycles but not for disease initiation.

Alternate hosts support the sexual stages of rust fungi and thereby are sources of new virulent forms and of early spring inoculum (aeciospores). It also may be possible for new races of rust fungi to develop apart from alternate hosts through mutation and parasexual mechanisms in the uredial stage. With mixed infections, nuclei or genetic characters may be exchanged or transferred between strains of rust fungi by incidental hyphal fusion.

Control

Rusts are best controlled by resistant wheat cultivars. The heritability of rust resistance has been known for 80 years and has been widely used by breeders to develop resistant wheats. Resistance can be narrowly or broadly based. Resistance based on a single host gene may soon be rendered ineffective by shifts in pathogen virulence (see Powdery Mildew and Melanism [Brown Necrosis]). Multigenic resistance (controlled by combinations of specific genes) may be stable for years. One successful form of resistance that appears to be multigenic is described as "slow rusting."

Destroying alternate hosts interrupts the life cycle of rust fungi, limits their diversity, indirectly increases the stability of resistant cultivars, and prevents the production of early spring inoculum (aeciospores). In the case of stem rust, an extensive federal program in the United States to eradicate susceptible barberries has partially achieved these goals.

Low-cost protective or eradicative fungicides are sometimes used for rust control. They are applied as foliar sprays where cost-benefit analyses show that they are profitable. Systemic foliar fungicides are used in some regions. Sulfur has been an

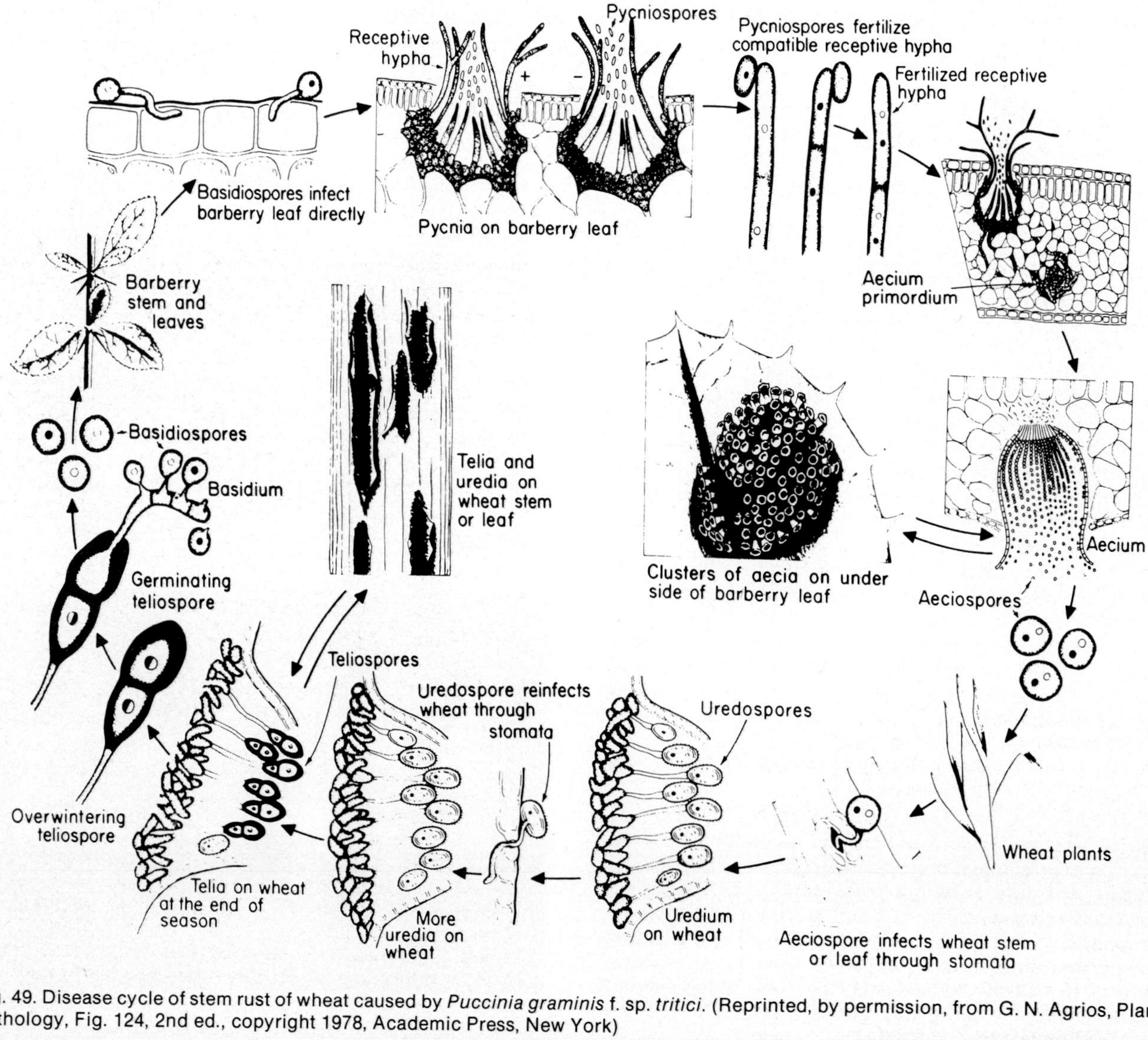

Fig. 49. Disease cycle of stem rust of wheat caused by *Puccinia graminis* f. sp. *tritici*. (Reprinted, by permission, from G. N. Agrios, Plant Pathology, Fig. 124, 2nd ed., copyright 1978, Academic Press, New York)

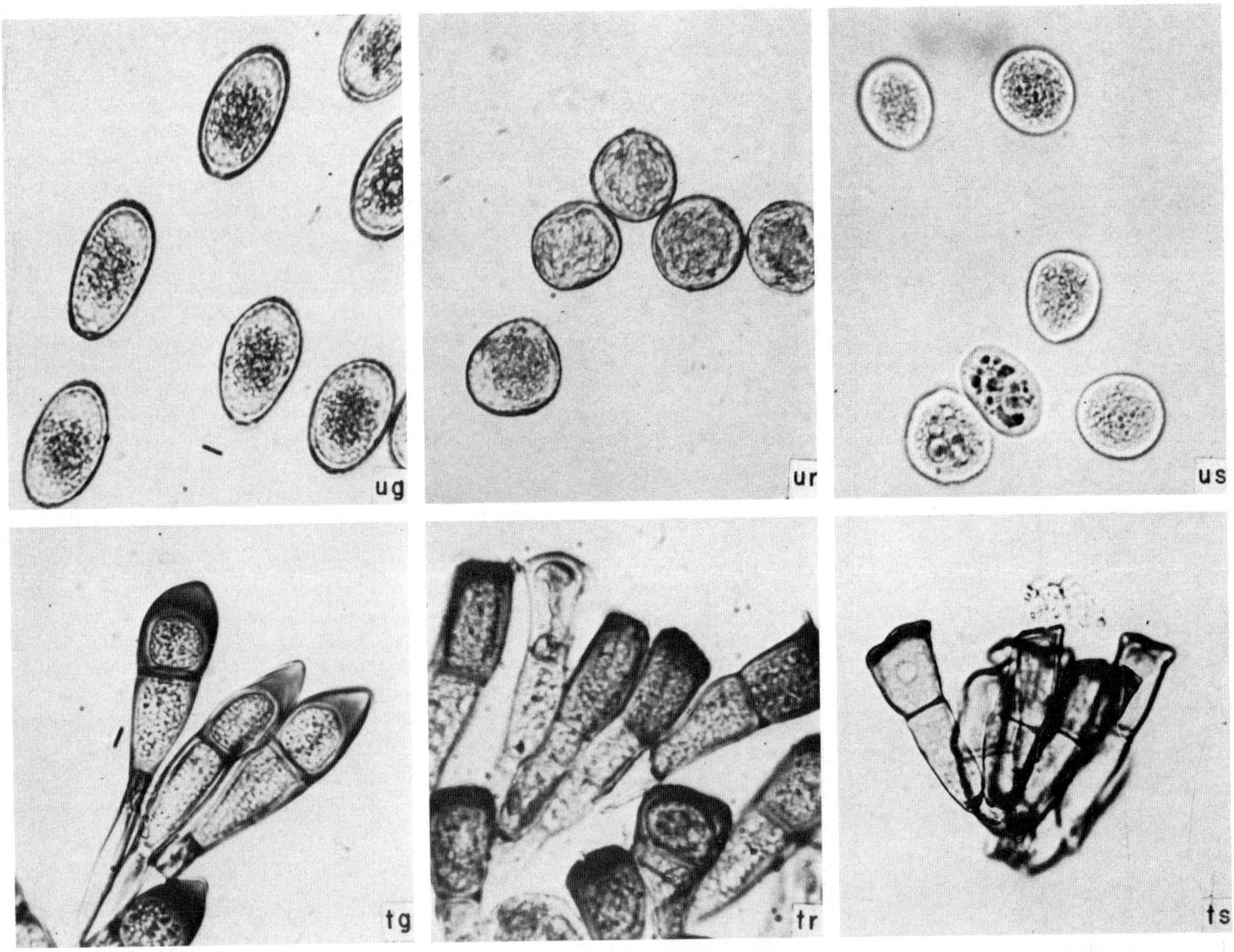

Fig. 50. Urediospores (u) and teliospores (t) of *Puccinia graminis* f. sp. *tritici* (g), *P. recondita* f. sp. *tritici* (r), and *P. striiformis* (s). (Courtesy W. Q. Loegering, USDA)

effective, nonspecific protective chemical, but its commercial use at present is uneconomical. Some systemic seed treatments show promise for controlling rust on seedlings.

Growing different cultivars within a production area restricts rust damage because the genetic diversity in the crop requires corresponding diversity in the rust population for epidemics to occur. This heterogeneity can be accomplished on a field basis by sowing different cultivars and on a plant basis by seeding mixtures of seed differing in genes for resistance (multilines).

Where possible, early-maturing cultivars should be grown. Spring wheats should be sown as early as possible and given adequate phosphorus to ensure early maturity and perhaps escape peak rust periods. Autumn rust infections can be reduced by late-autumn seeding. Early-autumn seeding should be avoided.

Fig. 51. Uredial sorus of *Puccinia recondita* f. sp. *tritici* erupting through the epidermis of a leaf. (Courtesy M. F. Brown and H. G. Brotzman)

Selected References

Borlaug, N. E. 1965. Wheat, rust, and people. Phytopathology 55:1088–1098.

Browder, L. E. 1985. Parasite:host:environment specificity in the cereal rusts. Annu. Rev. Phytopathol. 23:201–222.

Buchenau, G. W. 1957. Relationship between yield loss and area under the wheat stem rust and leaf rust progress curves. Phytopathology 65:1317–1318.

Burleigh, J. R., Schulze, A. A., and Eversmeyer, M. G. 1969. Some aspects of the summer and winter ecology of wheat rust fungi. Plant Dis. Rep. 53:648–651.

Bushnell, W. R., and Roelfs, A. P., eds. 1984. The Cereal Rusts. Vol. I. Academic Press, Orlando, FL. 546 pp.

Bushnell, W. R., and Stewart, D. M. 1971. Development of American isolates of *Puccinia graminis* f. sp. *tritici* on an artificial medium. Phytopathology 61:376–379.

Caldwell, R. M. 1968. Breeding for general and/or specific plant disease resistance. Pages 263–272 in: Proc. Int. Wheat Genet. Symp. 3rd, Canberra, Australia. K. W. Finlay and K. W. Shepherd, eds. Butterworth and Co., Sydney.

Calpouzos, L., Roelfs, A. P., Madison, M. E., Martin, F. B., Walsh, J. R., and Wilcoxson, R. D. 1976. A new model to measure yield losses caused by stem rust in spring wheat. Univ. Minn. Agric. Exp. Stn. Tech. Bull. 307. 23 pp.

Coakley, S. M., Line, R. F., and Boyd, W. S. 1983. Regional models for predicting stripe rust on winter wheat in the Pacific Northwest. Phytopathology 73:1382–1385.

Eversmeyer, M. G., King, C. L., and Willis, W. G. 1975. Control of leaf and stem rust of wheat by aerial application of a zinc and maneb fungicide in Kansas in 1974. Plant Dis. Rep. 59:607–609.

Johnston, T., Green, G. J., and Samborski, D. J. 1967. The world situation of the cereal rusts. Annu. Rev. Phytopathol. 5:183–200.

Loegering, W. Q., Hendrix, J. W., and Browder, L. E. 1967. The rust diseases of wheat. U.S. Dep. Agric. Agric. Handb. 334.

MacKenzie, D. R. 1976. Application of two epidemiological models for the identification of slow stem rusting in wheat. Phytopathology 66:55–59.

Nagarajan, S., Singh, H., Joshi, L. M., and Saari, E. E. 1976. Meteorological conditions associated with long-distance dissemination and deposition of *Puccinia graminis tritici* uredospores in India. Phytopathology 66:198–203.

Rakotondradona, R., and Line, R. F. 1984. Control of stripe rust and leaf rust of wheat with seed treatments and effects of treatments on the host. Plant Dis. 68:112–117.

Roelfs, A. P., and Bushnell, W. R. 1985. The Cereal Rusts. Vol. II. Academic Press, Orlando, FL. 606 pp.

Stem Rust

Stem rust was recognized in Roman times as the "greatest of plant diseases"; however, detailed characterization of its disease cycle did not begin until 1767. Stem rust of wheat (also called black rust and black stem rust) is caused by *Puccinia graminis* Pers. f. sp. *tritici* Eriks. & Henn., which also parasitizes certain barley and rye cultivars and some grasses, especially wild barley (*Hordeum jubatum* L.) and goatgrass (*Aegilops* spp.). Apart from wheat and its alternate hosts, the stem rust fungus is a relatively weak pathogen, but some grasses are important sources of primary inoculum. In addition to wheat, the fungus completes part of its life cycle on alternate hosts, especially barberries (*Berberis vulgaris* L. and *B. canadensis* Mill.) and certain species of *Mahonia*.

B. vulgaris is an erect woody shrub that reaches 3 m in height. It bears prominent spines and clusters of two to six leaves at stem joints (Fig. 52). Its bark is gray and its flowers are yellow and inconspicuous in long, drooping clusters. Red berries, formed in late summer, overwinter on the plant. Barberries are cultivated as ornamentals or grow in wild clusters on uncultivated land. The Japanese barberry, *B. thunbergii* DC., appears immune to stem rust.

Uredia of *P. graminis* f. sp. *tritici* occur on wheat stems, leaves, and leaf sheaths and occasionally on glumes, awns, and even seed. Uredial pustules are conspicuously erumpent, with tattered epidermal tissues at their margins (Plate 22). They may erupt through both leaf surfaces and tend to be larger on the underside. The pustules are oval, elongate, or spindle-shaped and up to 3×10 mm in size. Numerous infections may weaken stems and cause plants to lodge. Urediospores are $15–24 \times 21–40$ μm, orange-red, dehiscent, and oval, oblong, or ellipsoid (Figs. 48 and 50). Four median germ pores indent their thick, spiny walls.

Host maturity and the aging of uredia initiate the formation of black-brown teliospores in uredial sori or in separate erumpent telial sori. Teliospores are ellipsoid to clavate, $15–20 \times 40–60$ μm, and two-celled. They are tapered at their apex and have smooth, thick walls and a slight constriction at their septum (Fig. 50). A germ pore is terminal in the upper and lateral in the lower cell. Germination normally follows several weeks of cold dormancy and yields a hyaline basidium (promycelium) on which four hyaline sporidia (basidiospores) develop on sterigmata.

Pycnia on barberry are small, flask-shaped, and sunken except for the ostiole (Fig. 53). Supporting leaf tissues are typically discolored yellow-red. Pycnia exude slender, hyaline spores (pycniospores) and receptive hyphae in small, sticky droplets that attract insects (see Ergot). Aecia on the underside of barberry leaves are yellow and hornlike, projecting up to 5 mm from the leaf surface (Figs. 52 and 53). Aeciospores in long, dry chains are subglobose, $15–19 \times 16–23$ μm, smooth, and light orange-yellow.

Stem rust develops optimally near 26° C and is seriously hampered below 15° C and above 40° C. Delayed crop maturity especially favors the disease.

Selected References

Browder, L. E. 1966. A rapid method of assaying pathogenic potential of populations of *Puccinia graminis tritici*. Plant Dis. Rep. 50:673–676.

Dunkle, L. D., and Allen, P. J. 1971. Infection structure differentiation by wheat stem rust uredospores in suspension. Phytopathology 61:649–652.

Fig. 52. Aecial stage of *Puccinia graminis* f. sp. *tritici* (cause of stem rust) on leaves of barberry. (Courtesy D. L. Long, USDA)

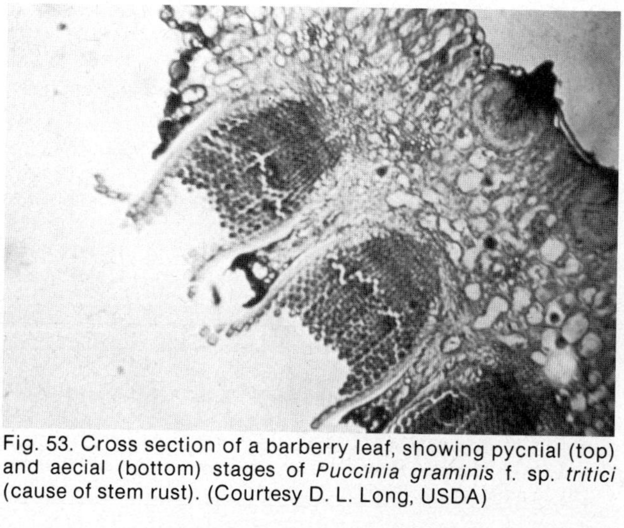

Fig. 53. Cross section of a barberry leaf, showing pycnial (top) and aecial (bottom) stages of *Puccinia graminis* f. sp. *tritici* (cause of stem rust). (Courtesy D. L. Long, USDA)

Green, G. J. 1971. Physiologic races of wheat stem rust in Canada from 1919 to 1969. Can. J. Bot. 49:1575–1588.
Roane, C. W., Stakman, E. C., Loegering, W. Q., Stewart, D. M., and Watson, W. M. 1960. Survival of physiologic races of *Puccinia graminis* var. *tritici* on wheat near barberry bushes. Phytopathology 50:40–44.
Roelfs, A. P. 1985. Wheat and rye stem rust. Pages 3–37 in: The Cereal Rusts. Vol. II. A. P. Roelfs and W. R. Bushnell, eds. Academic Press, Orlando, FL.
Roelfs, A. P., and McVey, D. V. 1979. Low infection types produced by *Puccinia graminis* f. sp. *tritici* and wheat lines with designated genes for resistance. Phytopathology 69:722–730.

Leaf Rust

Leaf rust (also called brown rust, dwarf rust, and orange rust) may be the most widely distributed wheat disease. Practically coexistent with the crop, it is most prevalent where wheat matures late, as in spring wheat regions. The causal fungus, *Puccinia recondita* Rob. ex Desm. f. sp. *tritici* (syns. *P. rubigo-vera* (DC.) Wint., *P. triticina* Eriks.), is a weak parasite on certain cultivars of barley and some *Aegilops* and *Agropyron* spp. These infections are relatively unimportant, as are infections on its alternate hosts. Pycnia and aecia are rare and occur primarily in Europe on meadow rue, *Thalictrum* spp. (Plate 25). Aecia also have been reported on species of *Anchusa, Anemonella, Clematis*, and *Isopyrum*.

Uredia of *P. recondita* f. sp. *tritici* are up to 1.5 mm in diameter and are scattered or clustered primarily on the upper surface of leaf blades (Fig. 51 and Plate 23). They are round to ovoid, orange-red, and erumpent, but without the conspicuously torn epidermal tissues at their margins seen with stem rust (Plates 22 and 23).

Urediospores are 15–30 μm in diameter, subgloboid, and red-brown, with three to eight germ pores scattered in their thick, echinulate walls (Fig. 51). Their temperature optimum for infection lies between those of *P. graminis* and *P. striiformis* urediospores.

Telial sori develop beneath the epidermis, principally on leaf sheaths and blades. They are the size of uredia, glossy black, and not erumpent. Teliospores are round or flattened at the apex like those of *P. striiformis* (Fig. 50). Teliospores may not be produced in some environments or where plants become infected near maturity. Like teliospores of the stem rust fungus, they are stimulated to germinate by cold treatment.

Leaf rust develops rapidly between 15 and 22°C when moisture is not limiting. *P. recondita* f. sp. *tritici* sometimes is synergistically damaging in combination with *Septoria* spp. (see Septoria Leaf and Glume Blotches).

Selected References

Burleigh, J. R., Roelfs, A. P., and Eversmeyer, M. G. 1972. Estimating damage to wheat caused by *Puccinia recondita tritici*. Phytopathology 62:944–946.
Jackson, A. O., and Young, H. C., Jr. 1967. Teliospore formation by *Puccinia recondita* f. sp. *tritici* on seedling wheat plants. Phytopathology 57:793–794.
Lathra, J. K., and Rao, M. V. 1979. Multiline cultivars—How their resistance influences leaf rust diseases. Euphytica 28:137–144.
Lee, T. S., and Shaner, G. 1984. Infection processes of *Puccinia recondita* in slow- and fast-rusting wheat cultivars. Phytopathology 74:1419–1423.
Mehta, Y. R., and Zadoks, J. C. 1970. Uredospore production and sporulation period of *Puccinia recondita* f. sp. *triticina* on primary leaves of wheat. Neth. J. Plant Pathol. 76:267–276.
Saari, E. E., Young, H. C., Jr., and Kernkamp, M. F. 1968. Infection of North American *Thalictrum* spp. with *Puccinia recondita* f. sp. *tritici*. Phytopathology 58:939–943.
Samborski, D. J. 1985. Wheat leaf rust. Pages 39–55 in: The Cereal Rusts. Vol. II. A. P. Roelfs and W. P. Bushnell, eds. Academic Press, Orlando, FL.

Stripe Rust

Stripe rust (also called yellow rust and glume rust) is confined to higher elevations and cooler climates and does not persist in many regions where leaf and stem rust occur. In North America, for example, it appears annually in western mountains and valleys from Canada to Mexico but is infrequent in the central plains. The pathogen is best sustained where night temperatures are cool.

Puccinia striiformis West. (syn. *P. glumarum* Eriks. & Henn.), the cause of stripe rust, is not known to have alternate hosts or sexual stages. It infects rye and over 18 genera of grasses in addition to wheat. Barley is damaged economically, and many perennial grasses are important reservoirs for the fungus.

Symptoms of stripe rust vary but usually appear earlier in spring than symptoms of leaf or stem rust. Uredia are yellow, appear principally on leaves and heads, and are often arranged into conspicuous stripes (Plate 24). Individual pustules measure 0.3–0.5 × 0.5–1 μm, but their linear orientation between vascular bundles and the development of runner hyphae can result in stripes as long as the leaf blade. Urediospores are 20–30 μm in diameter, yellow-orange, and spherical (Fig. 50). They have thick, echinulate walls and six to 12 scattered germ pores. In wheat heads, uredia normally occur on the ventral surface of the glumes, and seed sometimes are infected.

Telial pustules, prevalent on leaf sheaths, are persistently subepidermal. They are dark brown and often form long, dark streaks. Teliospores resemble those of *P. recondita* f. sp. *tritici* (Fig. 50) but are germinable without cold treatment.

Stripe rust of wheat originates from mycelium that overwinters in leaf tissues and especially from urediospores that survive locally or are windborne from distant hosts. Wheat at lower elevations sometimes is infected by urediospores from grasses at higher altitudes. Infections may occur throughout autumn and winter because mycelium remains viable to −5° C. Urediospores lose viability rapidly at temperatures above 15° C. They germinate optimally between 3 and 15° C, with limits near 0 and 21° C. Disease development is most rapid between 10 and 15° C with intermittent rain or dew. Infection types normally are reduced and may resemble resistant reactions if night temperatures exceed 18° C.

Selected References

Coakley, S. M., and Line, R. F. 1981. Quantitative relationships between climatic variables and stripe rust epidemics on winter wheat. Phytopathology 71:461–467.
Coakley, S. M., Line, R. F., and Boyd, W. S. 1983. Regional models for predicting stripe rust on winter wheat in the Pacific Northwest. Phytopathology 73:1382–1385.
Hendrix, J. W., and Fuchs, E. 1970. Influence of fall stripe rust infection on tillering and yield of wheat. Plant Dis. Rep. 54:347–349.
Mares, D. J. 1979. Light and electron microscope study of the interaction of yellow rust with a susceptible wheat cultivar. Ann. Bot. 43:183–189.
Milus, E. A., and Line, R. F. 1986. Number of genes controlling high-temperature, adult-plant resistance to stripe rust in wheat. Phytopathology 76:93–96.
Mulder, J. L., and Booth, C. 1971. *Puccinia striiformis*. Descriptions of Pathogenic Fungi and Bacteria, No. 291. Commonwealth Mycological Institute, Association of Applied Biologists, Kew, Surrey, England.
Rapilly, F. 1979. Yellow rust epidemiology. Annu. Rev. Phytopathol. 17:59–73.
Shaner, G., and Powelson, R. L. 1973. The oversummering and dispersal of inoculum of *Puccinia striiformis* in Oregon. Phytopathology 63:13–17.
Sharp, E. L. 1965. Prepenetration and postpenetration environment and development of *Puccinia striiformis* on wheat. Phytopathology 55:198–203.
Sharp, E. L., and Hehn, E. R. 1963. Overwintering of stripe rust in winter wheat in Montana. Phytopathology 53:1239–1240.
Stubbs, R. W. 1985. Stripe rust. Pages 61–91 in: The Cereal Rusts.

Vol. II. A. P. Roelfs and W. R. Bushnell, eds. Academic Press, Orlando, FL.

Tu, J. C., and Hendrix, J. W. 1967. The summer biology of *Puccinia striiformis* in southeastern Washington. II. Natural infection during the summer. Plant Dis. Rep. 54:384–386.

Volin, R. B., and Sharp, E. L. 1973. Physiologic specialization and pathogen aggressiveness in stripe rust. Phytopathology 63:699–703.

Tan Spot (Yellow Leaf Spot)

Tan spot (also called yellow leaf spot) occurs the world over on many species of Gramineae. It can be serious by itself, and it also frequently contributes to leaf-spotting complexes. Wheat, bromegrass (*Bromus* spp.), wheatgrass (*Agropyron* spp.), and rye are susceptible. Barley is resistant, and oats appear immune.

Symptoms

Tan spot develops on both upper and lower leaf surfaces during spring and summer. The spots appear initially as tan-brown flecks, which expand into lens-shaped, tan blotches up to 12 mm long, often with yellow borders (Plate 26 and Fig. 54). Large lesions coalesce and become darker brown at their centers because of outgrowths of conidia and conidiophores of the causal fungus. Spot blotch (*Bipolaris sorokiniana*) lesions are similar but are less associated with diffuse yellow borders and tend to blacken at the center.

Erumpent pseudothecia appear as dark, raised specks on wheat straw (Fig. 54).

Causal Organism

Pyrenophora tritici-repentis (Died.) Drechs. (syn. *P. trichostoma* (Fr.) Fckl.), anamorph *Drechslera tritici-repentis* (Died.) Shoem. (syn. *Helminthosporium tritici-repentis* Died.), causes tan spot. It grows on potato-dextrose agar and other conventional media, where it produces gray-white mycelium and columns of black hyphae. The latter are common on the side walls of culture plates. After one or more months at 16° C, the fungus produces pseudothecia and ascospores on autoclaved wheat and barley seed incubated on water agar. Large leaf spots placed in moist chambers at 16° C produce conidiophores after 12 hr in light and conidia after an additional 12 hr in the dark.

Erect, simple, olive-black conidiophores in leaf spots and on straw measure 7–8 × 100–300 μm and have a swollen base. Conidia are subhyaline, cylindrical, mostly four- to seven-septate, and 12–21 × 45–200 μm. Their conically tapered basal cell (Fig. 55) is diagnostic.

Pseudothecia on wheat or grass stubble are black, 200–350 μm in diameter, and sometimes beaked (Fig. 54). Ascospores are oval to globose, brown, and 18–28 × 45–70 μm. They are three-septate transversely, with slight septal constrictions. Their median cells may have a longitudinal septation. Normally each clavate ascus bears eight spores (Fig. 55).

Fig. 54. Tan spot lesions on a leaf (left) and dark pseudothecia of *Pyrenophora tritici-repentis* on straw (right). (Courtesy R. L. Forster)

Disease Cycle

P. tritici-repentis persists saprophytically on host debris. Pseudothecia mature on wheat straw during fall and winter and release ascospores as primary inoculum throughout the growing season. Ascospores are dispersed by wind, as are conidia, which appear in spring through late summer. Wheat infections are most numerous in proximity to host residues. Infections require a 6- to 48-hr wet period and occur throughout the growing season. Yield losses result especially when infections develop during heading and damage flag leaves.

Conidia formed in maturing lesions serve as secondary inoculum. Their production and release are promoted by rains or dew, as is lesion enlargement. In prolonged wet periods, even relatively resistant wheats develop leaf spots. Seedborne inoculum appears insignificant.

Control

Fungicides and cultural practices described for control of Septoria leaf and glume blotches are also effective against tan spot. The destruction of wheat stubble and crop residues limits

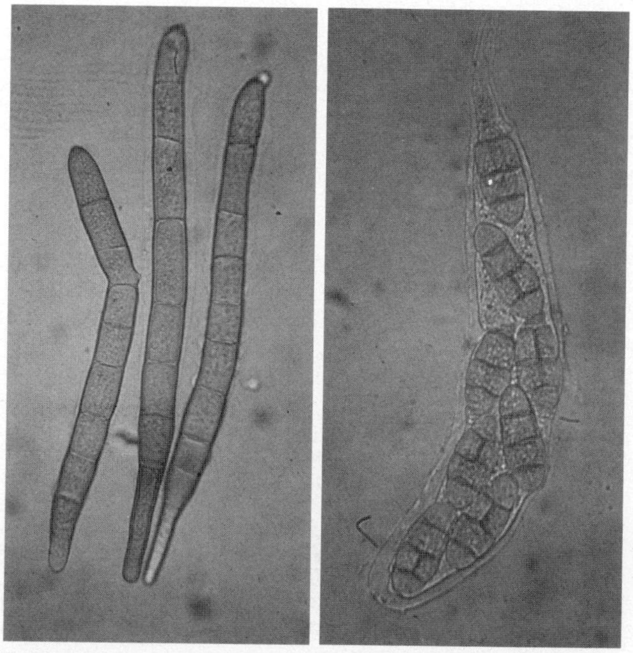

Fig. 55. Conidia (left) and ascus and ascospores (right) of *Pyrenophora tritici-repentis* (anamorph *Drechslera tritici-repentis*). (Courtesy G. N. Odvody)

multiplication and primary inoculum of *P. tritici-repentis*. Crop rotation with nonhosts also is advised.

Wheat cultivars tolerant or resistant to localized forms of *P. tritici-repentis* include Duri, Eklund, and Wells in North America, Red Chief in Australia, BH1146 in Brazil, and Gabo and Norteno 67 in India.

Selected References

Hosford, R. M., Jr. 1971. A form of *Pyrenophora trichostoma* pathogenic to wheat and other grasses. Phytopathology 61:28–32.

Hosford, R. M., Jr. 1972. Propagules of *Pyrenophora trichostoma*. Phytopathology 62:627–629.

Hosford, R. M., Jr., ed. 1982. Tan Spot of Wheat and Related Diseases. N.D. Agric. Exp. Stn. Publ. 116 pp.

Hosford, R. M., Jr., and Busch, R. H. 1974. Losses in wheat caused by *Pyrenophora trichostoma* and *Leptosphaeria avenaria* f. sp. *triticea*. Phytopathology 64:184–187.

Krupinsky, J. M. 1982. Observations on the host range of isolates of *Pyrenophora trichostoma*. Can. J. Plant Pathol. 4:42–46.

Lee, T. S., and Gough, F. J. 1984. Inheritance of Septoria leaf blotch (*S. tritici*) and Pyrenophora tan spot (*P. tritici-repentis*) resistance in *Triticum aestivum* cv. Carifen 12. Plant Dis. 68:848–851.

Luz, W. C. da, and Hosford, R. M., Jr. 1980. Twelve *Pyrenophora trichostoma* races for virulence to wheat in the Central Plains of North America. Phytopathology 70:1193–1196.

Mehta, Y. R. 1975. Mancha foliar do trigo causada por *Pyrenophora trichostoma*. (A leaf spot disease of wheat caused by *Pyrenophora trichostoma*.) Summa Phytopathol. 1:283–288.

Raymond, P. J., Bockus, W. W., and Norman, B. L. 1985. Tan spot of winter wheat: Procedures to determine host response. Phytopathology 75:686–690.

Rees, R. G., and Platz, G. J. 1983. Effects of yellow spot on wheat: Comparison of epidemics at different stages of crop development. Aust. J. Agric. Res. 34:39–46.

Sharp, E. L., Sally, B. K., and McNeal, F. H. 1976. Effect of Pyrenophora wheat leaf blight on the thousand kernel weight of 30 spring wheat cultivars. Plant Dis. Rep. 60:135–138.

Halo Spot

Halo spot affects many wild and cultivated grasses. In the United States it was widespread on Gaines wheat in the Pacific Northwest in the early 1960s. The disease also occurs on oats and rye, while barley is relatively unaffected except in certain Scandinavian countries. In addition to the United States and the United Kingdom, halo spot occurs in western Europe, Australia, and New Zealand.

Symptoms

Halo spots are elliptical and usually less than 10 mm long. They are sometimes so numerous that they cover the entire leaf blade. The spots, which may also occur on leaf sheaths and culms, derive their name from the prominent yellow halo that surrounds the dark border of the lesions. Lesions are tan to brownish gray with obscure pycnidia (Plate 27). Halo spots may fade with age, leaving pycnidia in unspotted host tissues.

Causal Organism

Halo spot is caused by *Selenophoma donacis* (Pass.) Sprague & Johns. (syn. *Septoria donacis* Pass.). Pycnidia in host tissues are erumpent, ostiolate, globose, 40–150 μm in diameter, and dark brown. Pycnidiospores are 2–4.5 × 18–35 μm, hyaline, and crescent-shaped, with pointed ends. Current-season spores are nonseptate, whereas overwintered spores may be uniseptate. The fungus grows and sporulates on nutrient media but may not always form pycnidia in culture.

Disease Cycle and Control

Pycnidiospores are exuded and dispersed in rainwater, in a manner similar to *Septoria* spp., and the disease cycle of halo spot is similar to that of speckled leaf spot (see Septoria Leaf and Glume Blotches).

Cultivars specifically resistant to halo spot have not been developed because the disease is relatively insignificant. However, the cultivar Gaines is susceptible in the United States. Older cultivars such as White Coin, Comanche, Elmar, and Kharkoff appear resistant. Fungicide treatments are not prescribed.

Selected References

Holton, C. S. 1965. Local epidemic outbreaks of fungus leaf spots on 'Gaines' wheat in 1964. Plant Dis. Rep. 49:242–243.

Holton, C. S., and Purdy, L. H. 1962. A severe outbreak of Selenophoma spot on 'Gaines' wheat. Plant Dis. Rep. 46:728.

Punithalingam, E., and Waller, J. 1973. *Selenophoma donacis*. Descriptions of Pathogenic Fungi and Bacteria, No. 400. Commonwealth Mycological Institute, Association of Applied Biologists, Kew, Surrey, England.

Sampson, G., and Clough, K. S. 1979. A Selenophoma leaf spot on cereals in the Maritimes. Can. Plant. Dis. Surv. 59:3.

Sprague, R. 1950. Diseases of Cereals and Grasses in North America. The Ronald Press Co., New York. 538 pp.

Septoria Leaf and Glume Blotches

Although many *Septoria* spp. parasitize the Gramineae, three ascomycetes with *Septoria* anamorphs are particularly virulent on wheat: *Leptosphaeria avenaria* f. sp. *triticea* (anamorph *S. avenae* f. sp. *triticea*), *L. nodorum* (anamorph *S. nodorum*), and *Mycosphaerella graminicola* (anamorph *S. tritici*). Diseases caused by *Septoria* spp. are economically significant in nearly all wheat-growing areas. In the last decade, the diseases appear to have become even more prevalent where wheat crops have dense foliage and receive heavy fertilization and where semidwarf and rust-resistant wheats are grown.

In most literature, *S. nodorum* is cited as the cause of glume blotch and *S. tritici* and *S. avenae* f. sp. *triticea* as causes of (speckled) leaf blotch. The diseases collectively remain best known as Septoria leaf spot, Septoria blotch, or the Septoria complex. The names Septoria tritici blotch, Septoria nodorum blotch, and Septoria avenae blotch are also recognized. Diseases caused by *Septoria* spp. may occur in different combinations within fields and on individual plants. The diseases are difficult to differentiate without careful microscopic examination of host tissues for pycnidiospore morphology.

The Septoria complex at present destroys nearly 2% of the world's wheat annually. Seed set is reduced, seed filling is impaired, and shriveled grain is lost with chaff at harvest. Some fungicide-protected fields yield 10–20% more grain than fields in which foliar diseases, principally Septoria leaf and glume blotches, are allowed to develop. Losses are greatest when epidemics develop before heading.

The three wheat pathogens are weakly virulent on hosts other than wheat. Barley, rye, and grasses, especially *Poa* and *Agrostis* spp., are susceptible, but lesions remain small and secondary sporulation is restricted. *S. nodorum* invades leaves but not heads on rye.

Symptoms

Symptoms develop throughout the growing season on all aerial plant parts. Initial symptoms are chlorotic flecks, usually on lower leaves in contact with soil. The flecks expand into irregular lesions, normally 1–5 × 4–15 mm (Plate 28). Lesions caused by *S. tritici* tend to be restricted laterally and assume parallel sides (Fig. 56). Those caused by *S. nodorum* and *S. avenae* f. sp. *triticea* are more lens-shaped (Plate 28). Lesions are initially water-soaked, then become dry, yellow, and finally, red-brown. Some develop gray-brown or ashen centers. Necrosis extends well beyond colonized cells, apparently

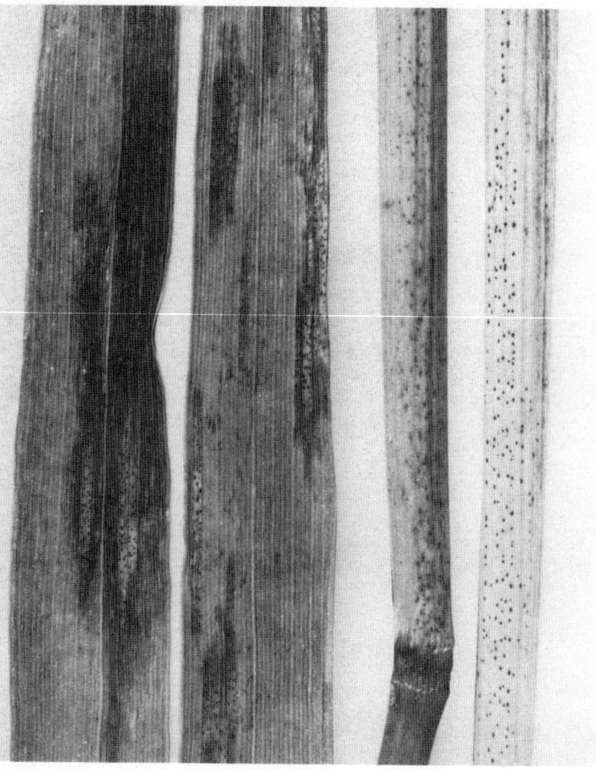

Fig. 56. Leaves and culms infected with *Mycosphaerella graminicola* (anamorph *Septoria tritici*) and speckled with pycnidia. (Courtesy Illinois Agricultural Experiment Station)

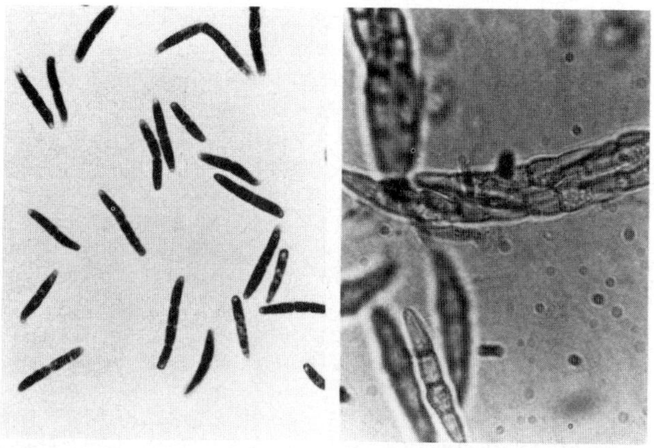

Fig. 57. Pycnidiospores (left) and ascus and ascospores (right) of *Leptosphaeria nodorum* (anamorph *Septoria nodorum*). (Courtesy A. L. Scharen)

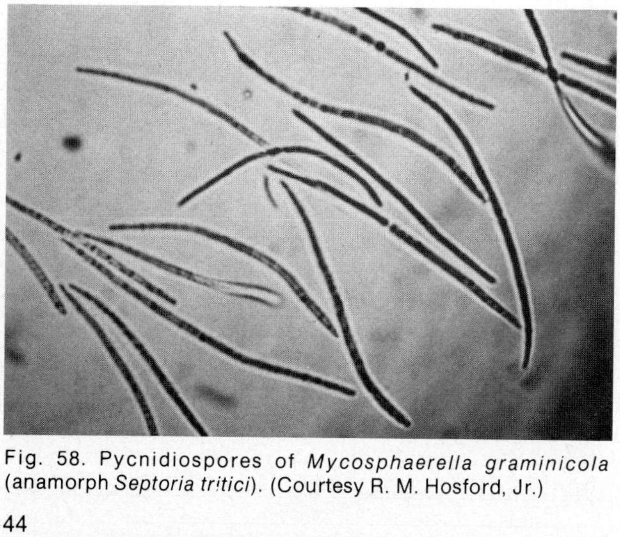

Fig. 58. Pycnidiospores of *Mycosphaerella graminicola* (anamorph *Septoria tritici*). (Courtesy R. M. Hosford, Jr.)

because of diffusible toxins. Lesions that originate at the base of the blade often kill the entire leaf.

Lesions are of greater diagnostic value when pycnidia mature within them. Pycnidia are gray-brown, globose, and 100–200 μm in diameter, with rugose walls. They exude masses of diagnostic spores in gelatinous droplets or columns (cirrhi) when moistened. Pycnidia often are spaced in orderly rows because they develop within substomatal cavities. Sometimes infected nodes are sunken and darkened or speckled with pycnidia.

Causal Organisms

Septoria spp. grow and sporulate on many artificial media. They are much more common in developing wheat plants than are their teleomorphs (perfect stages), which are typically found on stubble. Isolations from infected plants normally yield only the mycelial and pycnidial stages. Field isolates, or even cultures derived from single conidia (pycnidiospores), can vary greatly in morphology and virulence. Colonies are white, yellow, olive-gray, or pink and are often sectored. Parasexual mechanisms may be responsible for such heterogeneity. Only a few physiologic races have been designated.

L. nodorum Müller (anamorph *Septoria nodorum* (Berk.) Berk.), causes blotches on coleoptiles, upper leaves, necks, and glumes (Plates 28 and 29). Its mycelium and pycnidia also appear in seed. The fungus occurs worldwide and is especially significant in the southeastern United States, South America, Europe, Africa, and East Asia. *S. nodorum* was described in 1845, and its ascomycetous teleomorph (*L. nodorum*) was described in 1961. *L. nodorum* resembles *L. avenaria* f. sp. *triticea* in morphology and habit (see below). Its ascospores measure 4–6 × 24–32 μm (Fig. 57).

S. nodorum pycnidiospores are hyaline, 2–4 × 15–32 μm, and one-, two-, or three-septate, with rounded ends (Fig. 57). Conidial masses in cirrhi are pink. Occasionally, bacilliform, nonseptate microspores, measuring 0.7–1.0 × 3–6 μm, appear. The fungus grows optimally at 20–24° C, with limits near 4 and 32° C.

M. graminicola (Fuckel) Schroeter (anamorph *S. tritici* Rob. in Desm.), like *L. nodorum*, is found in over 50 countries. It thrives in cooler climates and is more active during cooler portions of the growing season. Wheat seedlings are as susceptible as adult plants, although the fungus is rarely seedborne. Pseudothecia and ascospores have been reported in New Zealand, Australia, the United Kingdom, Chile, and the United States.

Perithecia of *M. graminicola* occur primarily on senescent leaf blades. They are subepidermal, globose, dark brown, and 68–114 μm in diameter. Asci measure 11–14 × 30–40 μm. Ascospores are hyaline, elliptical, and 2.5–4 × 9–16 μm, with two cells of unequal size.

S. tritici develops hyaline, threadlike pycnidiospores that measure 1.7–3.4 × 39–86 μm, with three to seven indistinct septations (Fig. 58). Germination of pycnidiospores can be lateral or terminal. Cirrhi are milky white to buff. Sometimes in culture nonseptate, hyaline microspores, measuring 1–1.3 × 5–9 μm, occur outside pycnidia by yeastlike budding. Pycnidia of *S. tritici* tend to be more conspicuous in host tissues (Fig. 56) than those of *S. nodorum* or *S. avenae* f. sp. *triticea*. The fungus grows optimally between 22 and 26° C, with limits at 3 and 32° C.

L. avenaria Weber f. sp. *triticea* Johnson (anamorph *S. avenae* Frank f. sp. *triticea* Johns.) has been reported in parts of Europe, Asia, South America, the United States, and Canada. First described in Canada in 1947, it is the most recent *Septoria* sp. to be characterized on wheat.

The teleomorph, *L. avenaria* f. sp. *triticea*, occurs on wheat stubble and occasionally on green leaves. Its pseudothecia resemble pycnidia but contain eight-spored asci that measure 8–11 × 40–80 μm. Ascospores are hyaline to yellow, measure 3.5–6 × 18–25 μm, and are three-septate (rarely four-septate). A slight constriction appears at each septum, and terminal cells are tapered.

Pycnidiospores of *S. avenae* f. sp. *triticea* are hyaline and measure 2.5–4.5 × 21–45 μm. They are intermediate in size between pycnidiospores of *S. tritici* and *S. nodorum*. They are mostly straight, with rounded ends and three or four septations.

Disease Cycle

Straw, seed, and overwintering and volunteer wheat are sources of primary inoculum. Undisturbed wheat stubble apparently supports conidial (pycnidiospore) production. Seed also can harbor *S. nodorum*, and perhaps *S. tritici*, for a year or more. Ascospores, often airborne for long distances, serve as primary inoculum, but mycelium in debris also is infectious.

Pycnidiospores remain viable for months between 2 and 10° C, and those of *S. nodorum* can tolerate temperatures above this range. The slime in which the spores are exuded protects them from radiation and desiccation and stimulates their germination. Conidia, produced during wet periods, are disseminated by splashing rain and initiate infections throughout the growing season. Ascospores are prevalent in air during late summer and autumn. Germ tubes of both spore types penetrate wheat directly and through stomata.

Infection requires at least 6 hr of wetness (up to 16 hr for *S. nodorum*), and secondary spores are generated within 10–20 days. Spore germination and infection are optimal between 15 and 25° C but can occur between 5 and 35° C. Septoria tritici blotch is favored by temperatures between 15 and 20° C, whereas Septoria nodorum blotch develops best between 20 and 27° C. Wet, windy weather favors epidemics; dry periods not only prevent infection but also halt the development of lesions and pycnidia.

Nonwheat hosts do not appear to play a major role in Septoria leaf and glume blotch epidemics on wheat. In addition to contributing ascospore inoculum, the teleomorphs may serve in generating new virulent forms.

Control

Wheat cultivars moderately resistant to one or more *Septoria* spp. are available, but none are highly resistant to all three pathogens. The cultivars Hill, Arthur, Oasis, and Coker 747 are resistant to one or more *Septoria* spp. in North America. In Oklahoma and California, Timpaw and Anza, respectively, become infected by *S. tritici* but do not support pycnidium production. In general, early-maturing cultivars are most susceptible.

Other controls include the use of pathogen-free seed and the destruction or avoidance of infested straw, stubble, and volunteer wheat. Infested stubble is burned occasionally in Europe and Asia. In the United States, stubble mulch and minimum tillage practices may increase disease risk. Rotations in which wheat or other cereals appear every third year minimize most carry-over inoculum.

Wide row spacing and adequate but not excessive fertilization limit disease development by limiting foliage density and relative humidity in the crop canopy. In cooler climates, late-autumn seeding delays crop development in summer so that most foliage develops during warmer periods less favorable to *S. tritici* and *S. avenae* f. sp. *triticea*. Low-cost, broad-spectrum fungicides, such as maneb, control Septoria leaf and glume blotches and other foliar diseases in parts of Asia, Africa, and North America. Fungicide seed treatment reduces seedborne inoculum and seedling blight (Plate 30). Chemicals such as CCC [(2-chloroethyl) trimethylammonium chloride], used to limit culm elongation, can favor disease development.

Selected References

Babadoost, M., and Hebert, T. T. 1984. Factors affecting infection of wheat seedlings by *Septoria nodorum*. Phytopathology 74:592–595.

Eyal, Z. 1981. Integrated control of Septoria diseases of wheat. Plant Dis. 65:763–768.

Eyal, Z., and Ziv, O. 1974. The relationship between epidemics of Septoria leaf blotch and yield losses in spring wheat. Phytopathology 64:1385–1389.

Holmes, S. J. I., and Colhoun, J. 1974. Infection of wheat by *Septoria nodorum* and *S. tritici* in relation to plant age, air temperature and relative humidity. Trans. Br. Mycol. Soc. 63:329–338.

Holmes, S. J. I., and Colhoun, J. 1975. Straw-borne inoculum of *Septoria nodorum* and *S. tritici* in relation to incidence of disease on wheat plants. Plant Pathol. 24:63–66.

Hosford, R. M., Jr., Hogenson, R. O., Huguelet, J. E., and Kiesling, R. L. 1969. Studies of *Leptosphaeria avenaria* f. sp. *triticea* on wheat in North Dakota. Plant Dis. Rep. 53:378–381.

King, J. E., Cook, R. J., and Melville, S. M. 1983. A review of Septoria diseases of wheat and barley. Ann. Appl. Biol. 103:345–373.

Mehta, Y. R. 1975. *Leptosphaeria nodorum* on wheat in Brazil and its importance. Plant Dis. Rep. 59:404–406.

Osborn, A. E., Scott, P. R., and Caten, C. E. 1986. The effects of host passaging on the adaptation of *Septoria nodorum* to wheat and barley. Plant Pathol. 35:135–145.

Richardson, M. J., and Noble, M. 1970. *Septoria* species on cereals—A note to aid their identification. Plant Pathol. 19:159–163.

Sanderson, F. R. 1976. *Mycosphaerella graminicola* (Fuckel) Sanderson comb. nov., the ascogenous state of *Septoria tritici* Rob. apud Desm. N.Z. J. Bot. 14:359–360.

Scharen, A. L. 1966. Cyclic production of pycnidia and spores in dead wheat tissue by *Septoria nodorum*. Phytopathology 56:580–581.

Scharen, A. L., ed. 1985. *Septoria* of Cereals. Proc. 1983 Workshop, Montana State Univ., Bozeman. U.S. Dep. Agric. Agric. Res. Serv. Publ. No. 12. 116 pp.

Scharen, A. L., and Eyal, Z. 1983. Analysis of symptoms on spring and winter wheat cultivars inoculated with different isolates of *Septoria nodorum*. Phytopathology 73:143–147.

Shaner, G., Finney, R. E., and Patterson, F. L. 1975. Expression and effectiveness of resistance in wheat to Septoria leaf blotch. Phytopathology 65:761–766.

Shipton, W. A., Boyd, W. R. J., Rosielle, A. A., and Shearer, B. I. 1971. The common Septoria diseases of wheat. Bot. Rev. 37:231–262.

Sutton, B. C., and Waterston, J. M. 1966a. *Leptosphaeria nodorum*. Descriptions of Pathogenic Fungi and Bacteria, No. 86. Commonwealth Mycological Institute, Association of Applied Biologists, Kew, Surrey, England.

Sutton, B. C., and Waterston, J. M. 1966b. *Septoria tritici*. Descriptions of Pathogenic Fungi and Bacteria, No. 90. Commonwealth Mycological Institute, Association of Applied Biologists, Kew, Surrey, England.

Tar Spot

Tar spot is found on most members of the Gramineae. It appears principally on perennial grasses in moist, shaded areas and is rare on wheat and other annuals.

The disease is readily recognized by its glossy black spots (clypei) on leaf blades and sheaths. The distinct, somewhat sunken clypei, 0.1–1 × 0.2–5 μm, are crusty stromata of the causal fungus, *Phyllachora graminis* (Pers.) Fckl., with embedded perithecia. Ascospores are hyaline, ovoid to elliptical, unicellular, and 4–7 × 9–14 μm.

Stromata overseason on host plants and their residues. Grasses are the usual sources of primary inoculum for wheat infection. Tar spot on wheat is not economically important, and controls are not specified.

Selected References

Orton, C. R. 1944. Graminicolous species of *Phyllachora* in North America. Mycologia 36:18–53.

Parbery, C. R. 1967. Studies on graminicolous species of *Phyllachora* Nke. in Fckl. V. A. Taxonomic Monograph. Aust. J. Bot. 15:271–375.

Platyspora Leaf Spot

The first report of *Platyspora pentamera* (Karst.) Wehm. on wheat appeared in 1971. The fungus previously was known only at high altitudes, where it persists on the stems and leaves of numerous monocotyledonous and dicotyledonous hosts.

Wheat and rye (but not barley, oats, triticale, or wheatgrass) are now known to be infected in the Great Plains of North America.

At present, wheat infections are infrequent and not economically important. They appear as isolated, irregular, yellow-brown spots without internal fructifications. In the greenhouse, symptoms appear four to six days after inoculation, and infection requires 24–72 hr of continuous wetness. Natural infections eventually produce diagnostic pseudothecia on wheat straw. The pseudothecia are dark and 166–230 μm in diameter. Ascospores (eight per ascus) are airborne, distinctly flattened, 6–8 × 12 × 12–30 μm, and eight-celled. Each has four transverse septa, and longitudinal septa divide the three median cells. Colonies developed from ascospores on potato-dextrose agar are dense and gray-green. The pathogen resembles *Pyrenophora tritici-repentis* (Figs. 54 and 55) in morphology and habit.

Controls are largely unnecessary; however, those described for Septoria leaf and glume blotches should also control Platyspora leaf spot. The cultivar Hercules is resistant in North Dakota.

Selected References

Hosford, R. M., Jr. 1971. *Platyspora pentamera* in the Great Plains on wheat. Mycologia 63:668–669.

Hosford, R. M., Jr. 1975. *Platyspora pentamera*, a pathogen of wheat. Phytopathology 65:499–500.

Phoma Spot

Phoma spp. invade wheat in India, Mexico, and parts of South America. The pathogens are widely distributed on numerous hosts but relatively rare on wheat. *P. insidiosa* Tassi invades wheat glumes and awns in India, causing oval, brown lesions 2–7 mm long, with pycnidia in their centers. Wheat and triticale in Mexico occasionally show dark brown lesions on leaf sheaths caused by *P. glomerata* (Corda) Wr. & Hochapf.

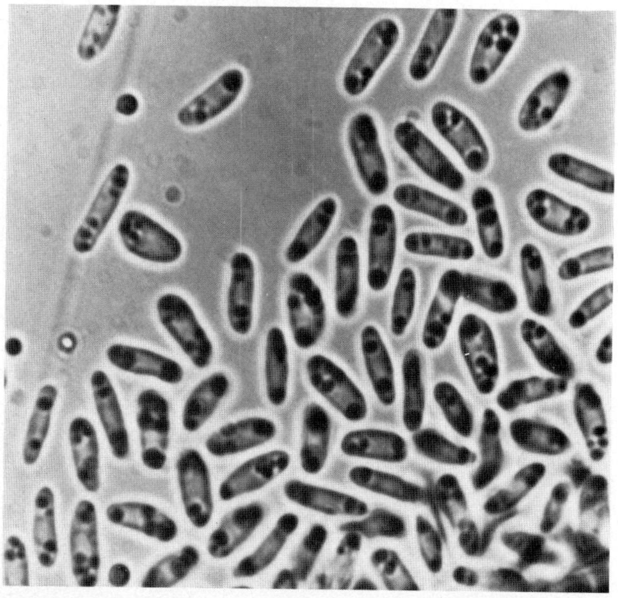

Fig. 59. Conidia (pycnidiospores) of *Phoma glomerata*. (Courtesy R. M. Hosford, Jr.)

Pycnidia and pycnidiospores are abundant within host lesions and on artificial media. Pycnidia are dark, globose, ostiolate, sunken in host tissues, and 100–200 μm in diameter. Conidia (pycnidiospores) are numerous, hyaline, ovoid to ellipsoid, one-celled, and approximately 3 × 6 μm (Fig. 59). *P. glomerata* also produces chains of chlamydospores.

Prolonged periods of continuous wetting are required for infection and symptom development. Controls are not prescribed, but cultivars ranging from highly resistant to highly susceptible have been identified in greenhouse tests.

Selected References

Hosford, R. M., Jr. 1975. *Phoma glomerata*, a new pathogen of wheat and triticales, cultivar resistance related to wet period. Phytopathology 65:1236–1239.

Morgan-Jones, G. 1967. *Phoma glomerata*. Descriptions of Pathogenic Fungi and Bacteria, No. 134. Commonwealth Mycological Institute, Association of Applied Biologists, Kew, Surrey, England.

Nema, K. G., Dave, G. S., and Khosla, H. K. 1971. A new glume blotch of wheat. Plant Dis. Rep. 55:95.

Punithalingam, E., and Holliday, P. 1972. *Phoma insidiosa*. Descriptions of Pathogenic Fungi and Bacteria, No. 333. Commonwealth Mycological Institute, Association of Applied Biologists, Kew, Surrey, England.

Leptosphaeria Leaf Spots

In the 1930s *Leptosphaeria herpotrichoides* de Not. was identified as saprophytic or weakly parasitic at the base of wheat plants. In the 1970s, the fungus was found producing mature pseudothecia on wheat stubble in North Dakota and Minnesota. *L. microscopica* Karst. (anamorph *Phaeoseptoria urvilleana* Sprague) also persists on wheat and grass stubble in Argentina, Europe, and the United States.

Both fungi are inconsequential wheat pathogens under usual growing conditions but are weak foliar pathogens under conditions of excessive and continuous moisture. When continuous wet periods last longer than 48 hr (and in some instances up to 121 hr), wheat leaves may develop yellow-brown blotches (Plate 31). Barley, rye, and oats are more resistant or immune.

L. herpotrichoides makes pseudothecia that measure 140–396 μm in diameter and ascospores that measure 3.3–7.6 × 21–24 μm, with two to nine septations. Pseudothecia of *L. microscopica* measure 98–224 μm in diameter and contain faintly colored ascospores that are three-septate and measure 5–7 × 21–26 μm. *P. urvilleana* makes dark pycnidia measuring 132 × 235 μm, with numerous pycnidiospores that measure 3–6 × 38–57 μm and have from zero to nine septations.

The fungi grow readily on potato-dextrose agar, where they form prostrate gray colonies, fruiting structures, ascospores, and conidia.

If necessary, controls described for tan spot and Septoria leaf and glume blotches should also control Leptosphaeria leaf spots.

Selected References

Hosford, R. M., Jr. 1978a. Effects of wetting period on resistance to leaf spotting of wheat, barley, and rye by *Leptosphaeria herpotrichoides*. Phytopathology 68:591–594.

Hosford, R. M., Jr. 1978b. Effects of wetting period on resistance to leaf spotting of wheat by *Leptosphaeria microscopica* with conidial stage *Phaeoseptoria urvilleana*. Phytopathology 68:908–912.

Fungal Diseases Principally Observed on Lower Stems and Roots

Foot Rot (Eyespot)

Foot rot, also called eyespot and strawbreaker, is an important wheat disease that is named for its effects on the base (foot) of the wheat plant. Its characterization began in France and in the United States in the early 1900s. Today, foot rot is recognized in the northern United States and in parts of South America, Europe, New Zealand, Australia, and Africa.

Wheat is more susceptible to the causal fungus (*Pseudocercosporella herpotrichoides*) than barley, rye, oats, and grasses, and winter cereals are damaged more often than spring cereals. The disease normally is more prevalent where wheat is repeatedly grown and in climates that are cool and moist.

Foot rot may kill entire plants outright. More frequently, however, it weakens or kills individual tillers, reduces kernel size and number, causes culms to lodge, and renders plants difficult to harvest. Mild and moderate infections are inconspicuous because individual damaged or lodged culms may not be noticed in dense plant populations.

Symptoms

Foot rot is diagnosed from lesions that are most distinctive at the plant base after jointing but also appear on younger plants. The lesions do not appear on roots and rarely develop more than 4 cm above or below soil level. The lesions begin development superficially on leaf sheaths at or near the soil surface and with time grow wider and longer and penetrate the culm (see Insects [billbugs] and Sharp Eyespot).

The elliptical or "eye" shape of the lesions is diagnostic (Plate 32). Lesions are distinct, white to tan-brown initially, and oriented lengthwise with the stem. They may eventually "girdle" the culm, develop fungus-darkened centers (see Sharp Eyespot and Anthracnose), and increase to 4 cm in length. Well-developed lesions are dark in color and cannot be removed by stripping off leaf sheaths. Sometimes a diagnostic weft of gray mycelium appears within the lumen of the culm beneath severe lesions. Lesions on maturing plants give the culm base a charred appearance, which, unlike the "dark stocking" of take-all, is not black and shiny. Diseased tillers tend to mature early and produce whiteheads that contain incompletely filled seed and, in humid climates, support "sooty" molds (Plate 9).

Foot rot lesions are firm at first but become sunken and brittle. They weaken the stem, so that diseased tillers begin to fall randomly (see Insects [boring insects]). When lesions are moderate or infrequent, many damaged plants are supported by their healthy neighbors. When lesions are severe and frequent, large areas of lodged plants occur (Fig. 60). Lodging is usually nondirectional and irreversible.

Causal Organism

Foot rot is caused by *Pseudocercosporella herpotrichoides* (Fron) Deighton (syn. *Cercosporella herpotrichoides* Fron), a fungus that produces mycelium that is vegetative, yellow-brown, linear, and branching and mycelium that is dark and stromalike. This latter dark mycelium accounts for the superficial charring on wheat stems. Conidiophores are simple or sparingly branched. Conidia, most abundant in early spring, are hyaline, slightly curved, mostly five- to seven-celled, and 1.5–3.5 × 35–70 μm. Sporulation in vitro tends to originate from loose sporodochia.

P. herpotrichoides grows on moist, sterile oat kernels and on a variety of simple agar media. Oat kernel cultures are convenient for inoculating soil used in greenhouse or field experiments. On potato-dextrose agar, young colonies are gray, compact, and mounded.

No teleomorph of *P. herpotrichoides* is known. Some early literature incorrectly lists *Leptosphaeria herpotrichoides* de Not. as the teleomorph of *P. herpotrichoides* and the cause of eyespot.

Disease Cycle

Eyespot infections occur on winter wheat throughout autumn, winter, and spring. Infections originate mainly from water-splashed conidia but also from mycelium persisting on host debris. The pathogen is dormant or least active in summer but sporulates efficiently on hosts and residues during periods of cool, wet weather. Conidial production is maximum when temperatures fluctuate near 10° C; it does not occur below 0 or above 20° C. Infection can occur within 15 hr between 6 and 15° C in a water-saturated atmosphere but is dramatically slowed or prevented above 16° C. Sometimes symptoms are not visible for two to three weeks after infection, and autumn infections may not be expressed until spring.

Conidia are distributed principally by splashing rain and function as primary inoculum. They germinate and penetrate coleoptiles and leaf sheaths directly or through stomata near ground level. Secondary conidia form on new lesions within four to 12 weeks. Like secondary inoculum from alternative grassy hosts, secondary infections appear inconsequential to the current epidemic. However, the secondary lesions may provide inoculum for subsequent crops.

Eyespot is favored by high soil moisture, a dense crop canopy, recurrent host crops, early seeding, and high humidity near soil level. Mild winters and cool springs prolong sporulation and infection periods. Plants may be predisposed to infection in spring by frosts and nitrogen fertilization. In hot, dry weather, diseased culms undergo extra moisture stress.

Control

Spring wheat and late-seeded winter wheat are less exposed to infection. Thinly seeded fields limit disease by maintaining

Fig. 60. Lodged wheat, an expression of foot rot (eyespot) caused by *Pseudocercosporella herpotrichoides*. (Courtesy J. Sitton)

lower relative humidity within the crop canopy but tend to lack the yield potential of thicker stands. Because *P. herpotrichoides* dies as infested residues decay, crop rotations in which susceptible cereals are not grown for two or more years are advised.

Foot rot is reduced when crop residues are maintained on the soil surface. The residues may limit inoculum dispersal via rain splash and/or retard seedling growth, making the crop canopy less favorable to the pathogen.

Resistant cultivars are not available, but some cultivars, because of stiffer straw, are tolerant. Cappelle-Desprez in Europe and Luke and Lewjain in the northwestern United States are damaged less than most cultivars. Resistance in certain species of *Aegilops* has recently been transferred to experimental wheat lines.

Chemicals like CCC [(2-chloroethyl) trimethylammonium chloride] stiffen straw and reduce lodging without influencing disease incidence. Fungicides such as benomyl, thiophanate-methyl, and thiabendazole reduce or prevent eyespot damage. Their use in problem areas is economically feasible.

Selected References

Booth, C., and Waller, J. M. 1973. *Pseudocercosporella herpotrichoides*. Descriptions of Pathogenic Fungi and Bacteria, No. 386. Commonwealth Mycological Institute, Association of Applied Biologists, Kew, Surrey, England.

Bruehl, G. W., and Machtmes, R. 1984. Effects of "dirting" on strawbreaker foot rot of winter wheat. Plant Dis. 68:868–870.

Bruehl, G. W., Machtmes, R., and Cook, R. J. 1982. Control of strawbreaker foot rot of winter wheat by fungicides in Washington. Plant Dis. 66:1056–1058.

Bruehl, G. W., Machtmes, R., and Murray, T. 1982. Importance of secondary inoculum in strawbreaker foot rot of winter wheat. Plant Dis. 66:845–847.

Bruehl, G. W., Peterson, C. J., Jr., and Machtmes, R. 1974. Influence of seeding date, resistance and benomyl on Cercosporella foot rot of winter wheat. Plant Dis. Rep. 58:554–558.

Chang, E. P., and Tyler, L. J. 1964. Sporulation by *Cercosporella herpotrichoides* on artificial media. Phytopathology 54:729–735.

Deacon, J. W. 1973. Behavior of *Cercosporella herpotrichoides* and *Ophiobolus graminis* on buried wheat plant tissues. Soil Biol. Biochem. 5:339–353.

Defosse, L., and Dekegel, D. 1974. Penetration of *Cercosporella herpotrichoides* Fron [*Pseudocercosporella herpotrichoides* (Fron) Deighton] into the coleoptile of wheat (*Triticum aestivum*) observed in electron microscopy. Ann. Phytopathol. 6:471–474.

Evans, M. E., and Rawlinson, C. J. 1975. A method for inoculating wheat with *Cercosporella herpotrichoides*. Ann. Appl. Biol. 80:339–341.

Fehrmann, H., and Schrodter, H. 1972. Ecological investigations on the epidemiology of *Cercosporella herpotrichoides*. IV. Elaboration of a practical method for the control of eyespot disease in wheat with systemic fungicides. Phytopathol. Z. 74:161–174.

Glynne, M. D. 1963. Eyespot (*Cercosporella herpotrichoides*) and other factors influencing yield of wheat in the six-course rotation experiment at Rothamsted (1930–60). Ann. Appl. Biol. 61:189–214.

Herrman, T., and Wiese, M. V. 1985. Influence of cultural practices on incidence of foot rot in winter wheat. Plant Dis. 69:948–950.

Murray, T. D., and Bruehl, G. W. 1983. Role of the hypodermis and secondary cell wall thickening in basal stem internodes in resistance to strawbreaker foot rot in winter wheat. Phytopathology 73:261–268.

Rowe, R. C., and Powelson, R. L. 1973. Epidemiology of Cercosporella footrot of wheat: Disease spread. Phytopathology 63:984–988.

Scott, P. R., and Hollins, T. W. 1974. Effects of eyespot on the yield of winter wheat. Ann. Appl. Biol. 78:269–279.

Scott, P. R., and Hollins, T. W. 1980. Pathogenic variation in *Pseudocercosporella herpotrichoides*. Ann. Appl. Biol. 94:297–300.

Aureobasidium Decay

Microdochium bolleyi (Sprague) de Hoog & Hermanides-Nijhof (syns. *Aureobasidium bolleyi, Gloeosporium bolleyi*) was described in North Dakota in 1913 as one of several fungi associated with darkened and rotted wheat roots. The fungus is a ubiquitous colonizer of roots, crowns, and seeds of wheat and other Gramineae. Isolates are largely nonpathogenic but occasionally cause mild necrosis of seedling roots.

Fig. 61. Dark chlamydospores of *Microdochium bolleyi* in root epidermal cells. (Courtesy W. W. Bockus)

Cultures on potato-dextrose agar are pink to red and darken with age. Conidia are numerous, hyaline, one-celled, and ovoid to bean-shaped. They are produced either on short hyphal denticles or on subglobose conidiogenous cells. Clumps of dark chlamydospores in epidermal cells of wheat roots are diagnostic (Fig. 61).

Selected References

De Hoog, G. S., and Hermanides-Nijhof, E. J. 1977. The black yeasts and allied hyphomycetes. Stud. Mycol. 15:178–221.

Murray, D. I. L., and Gadd, G. M. 1981. Preliminary studies on *Microdochium bolleyi* with special reference to colonization of barley. Trans. Br. Mycol. Soc. 76:397–403.

Sprague, R. 1948. Gloeosporium decay in Gramineae. Phytopathology 38:131–136.

Root-Associated Agaricales

Between 1920 and 1940, basidiocarp gill fungi (toadstools) were reported in association with wheat roots in the United States. The fungi, all members of the Agaricales, are presumed parasitic but appear inconsequential on wheat. *Naucoria cerealis* Boewe, *Pholiota dura* (Bolt.) Fries (syn. *Agrocybe dura* Bolt.), *P. praecox* (Pers.) Fries, and *Marasmius tritici* Young frequently occur in association with cereals. Members of the last-named genus (recently placed under *Marasmiellus*) are pathogenic on beach grass and are associated with "fairy rings" in turfgrass.

Selected References

Boewe, G. H. 1938. *Naucoria* on small grains in Illinois. Phytopathology 28:852–855.

Sprague, R. 1938. Gill fungi associated with the roots of cereals. (Note) Phytopathology 28:78–79.

Tehon, L. R. 1924. *Marasmius* on wheat. Mycologia 16:132–133.

Young, P. A. 1925. A *Marasmius* parasitic on small grains in Illinois. Phytopathology 15:115–118.

Rhizoctonia Root Rot

Rhizoctonia root rot affects wheat in all temperate regions. However, like many other root diseases, it tends to go unnoticed unless roots (especially root tips) are carefully examined under magnification. *Rhizoctonia solani*, the causal fungus, is active in the top 10–15 cm of soil, where it girdles and eventually severs (prunes) individual roots and rootlets. Plants with a few pruned roots grow and yield normally. However, when root pruning is severe, host plants are severely stunted and exhibit symptoms of drought or nutrient deficiency.

Severe Rhizoctonia root rot occurs in Australia, parts of Europe, and the Pacific Northwest of the United States. Because the disease tends to be localized in wheat fields, it is sometimes referred to as "bare patch" or "purple patch." In Scotland, severe Rhizoctonia root rot of barley is called "barley stunt disorder."

Symptoms

Distinct patches of stunted, lodged, or white-headed plants are indicative of Rhizoctonia root rot. Seedlings occasionally are killed outright ("bare patch") (Fig. 62). Most infected plants that are not winter-killed are able to tolerate root browning and pruning and may outgrow the disease by producing new roots. "Purple patch" denotes areas in the field where the leaves on diseased plants appear stiff and dull blue-gray. Maturity is delayed, as with Pythium root rot, rather than hastened, as with crown and culm infections caused by most other soilborne wheat pathogens.

Under low magnification, root ends appear reddish brown and taper to a diagnostic fine point (spear tip) (Fig. 63). Root lesions caused by *R. solani* are mostly small, often only 2–3 mm long. Usually only half of a lesion is evident at the end of a root because the remainder is easily severed when plants are dug from soil.

Causal Organism

R. solani Kühn, like *R. cerealis*, is identified only from mycelial characteristics because it produces no spores. Its hyphal cells are multinucleate (Fig. 64), whereas those of *R. cerealis* (the cause of sharp eyespot) are binucleate. This distinction is useful, but the location of the two species on wheat plants is not absolute. Occasionally, multinucleate isolates occur on culms and binucleate strains are isolated from roots.

R. solani isolates can be further subdivided into anastomosis groups (AG) based on hyphal fusion between compatible strains. Although strains of *R. solani* belonging to several anastomosis groups have been isolated from wheat roots, most isolates responsible for bare patch belong to AG 8.

The teleomorph of *R. solani* occurs in nature but is rarely found in association with wheat. It is *Thanatephorus cucumeris* (Frank) Donk (syns. *Pellicularia filamentosa* (Pat.) Rog., *Corticium solani* Prill. & Delacr.), a basidiomycete. *T. cucumeris* forms club-shaped basidia and four apical sterigmata on which oval, hyaline sporidia are borne.

R. solani grows on many artificial media and on plant residues in soil. Its mycelium is white to deep brown, and hyphae, 4–15 μm wide, tend to branch at right angles. A septum near each hyphal branch and a slight constriction at the branch base are diagnostic (Fig. 64). The mycelium readily differentiates into brown-black, irregularly shaped sclerotia in culture and on host plants.

Disease Cycle

R. solani persists in soil using soil nutrients and fragments of crop residue as energy sources. It is virtually omnipresent

Fig. 62. Bare patch (Rhizoctonia root rot) caused by *Rhizoctonia solani*. (Courtesy R. D. Price)

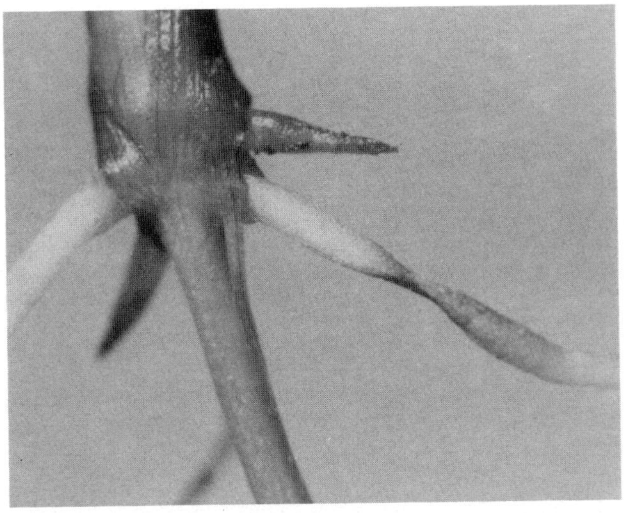

Fig. 63. Roots rotted and spear-tipped by *Rhizoctonia solani*. (Courtesy D. M. Weller, USDA)

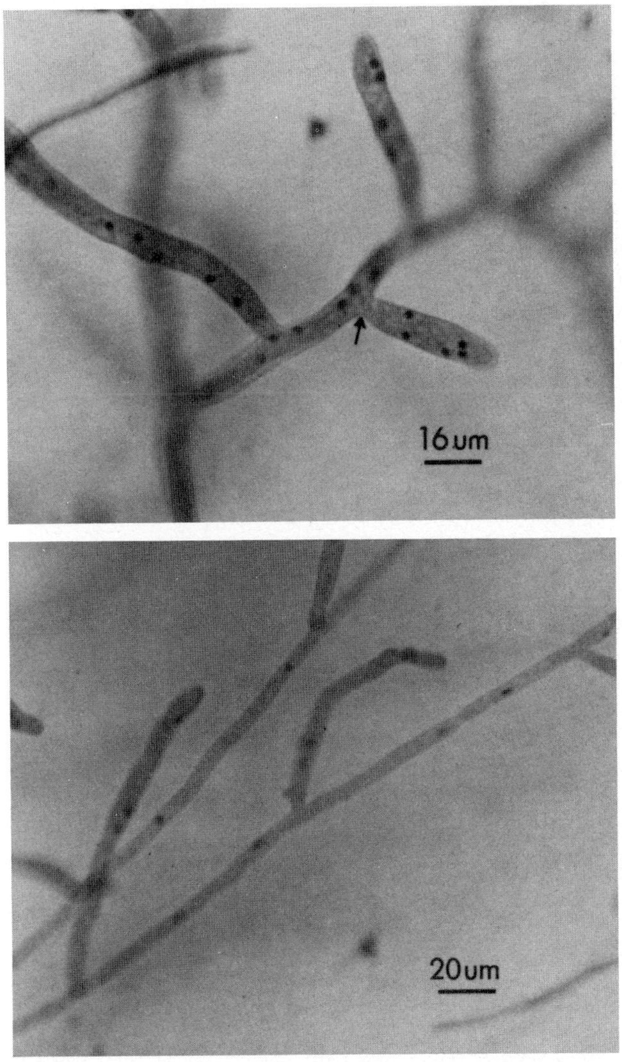

Fig. 64. Multinucleate hyphae of *Rhizoctonia solani* (top), cause of Rhizoctonia root rot, and binucleate hyphae of *R. cerealis* (bottom), cause of sharp eyespot. (Courtesy C. Castro)

as mycelium and sclerotia, and virtually all wheat is exposed to attack. However, damage depends greatly on environment. Root infections can occur at any time during the growing season but are most stressful before seed set. The occurrence of severe disease in patches may reflect the ability of the fungus to establish a network of hyphae on crop residues near the soil surface.

Control

Truly resistant wheat cultivars are not available. However, crop rotation may provide some benefit in spite of the broad host range of *R. solani* and its capacity to persist in soil. Late, shallow seeding and conditions that promote root growth may limit the effects of the disease. A practical control measure is soil tillage to disturb the network of mycelium and promote decay of crop residue. Root rot damage is likely to be greatest when soil is untilled and then kept moist during seedling development. In both the Pacific Northwest and Australia, reduced tillage favors the disease. Delaying seeding a few days after final tillage also appears advantageous. Some chemical seed treatments partially control the disease.

Selected References

MacNish, G. C. 1985. Methods of reducing rhizoctonia patch of cereals in Western Australia. Plant Pathol. 34:175–181.

Mordue, J. E. M. 1974. *Thanatephorus cucumeris*. Descriptions of Pathogenic Fungi and Bacteria, No. 406. Commonwealth Mycological Institute, Association of Applied Biologists, Kew, Surrey, England.

Murray, D. I. L. 1981. *Rhizoctonia solani* causing barley stunt disorder. Trans. Br. Mycol. Soc. 76:383–395.

Neate, S. M., and Warcup, J. H. 1985. Anastomosis groupings of some isolates of *Thanatephorus cucumeris* from agricultural soils in South Australia. Trans. Br. Mycol. Soc. 85:615–620.

Ogoshi, A., and Ui, T. 1985. Anastomosis groups of *Rhizoctonia solani* and binucleate *Rhizoctonia*. Pages 57–58 in: Ecology and Management of Soilborne Plant Pathogens. C. A. Parker, A. D. Rovira, K. J. Moore, P. T. W. Wong, and J. F. Kollmorgen, eds. American Phytopathological Society, St. Paul, MN. 358 pp.

Parmeter, J. R. 1970. *Rhizoctonia solani*, Biology and Pathology. University of California Press, Berkeley. 255 pp.

Rovira, A. D., and Venn, N. R. 1985. Effect of rotation and tillage on take-all and Rhizoctonia root rot in wheat. Pages 255–258 in: Ecology and Management of Soilborne Plant Pathogens. C. A. Parker, A. D. Rovira, K. J. Moore, P. T. W. Wong, and J. F. Kollmorgen, eds. American Phytopathological Society, St. Paul, MN. 358 pp.

Weller, D. M., Cook, R. J., MacNish, G. E., Bassett, N., Powelson, R. L., and Petersen, R. R. 1986. Rhizoctonia root rot of small grains favored by reduced tillage in the Pacific Northwest. Plant Dis. 70:70–73.

Sharp Eyespot

Sharp eyespot occurs on wheat, barley, and rye in Europe, North America, and most other temperate wheat-growing regions. Oats are less susceptible than other cereals. Strains of the sharp eyespot pathogen, *Rhizoctonia cerealis*, also cause "yellow patch" in turfgrass.

Severe sharp eyespot in wheat causes premature ripening ("whiteheads") and lodging. More often, however, sharp eyespot lesions are superficial and inconsequential, and control measures have not been justified. With the increased control of foot rot (caused by *Pseudocercosporella herpotrichoides*), however, particularly in Europe, sharp eyespot has become more frequent.

Symptoms

Sharp eyespot, like foot rot, usually begins on an outer leaf sheath near the base of the plant. Leaf sheaths develop circular or elliptical, light brown areas circumscribed by a thin, necrotic, dark brown border. Affected leaf sheath tissues rot, leaving a characteristic hole rather than a fibrous net as with foot rot. Several lesions ranging up to 1 cm in diameter may occur at the base of the same plant. Severe infections cause seedling blight, but most infected plants survive to maturity.

Lesions on culms are light brown or straw-colored and are surrounded by a sharply defined, dark brown border (Plate 33). The lesions may resemble the eyespots of foot rot but are more superficial, less lens-shaped, and more sharply delineated. They normally develop later than foot rot lesions and can occur up to 30 cm above the soil line. On maturing culms, mycelium beneath lesions is often abundant and ash white.

Severe sharp eyespot causes premature ripening and lodging of infected plants. Lodged culms frequently bend at the second or third internode. *R. cerealis* is less able to mechanically weaken infected culms than *P. herpotrichoides*.

Causal Organism

R. cerealis van der Hoeven is a widespread soilborne plant pathogen. It forms no spores and produces mycelium that is characteristically binucleate (Fig. 64). Its hyphae are white to gray-brown and 4–15 μm wide and tend to branch at right angles (see Rhizoctonia Root Rot). The teleomorph of *R. cerealis* is *Ceratobasidium cereale* Murray & Burpee, a basidiomycete that is rare in nature and is not observed in association with sharp eyespot symptoms. *R. cerealis* mycelium grows on many artificial media and forms abundant sclerotia. Sclerotia are irregular in shape and dark in color and lack a distinct rind. They often occur within the culm or between the leaf sheaths of wheat.

Disease Cycle

Culm infections apparently originate from soilborne sclerotia or from mycelium in host debris. Infection depends on cool, moist conditions near the base of the plant. Thereafter, lesion development is favored by light and well-drained soils. The disease especially affects cereals grown continuously on the same land. After invading seedling leaf sheaths, the pathogen spreads by mycelial growth within and on the plant during the growing season. Sclerotia apparently develop during the summer and are a principal source of inoculum returned to the soil after harvest.

The importance of the *C. cereale* teleomorph to the epidemiology of sharp eyespot is unknown.

Control

No resistant cultivars are available to control sharp eyespot. Late-autumn seeding of winter wheat results in less disease but may also lower crop yield potential. Oats are usually less susceptible than wheat, barley, and rye, and rotation with legume or other nonhost crops is beneficial. No effective and economical chemical controls are available.

Selected References

Burpee, L. L. 1980. *Rhizoctonia cerealis* causes yellow patch of turfgrasses. Plant Dis. 64:1114–1116.

Clarkson, J. D. S., and Cook, R. J. 1983. Effect of sharp eyespot (*Rhizoctonia cerealis*) on yield loss in winter wheat. Plant Pathol. 32:421–428.

Clarkson, J. D. S., and Griffin, M. J. 1977. Sclerotia of *Rhizoctonia solani* in wheat stems with sharp eyespot. Plant Pathol. 26:98.

Hollins, T. W., and Scott, P. R. 1985. Differences between wheat cultivars in resistance to sharp eyespot caused by *Rhizoctonia cerealis*. Tests of Agrochemicals and Cultivars, No. 6. Ann. Appl. Biol. 106(Suppl.):166–167.

Lipps, P. E., and Herr, L. J. 1982. Etiology of *Rhizoctonia cerealis* in sharp eyespot of wheat. Phytopathology 72:1574–1577.

Murray, D. I. L., and Burpee, L. L. 1984. *Ceratobasidium cereale* sp. nov., the teleomorph of *Rhizoctonia cerealis*. Trans. Br. Mycol. Soc. 82:170–172.

Ogoshi, A., and Ui, T. 1985. Anastomosis groups of *Rhizoctonia solani* and binucleate *Rhizoctonia*. Pages 57–58 in: Ecology and Management of Soilborne Plant Pathogens. C. A. Parker, A. D. Rovira, K. J. Moore, P. T. W. Wong, and J. F. Kollmorgen, eds. American Phytopathological Society, St. Paul, MN. 358 pp.

Richardson, M. J., and Cook, R. J. 1985. *Rhizoctonia* on small-grain cereals in Great Britain. Pages 63–65 in: Ecology and Management of Soilborne Plant Pathogens. C. A. Parker, A. D. Rovira, K. J. Moore, P. T. W. Wong, and J. F. Kollmorgen, eds. American Phytopathological Society, St. Paul, MN. 358 pp.

Sterne, R. E., and Jones, J. P. 1978. Sharp eyespot of wheat in Arkansas caused by *Rhizoctonia solani*. Plant Dis. Rep. 62:56–60.

Take-All

The term "take-all" originated in Australia more than 100 years ago. It was used to describe a severe seedling blight now known to be caused by the soilborne fungus *Gaeumannomyces graminis* var. *tritici*. Today take-all on wheat is recognized as a disease of the roots, crown, and basal stem (foot). Take-all is most important in temperate climates where wheat and grass culture is intensive, soil pH is neutral or alkaline, and moisture is plentiful. It primarily affects autumn-seeded wheat and is widely distributed in Australia, Europe, South Africa, Japan, Brazil, Chile, Argentina, the coastal region of China, and most of North America.

The take-all fungus shows some specialization for wheat but also invades bromegrass (*Bromus* spp.), wheatgrass, and quackgrass (*Agropyron* spp.). Damage to wheat is related to the extent and time of colonization of roots and culm bases. When take-all symptoms become obvious, seed yields normally are less than half those of healthy plants.

Symptoms

Most plants withstand mild root infections and appear symptomless. Severely infected plants are stunted and ripen prematurely. Take-all symptoms are most obvious near heading, when plants appear uneven in height, begin to die prematurely, and exhibit white heads. Earlier, plants with take-all are stunted and mildly chlorotic and have fewer tillers. When tillers are killed prematurely, their heads are distinctly bleached (white-headed) and sterile (Plate 34). Early maturation causes heads to contain only shriveled grain and makes them subject to darkening by "sooty" molds (Plate 9). Diseased plants typically pull up easily or break off near the soil line. On close examination, their roots appear sparse, blackened, and brittle (Plate 35).

Symptoms may be limited to blackened roots if soil moisture is limited early in the development of the crop. If soil moisture is sufficient throughout the growing season, the black-brown rot extends into the crown and up the culm base, where a superficial, dark, shiny mycelial plate surrounding the culm beneath the lowest leaf sheath is diagnostic (Plate 35). Under prolonged moist conditions, the leaf sheath surrounding this mycelial plate may be speckled by dark, erumpent perithecia. Diseased culms weakened at their base may occasionally lean and fall nondirectionally (see Foot Rot [Fig. 60]).

When perithecia and blackened lower stems are not produced, take-all diagnosis depends on close examination of roots for internal and superficial dark mycelium and runner hyphae. Coarse runner hyphae of the pathogen are often grouped in strands several millimeters long. They are especially conspicuous in and on transparent intact roots (Plate 36).

Causal Organism

The take-all fungus was long included under *Ophiobolus graminis* Sacc. In 1952, however, when infection pads (simple hyphopodia) and unitunicate asci were identified, it was transferred to *Gaeumannomyces graminis* (Sacc.) von Arx & Olivier. Isolates from wheat are designated *G. graminis* var. *tritici* Walker and have simple hyphopodia. The closely related pathogen *G. graminis* var. *avenae*, on oats and bentgrass (*Agrostis* spp.), also has simple hyphopodia, but its ascospores are longer (110–130 μm, compared to 70–100 μm for *G. graminis* var. *tritici*). Another closely related fungus, *G. graminis* var. *graminis*, on grasses, has ascospores similar in length to those of *G. graminis* var. *tritici* but has lobed rather than simple hyphopodia. *G. graminis* var. *graminis*, normally only weakly pathogenic on wheat, has been reported as a take-all pathogen on wheat in China.

O. herpotrichus (Fr.) Sacc. and *Wojnowicia graminis* (McAlp.) Sacc. & Sacc. (syn. *Hendersonia crastophila* Sacc.) are occasionally confused with the take-all pathogen. *O. herpotrichus* resembles *G. graminis* var. *tritici* morphologically but is not virulent on wheat. *W. graminis* invades wheat and causes a mild darkening of the basal stem but produces pycnidia rather than perithecia. It was once incorrectly assumed to be the anamorph of *O. graminis*. *Phialophora graminicola* (teleomorph *G. cylindrosporus*) also produces dark brown runner hyphae on wheat roots and may be confused with the take-all fungus. However, it is a much weaker parasite than *G. graminis* var. *tritici* and becomes damaging on wheat and grasses only when soil temperatures are high (approaching 30° C).

The simple hyphopodia of *G. graminis* var. *tritici* develop in contact with the host surface, somewhat like appressoria, and have a transparent pore through which the host is penetrated via a hyphal peg. Hyphopodia are flattened, 9–15 μm in diameter, and slightly darker than the hyphae that bear them. Runner hyphae are septate and 4–7 μm wide and are frequently grouped into strands.

Perithecia erupt through leaf sheaths and are black and 200–400 μm in diameter, with beaks 150–300 μm long. Asci measure 10–15 × 80–130 μm, and each has a distinct apical ring 2–3 μm in diameter. Ascospores (eight per ascus) are hyaline, slender, 3–4 × 70–100 μm, and three- to seven-septate.

The take-all fungus is best cultured from ascospores or by baiting the fungus from infected roots onto seedlings. It grows vegetatively and rarely forms perithecia on nutrient media. Instead, hyaline microspores (phialospores), 1–1.5 × 4–7 μm, may develop from sporogenous terminal hyphae.

Disease Cycle

G. graminis var. *tritici* persists in infected host plants and in host debris. Both hyphae and ascospores can serve as inoculum, although the latter are less important epidemiologically. Roots become infected as they grow through soil near infested debris. Roots are first colonized superficially by pigmented (runner) hyphae (Plate 36) and then are penetrated directly by hyaline (feeder) hyphae beneath hyphopodia. Infection occurs throughout the growing season, but temperatures between 10 and 20° C are optimal. Root infections in autumn or early spring are most likely to progress to the crown and foot.

Most plant-to-plant spread of the pathogen occurs via runner hyphae advancing across "root bridges." Pathogen dispersal otherwise depends on the movement of infested soil and host debris. Ascospores are also dispersed by splashing rain and to some extent by wind. Ascospores are thought to have introduced the take-all pathogen into new polder soils in The Netherlands.

Take-all is favored by neutral to alkaline, infertile (especially nitrogen- and phosphorus-deficient), and poorly drained soils. Like many other diseases caused by soilborne pathogens, it is favored by continuous host-cropping. Because the fungus is most active in moist soil, the disease is most severe in wet areas or years and in irrigated fields. Host plants ripen prematurely (Plate 34) when water transport through roots and culms does not keep pace with water loss through leaves. The sudden development of whiteheads after a period of hot, dry weather gives the impression that the disease develops late in the season and is favored by hot, dry conditions. However, root and basal stem damage occurs much earlier, and hot, dry weather serves only to accelerate ripening and water stress. The pathogen rapidly becomes inactive at any stage during the disease cycle if conditions turn dry.

Control

Crop rotation is the best way to control take-all. Usually a one-year break from wheat or barley is sufficient to reduce inoculum to inconsequential levels. Oats, corn, and dicotyledonous crops such as legumes are suitable alternative crops in most soils. Grassy weeds and volunteer wheat plants can harbor the pathogen and should be destroyed. Take-all may be severe in wheat following alfalfa where grassy weeds were common and has been known to occur in wheat after soybeans. No wheat or barley cultivars resistant to take-all are available.

Because the fungus depends on infested crop residue for survival apart from host plants, tillage can be beneficial by fragmenting and accelerating the decomposition of the infested residue. In North America and Australia, frequent and thorough tillage of the soil limits take-all in consecutive wheat crops. However, in England and parts of the United States (Georgia), take-all is sometimes less severe with direct drilling (no-till).

Applications of lime generally favor take-all, especially if soil pH is elevated above 6.0. Ammoniacal and slow-release forms of nitrogen may suppress take-all, in contrast to nitrates, which favor the disease. Equally important is an adequate supply of other essential nutrients to promote root growth. Wheat should not be nutritionally stressed for lack of nitrogen, phosphorus, potassium, or trace nutrients at any time during the growing season. Where soils are slightly acid (i.e., pH 5.5–6.0), the use of chloride in combination with ammonium in the seed furrow provides some control.

A phenomenon called "take-all decline" can be effective in successive wheat crops, especially in slightly acid soils. After increasing in incidence and severity in the first two to five consecutive wheat crops, take-all typically becomes less severe in subsequent crops. The decline is a form of biological control caused by a buildup of microorganisms antagonistic to *G. graminis* var. *tritici*. Take-all decline occurs in most wheat-growing regions of the temperate world.

Selected References

Asher, M. J. C., and Shipton, P. J., eds. 1981. Biology and Control of Take-All. Academic Press, New York. 538 pp.

Christensen, N. W., Powelson, R. L., and Brett, M. 1987. Epidemiology of wheat take-all as influenced by soil pH and temporal changes in organic soil N. Plant Soil 98:221–230.

Cook, R. J. 1981. The influence of rotation crops on take-all decline phenomenon. Phytopathology 71:189–192.

Hornby, D. 1969. Methods of investigating populations of the take-all fungus (*Ophiobolus graminis*) in soil. Ann. Appl. Biol. 35:435–442.

Huber, D. M., and Watson, R. D. 1974. Nitrogen form and plant disease. Annu. Rev. Phytopathol. 12:139–165.

Moore, K. J., and Cook, R. J. 1984. Increased take-all of wheat with direct drilling in the Pacific Northwest. Phytopathology 74:1044–1049.

Nilsson, H. E. 1973. Varietal differences in resistance to take-all disease of winter wheat. Swed. J. Agric. Res. 3:89–93.

Reis, E. M., Cook, R. J., and McNeal, B. L. 1982. Effect of mineral nutrition on take-all of wheat. Phytopathology 72:224–229.

Shipton, P. J., Cook, R. J., and Sitton, J. W. 1973. Occurrence and transfer of a biological factor in soil that suppresses take-all of wheat in eastern Washington. Phytopathology 63:511–517.

Taylor, R. G., Jackson, T. L., Powelson, R. L., and Christensen, N. W. 1983. Chloride, nitrogen form, lime, and planting date effects on take-all root rot of winter wheat. Plant Dis. 67:1116–1120.

Walker, J. 1973. *Gaeumannomyces graminis* var. *tritici*. Descriptions of Pathogenic Fungi and Bacteria, No. 383. Commonwealth Mycological Institute, Association of Applied Biologists, Kew, Surrey, England.

Walker, J. 1975. Take-all disease of Gramineae: A review of recent work. Rev. Plant Pathol. 54:113–144.

Weller, D. M., and Cook, R. J. 1983. Suppression of take-all of wheat by seed treatments with fluorescent pseudomonads. Phytopathology 73:463–469.

Pythium Root Rot

Pythium spp. inhabit agricultural soils the world over and parasitize the roots of numerous higher plants. Symptoms of *Pythium* infection are subtle and difficult to diagnose. All spring- and autumn-sown small grains and forage grasses can be thinned and stunted by Pythium root rot (also called browning root rot). The disease is caused by one or more *Pythium* spp. acting singly or in combination.

Symptoms of Pythium root rot tend to be more uniformly distributed in wheat fields than are symptoms of other diseases caused by soilborne pathogens, such as take-all and Rhizoctonia root rot. Because many or most plants can be affected to the same degree, the disease may go unnoticed. However, several studies suggest that the increased wheat growth associated with soil fumigation results mainly from the control of *Pythium* spp.

Symptoms

Pythium root rot is difficult to diagnose without comparing diseased plants with plants known to be free of *Pythium* (grown in chemical- or heat-treated soil). Severe *Pythium* damage to wheat appears as missing, stunted, and poorly tillered plants (Plate 37). On seedlings, the first true leaf is often noticeably shorter than a normal, healthy first leaf, perhaps as a result of early embryo infections in moist soil. Adult plants affected by *Pythium* may appear stunted, chlorotic, and nitrogen-deficient. They often show delayed heading and maturity and develop heads that are small and poorly filled.

Pythium-damaged fine roots are difficult to recover from soil. They have yellow-brown to necrotic root tips and few root hairs. A small root system with brown lateral roots and root cortical tissue is typical (browning root rot). Severe *Pythium* infection results in general root necrosis.

Causal Organisms

To date, at least 19 *Pythium* spp. exhibit some degree of pathogenicity on wheat roots. Species best documented as the cause of Pythium root rot are *P. arrhenomanes* Drechs., *P. graminicola* Subr., *P. aphanidermatum* (Edson) Fitz., *P. volutum* Vant. & Tru., *P. myriotylum* Drechs., *P. ultimum* var. *sporangiiferum*, *P. aristosporum*, and *P. irregulare*. Other species associated with root rot symptoms include *P. torulosum*, *P. sylvaticum*, and *P. heterothallicum*.

Pythium spp. are identified by their morphological characteristics on a variety of special media. They have aseptate mycelium and grow best between 22 and 30°C (see Downy Mildew). However, some root-infecting species can grow at 0–3°C (see Snow Rot), and others grow at 33–35°C.

Pythium spp. produce sporangia that are spherical to lobate, 3–5 μm wide, and up to 150 μm long, with prominent exit tubes. Zoospores, approximately 12 μm in diameter, number 10–40 per sporangium and are propelled by two flagella. Oospores are smooth, spherical to oblong, and 27–40 μm in diameter, smaller than those of *Sclerophthora macrospora* (Fig. 46).

Disease Cycle

The pathogens exist mainly in the top 10–15 cm of soil, where seed is sown, where fine wheat roots are abundant, and where crop residues, especially wheat chaff and straw, are deposited. Such crop residues serve as a source of nutrients and moisture for *Pythium* spp.

Oospores in and on residues serve as primary inoculum (Fig. 65). They can persist for years free in soil or embedded in fragments of plant refuse. They germinate and initiate infections directly or indirectly from released zoospores. Initial infections often begin in the embryo of germinating wheat seeds a few days after seeding. Thereafter, new roots, and especially fine lateral roots and root hairs, become infected.

New oospores are produced in parasitized roots, and some saprophytic multiplication also may occur on wheat straw. Because *Pythium* spp. are primary colonists, saprophytic colonization is limited to clean, unweathered straw, chaff, and debris not already colonized by other fungi.

Straw left on the soil surface or incorporated only slightly, by reduced or conservation tillage, favors Pythium root rot. Poor

1. Healthy wheat plant (left) and plant infected with aster yellows mycoplasma (right). (Courtesy C. C. Gill)

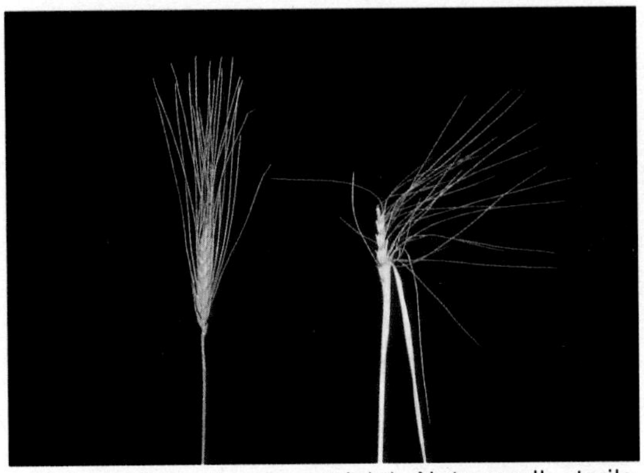

2. Aster yellows in wheat (right). Note small, sterile head with distorted awns compared to healthy plant (left). (Courtesy L. N. Chiykowski)

3. Healthy leaf (left) and leaves with bacterial mosaic seven days after inoculation with *Clavibacter michiganense* subsp. *tessellarius* by vacuum infiltration. (Courtesy A. K. Vidaver)

4. Symptoms of black chaff (caused by *Xanthomonas campestris* pv. *translucens*) on glumes. (Courtesy M. G. Boosalis)

5. Symptoms of black chaff (caused by *Xanthomonas campestris* pv. *translucens*) on leaves. (Courtesy R. A. Kilpatrick)

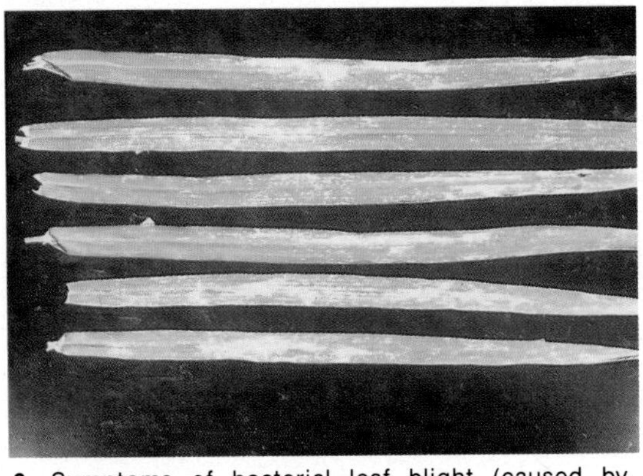

6. Symptoms of bacterial leaf blight (caused by *Pseudomonas syringae* pv. *syringae*) on fully expanded leaves. (Courtesy J. D. Otta)

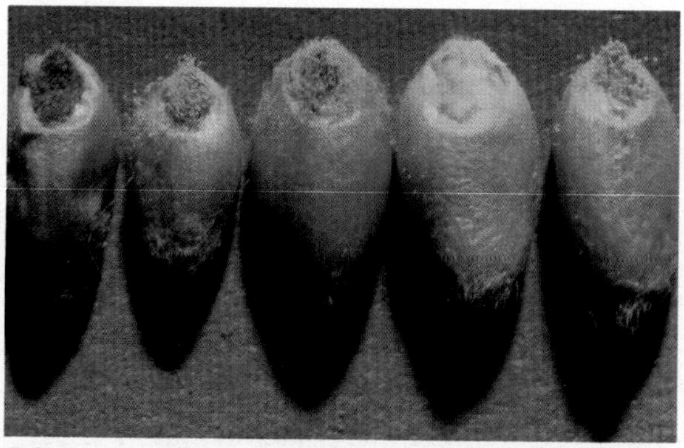

7. Storage fungi emerging from embryos of seed incubated in a moist atmosphere. (Courtesy C. M. Christensen)

8. Wheat seed variously discolored and black-pointed by *Helminthosporium sativum* and *Alternaria* spp. (Courtesy J. E. Huguelet)

9. Black (sooty) head mold (caused by *Cladosporium* sp.). (Courtesy M. V. Wiese)

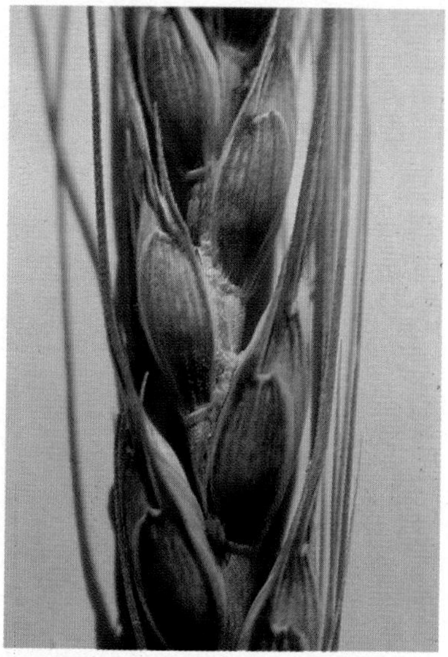

10. Spikelet with mycelium of *Fusarium* sp., a sign of scab. (Courtesy C. R. Pierobom and G. C. Luzzardi)

11. Teliospore cloud of *Tilletia controversa* released during harvest of plants affected by dwarf bunt. (Courtesy J. A. Hoffmann)

12. Glumes spread by sori (bunt balls) of common bunt (caused by *Tilletia tritici* and *T. laevis*). (Courtesy Plant Pathology Department, Washington State University)

13. Kernels damaged by Karnal (partial) bunt (left) and a teliospore of *Tilletia indica* (right). (Courtesy H. J. Dubin)

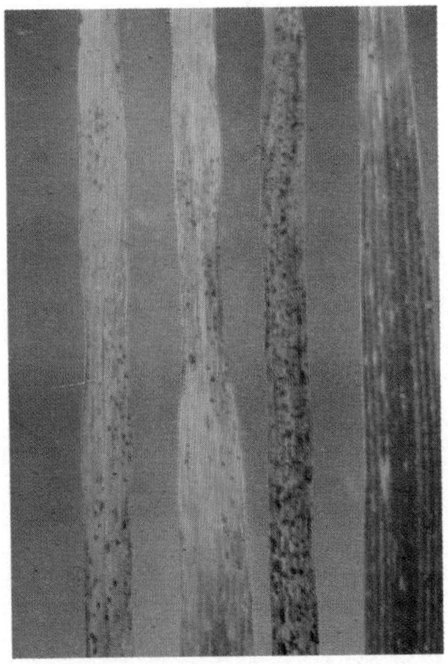

14. Leaves with symptoms of Ascochyta leaf spot, caused by *Ascochyta tritici*. (British Crown Copyright. Used by permission of the Controller of Her Britannic Majesty's Stationery Office)

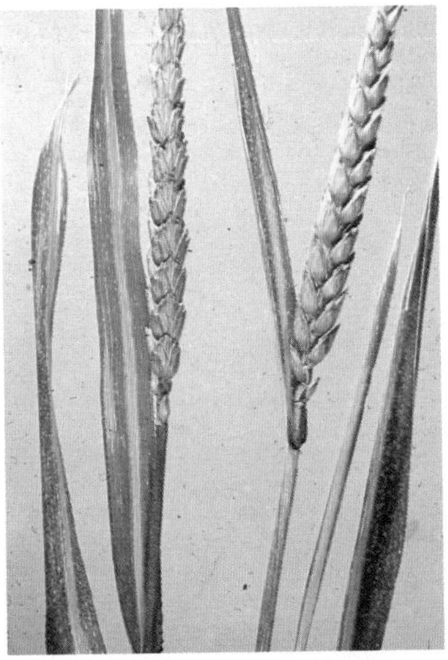

15. Leaves with Cephalosporium stripe, caused by *Cephalosporium gramineum*. (British Crown Copyright. Used by permission of the Controller of Her Britannic Majesty's Stationery Office)

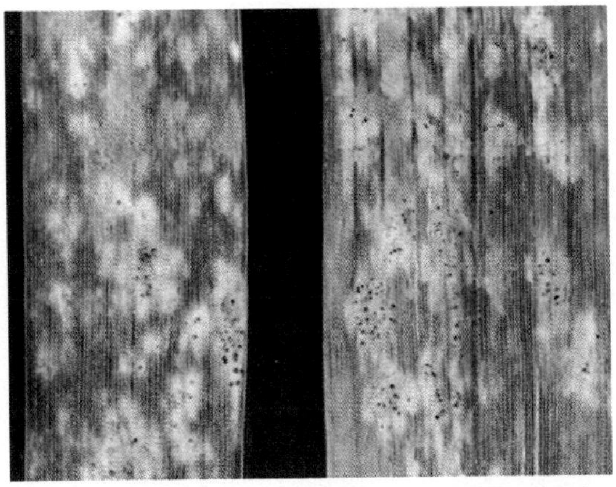

16. Aging colonies of powdery mildew (caused by *Erysiphe graminis* f. sp. *tritici*) speckled with cleistothecia on maturing leaves. (Courtesy Clemson University Extension Service)

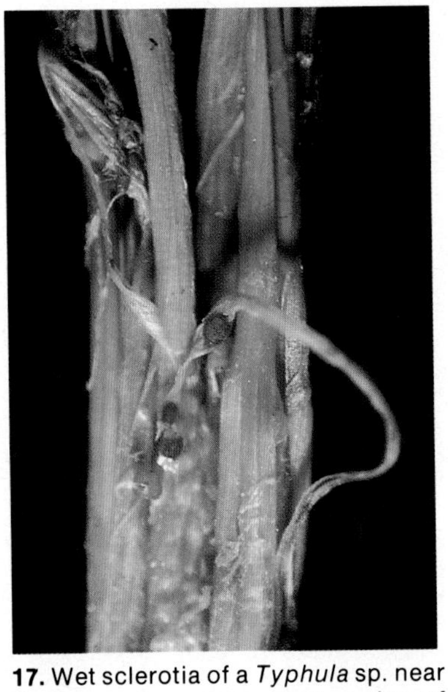

17. Wet sclerotia of a *Typhula* sp. near the base of a young plant, a sign of speckled snow mold. (Courtesy Plant Pathology Department, Washington State University)

18. Young plants killed by Sclerotinia snow mold. Note dark sclerotia of *Sclerotinia borealis* (arrows). (Courtesy J. D. Smith)

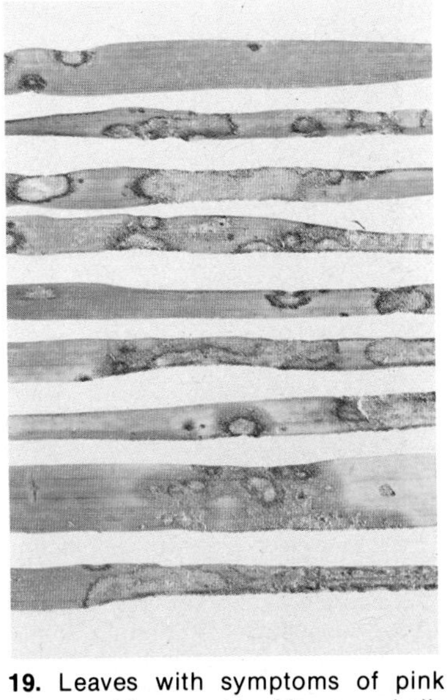

19. Leaves with symptoms of pink snow mold, caused by *Monographella nivalis* (anamorph *Microdochium nivale* [*Fusarium nivale*]). (Courtesy J. D. Smith)

20. Early spring stand of winter wheat thinned by pink snow mold, caused by *Monographella nivalis* (anamorph *Microdochium nivale* [*Fusarium nivale*]). (Courtesy Plant Pathology Department, Washington State University)

21. Plants killed by snow rot (caused by *Pythium* spp.) within standing and running water from snow melt. (Courtesy P. E. Lipps)

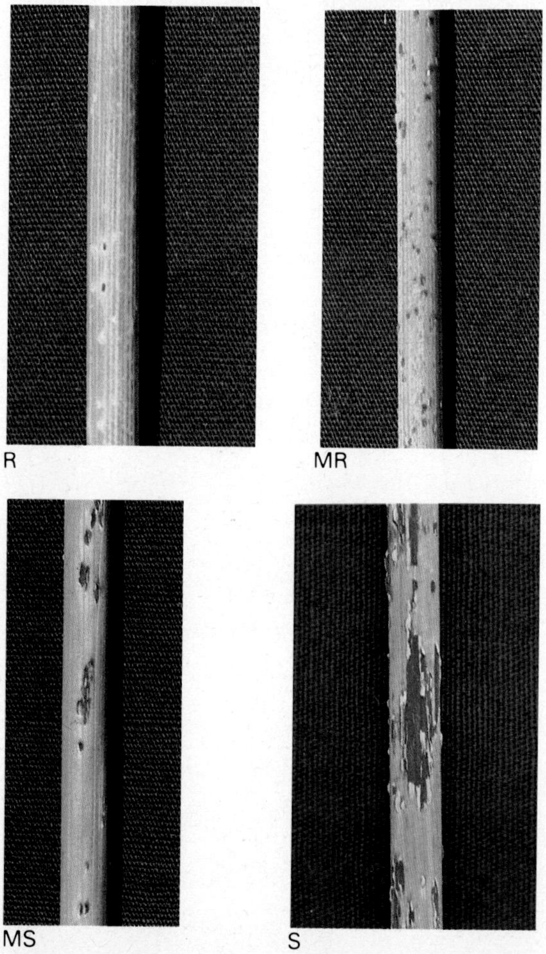

22. Stem rust (caused by *Puccinia graminis* f. sp. *tritici*) infection types on wheat culms: R = resistant, MR = moderately resistant, MS = moderately susceptible, and S = susceptible. (Reprinted from Rust Scoring Guide, by permission of the Research Institute for Plant Protection, Wageningen, The Netherlands, and the International Maize and Wheat Improvement Center, 06600 Mexico, D.F., Mexico)

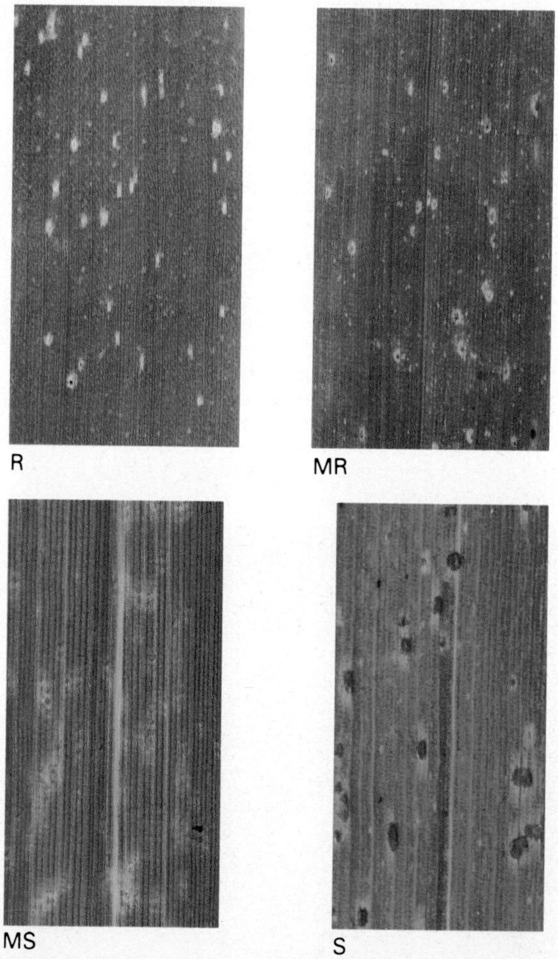

23. Leaf rust (caused by *Puccinia recondita* f. sp. *tritici*) infection types on leaves: R = resistant, MR = moderately resistant, MS = moderately susceptible, and S = susceptible. (Reprinted from Rust Scoring Guide, by permission of the Research Institute for Plant Protection, Wageningen, The Netherlands, and the International Maize and Wheat Improvement Center, 06600 Mexico, D.F., Mexico)

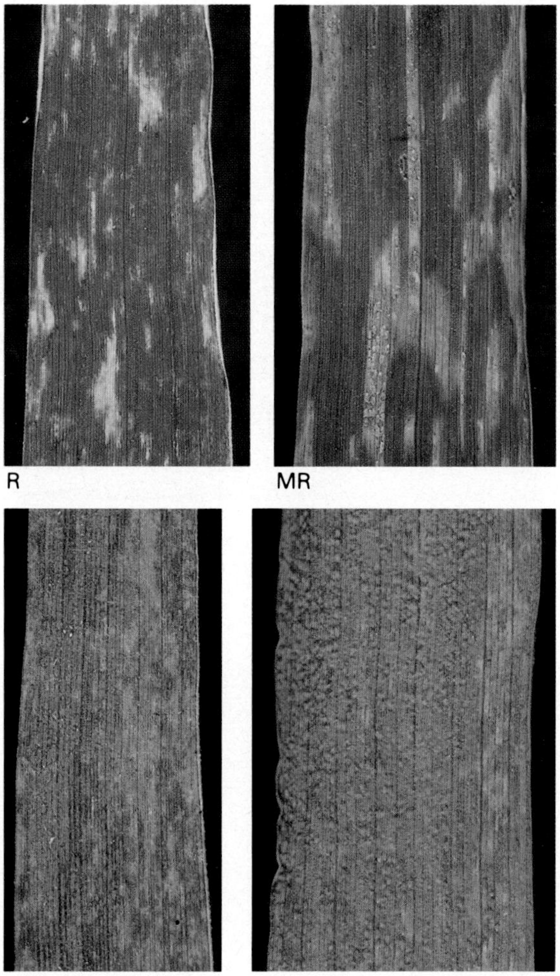

24. Stripe rust (caused by *Puccinia striiformis*) infection types on leaves: R = resistant, MR = moderately resistant, MS = moderately susceptible, and S = susceptible. (Reprinted from Rust Scoring Guide, by permission of the Research Institute for Plant Protection, Wageningen, The Netherlands, and the International Maize and Wheat Improvement Center, 06600 Mexico, D.F., Mexico)

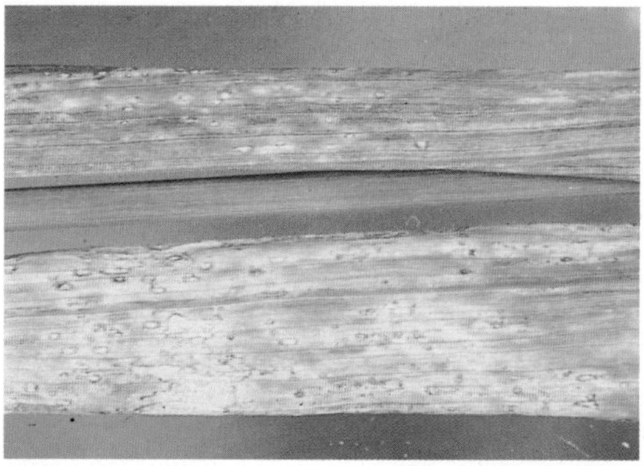

27. Leaves with halo spot (caused by *Selenophoma donacis*). (British Crown Copyright. Used by permission of the Controller of Her Britannic Majesty's Stationery Office)

25. Aecial stage of *Puccinia recondita* f. sp. *tritici* (cause of leaf rust) on the underside of a leaf of meadow rue (*Thalictrum* sp.). (Courtesy D. J. Samborski)

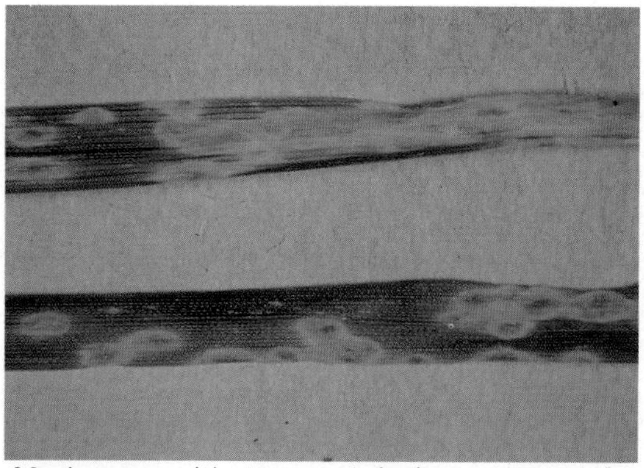

26. Leaves with tan spot lesions, caused by *Pyrenophora tritici-repentis*. (Courtesy E. L. Sharp)

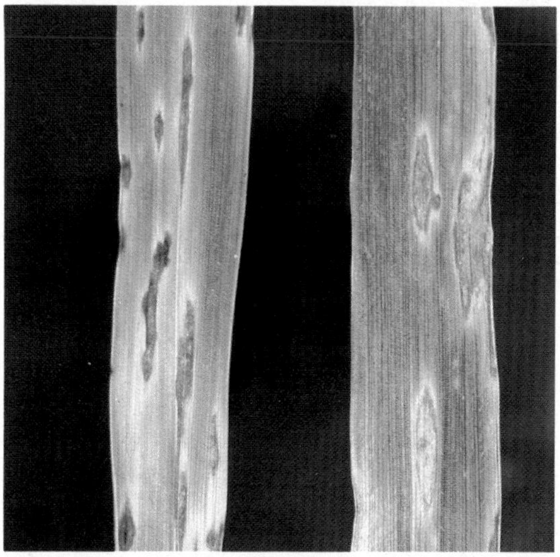

28. Septoria leaf blotch caused by *Leptosphaeria nodorum* (anamorph *Septoria nodorum*). (Courtesy Clemson University Extension Service)

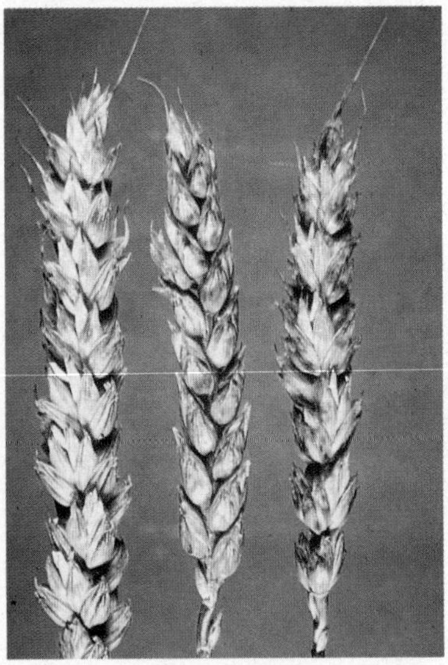

29. Glume blotch caused by *Leptosphaeria nodorum* (anamorph *Septoria nodorum*). (British Crown Copyright. Used by permission of the Controller of Her Britannic Majesty's Stationery Office)

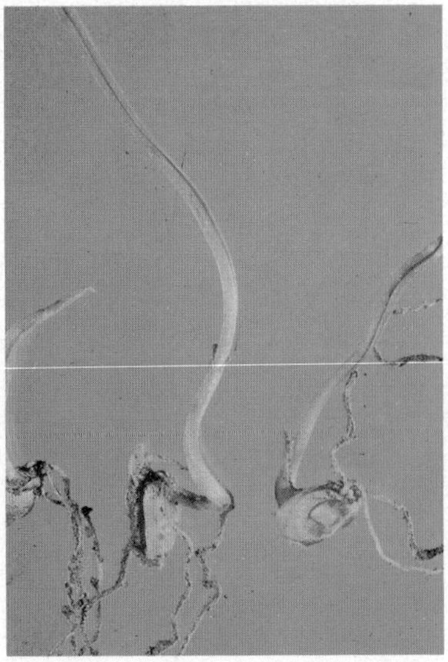

30. Seedling blight caused by *Leptosphaeria nodorum* (anamorph *Septoria nodorum*). (Courtesy BASF)

31. Leptosphaeria leaf spot caused by *Leptosphaeria herpotrichoides*. (Courtesy R. M. Hosford, Jr.)

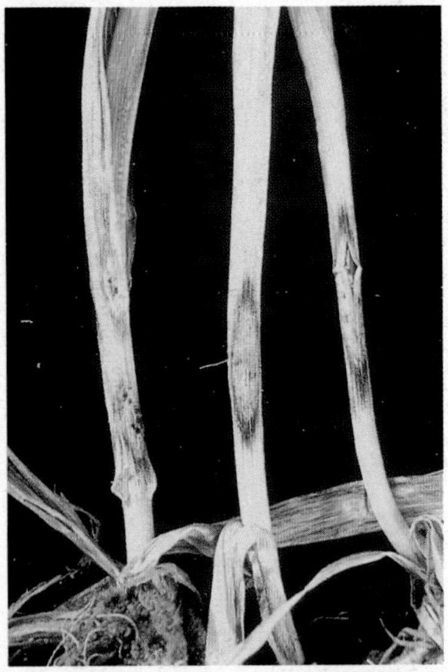

32. Culms with foot rot (eyespot) lesions caused by *Pseudocercosporella herpotrichoides*. (British Crown Copyright. Used by permission of the Controller of Her Britannic Majesty's Stationery Office)

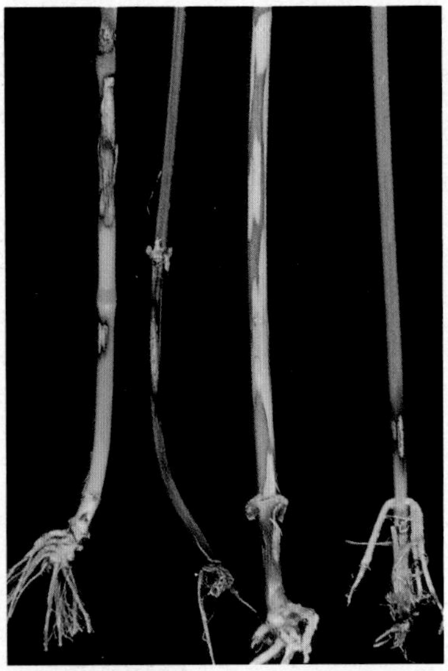

33. Sharp eyespot on culms, caused by *Rhizoctonia cerealis*. (Copyright Plant Breeding Institute, Cambridge, England. Used by permission. Contributed by T. W. Hollins via R. J. Cook)

34. Premature white heads resulting from plant stress associated with severe root or foot disease. (Courtesy BASF)

35. Degenerate roots and blackened culm bases indicative of take-all, caused by *Gaeumannomyces graminis* var. *tritici*. (Courtesy M. V. Wiese)

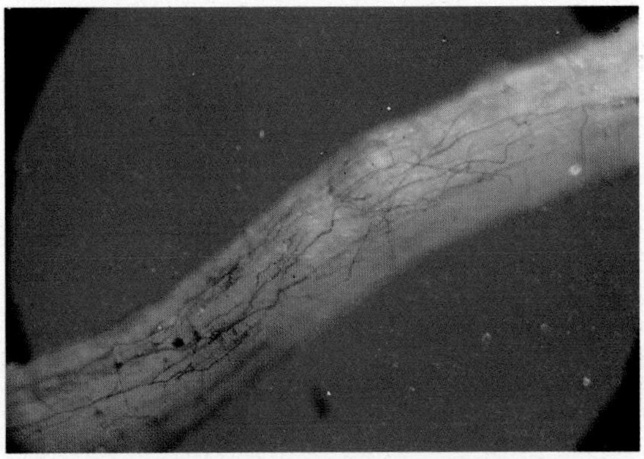

36. Dark runner hyphae of *Gaeumannomyces graminis* var. *tritici* on a root segment. (Courtesy BASF)

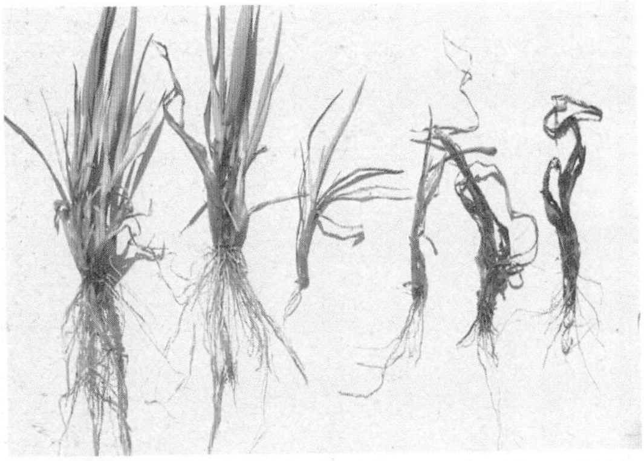

37. Seedlings damaged by Pythium root rot in wet soil. (Courtesy G. W. Bruehl)

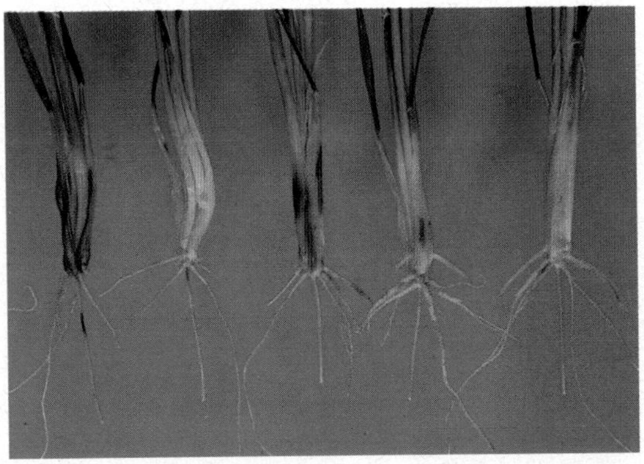

38. Young plants with early symptoms of common root rot (left) caused by *Bipolaris sorokiniana* (syn. *Helminthosporium sativum*) and a healthy plant (right). (Courtesy R. D. Tinline, Agriculture Canada, via R. J. Cook)

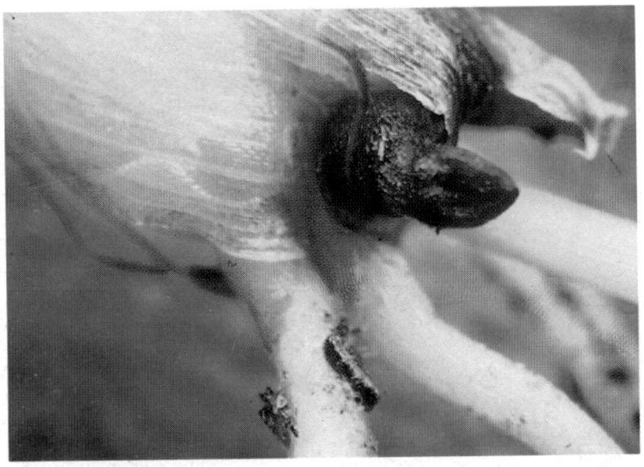

39. Secondary root emerging from crown, showing damage by *Fusarium culmorum*. (Courtesy R. J. Cook, USDA)

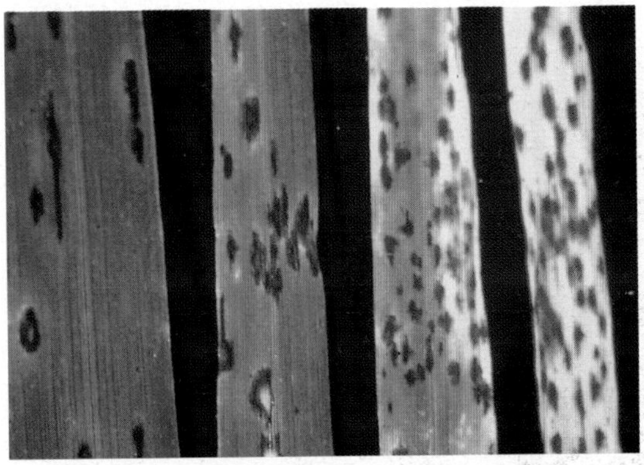

40. Spot blotch on leaves, caused by *Bipolaris sorokiniana* (syn. *Helminthosporium sativum*). (Courtesy R. V. Sturgeon)

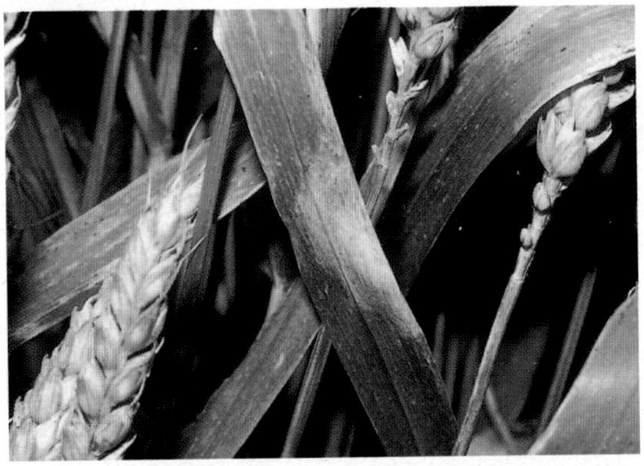

41. Fusarium leaf spot. (Courtesy R. J. Cook, Ministry of Agriculture, Fisheries and Food, United Kingdom)

42. Subcrown internodes darkened by *Bipolaris sorokiniana* (syn. *Helminthosporium sativum*) (left) and healthy plants (right). (Courtesy R. D. Tinline, Agriculture Canada, via R. J. Cook)

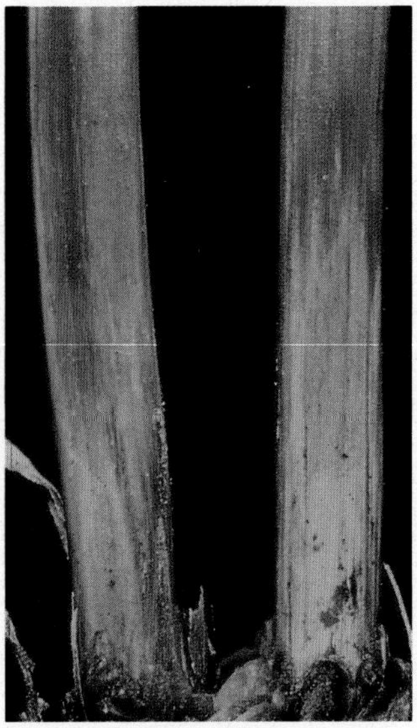

43. Severe Fusarium (dryland) root rot that has progressed through the crown and up the culm base. (Courtesy BASF)

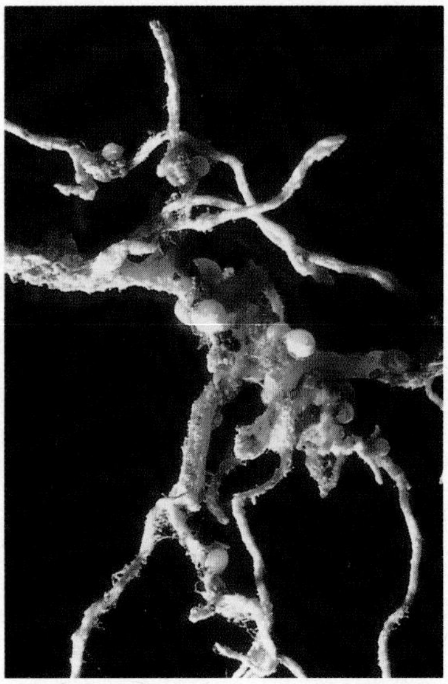

44. Cysts of the cereal cyst nematode, *Heterodera avenae*, on roots. (Courtesy R. H. Brown)

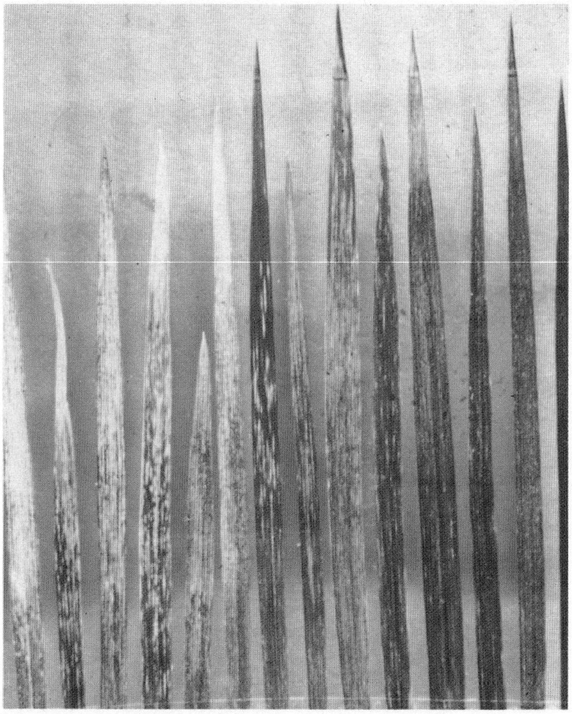

45. Array of leaf symptoms exhibited by virus-infected wheat plants. (Courtesy W. H. Sill)

46. Cluster of plants infected with barley yellow dwarf virus. (British Crown Copyright. Used by permission of the Controller of Her Britannic Majesty's Stationery Office)

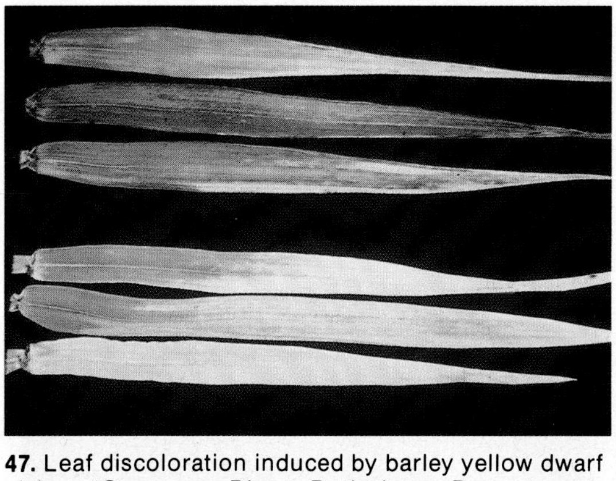

47. Leaf discoloration induced by barley yellow dwarf virus. (Courtesy Plant Pathology Department, Washington State University)

48. Enations on leaves with enanismo. (Courtesy G. E. Galvez)

49. Leaf segment supporting larvae of the wheat curl mite (*Aceria tulipae*). (Courtesy Plant Pathology Department, University of Nebraska, Lincoln)

50. Plants infected with wheat yellow mosaic virus (wheat spindle streak mosaic virus). (Courtesy M. V. Wiese)

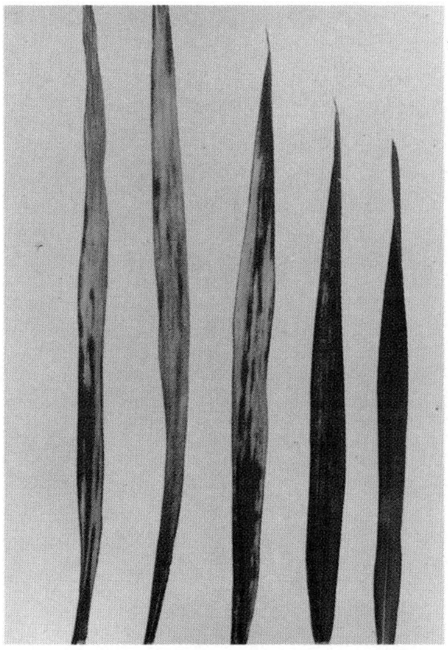

51. A range of leaf symptoms of seedborne wheat yellows. (Courtesy Xiaojie Wu, via R. J. Cook)

52. Pale western cutworm pupa (left), feeding larva (center), and adult (right). (Courtesy Department of Plant, Soil and Entomological Sciences, University of Idaho)

53. Leaves damaged by feeding larvae of the cereal leaf beetle. (Courtesy J. L. Clayton)

54. Culms defoliated and broken by grasshoppers. (Courtesy Department of Plant, Soil and Entomological Sciences, University of Idaho)

55. Wireworm (arrow) foraging at the base of young plants. The center plant has been severed. (Courtesy M. V. Wiese)

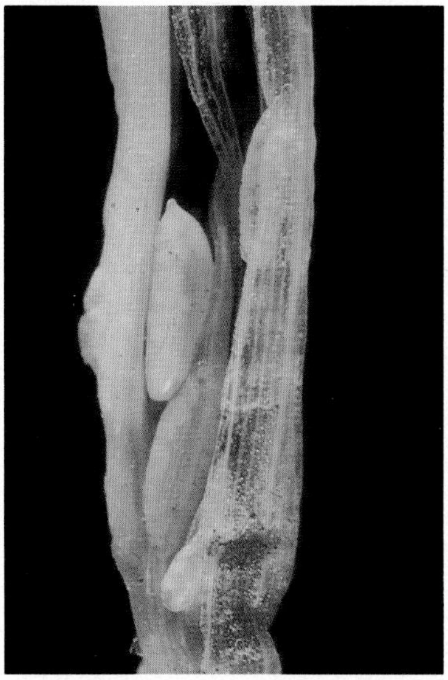

56. Hessian fly larvae attached to culm within a lower leaf sheath. (Courtesy Department of Plant, Soil and Entomological Sciences, University of Idaho)

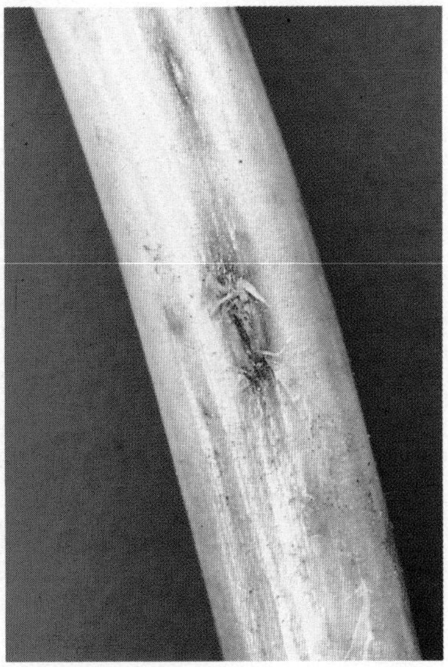

57. Billbug injury at base of culm. (Courtesy N. A. Smith)

58. Grain weevils (*Sitophilus* sp.) and weevil-damaged wheat kernels. (Courtesy Oregon State University Extension Service, in cooperation with the Environmental Protection Agency)

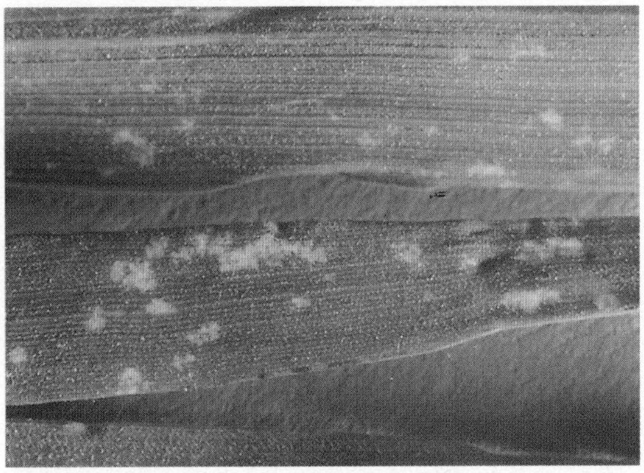

59. Symptoms of leafhopper feeding on leaves. (Courtesy Oregon State University Extension Service, in cooperation with the Environmental Protection Agency)

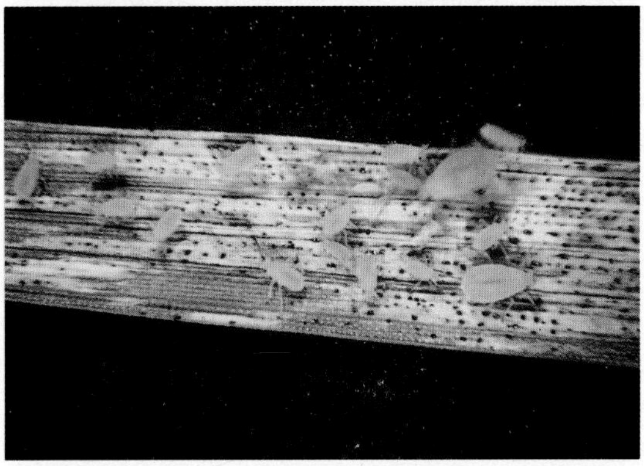

60. Leaf with greenbugs (*Schizaphis graminum*), showing injury. (Courtesy Entomology Department, University of Nebraska, Lincoln)

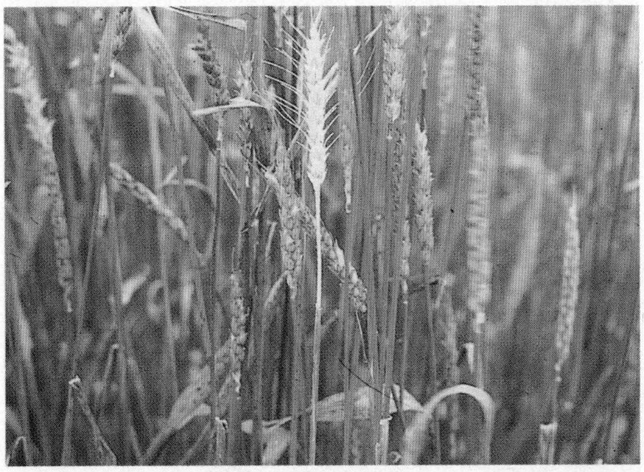

61. Loose (white) ear—a head that has been severed by the wheat stem maggot or wheat stem sawfly, causing it to be prematurely ripe and bleached. (Courtesy Plant Pathology Department, University of Nebraska, Lincoln)

62. Leaves injured by chlorine gas. (Courtesy I. J. Hindawi)

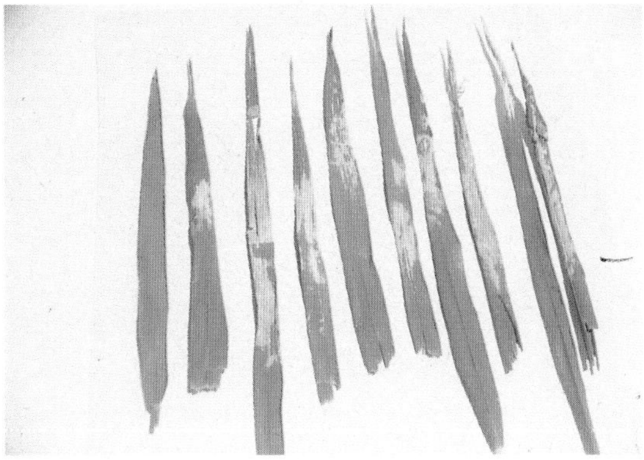

63. Leaves injured by sulfur dioxide and a healthy leaf (left). (Courtesy T. W. Barnett)

64. Leaf injured by chronic exposure to ozone (right) and a healthy leaf (left). (Courtesy A. S. Heagle)

65. Localized injury to leaves from chemical pesticides and/or fertilizers concentrated and applied in droplets. (Courtesy M. V. Wiese)

66. Color-banded seedlings. (Courtesy G. W. Bruehl)

67. Heads injured by frost. (Courtesy G. W. Bruehl)

68. Seedlings with frost-induced albinism. Bleached tillers often recover their green color and resume growth. (Courtesy R. H. Callihan)

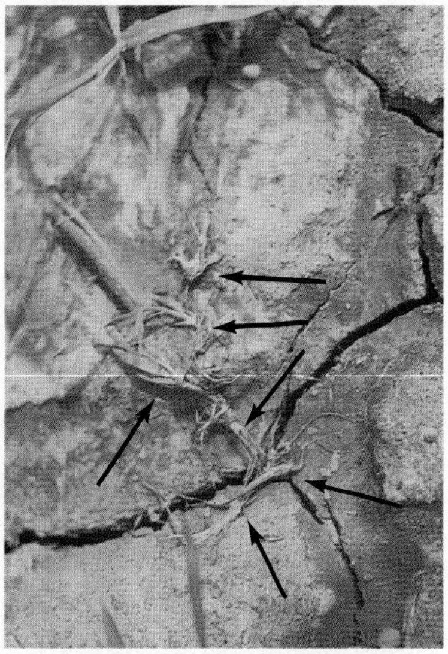

69. Crowns of overwintering plants (arrows) uprooted by mechanical forces from frost and ice formation in soil. (Courtesy M. C. Shurtleff)

70. Mature kernels sprouting in heads before harvest. (Courtesy A. M. Whitte)

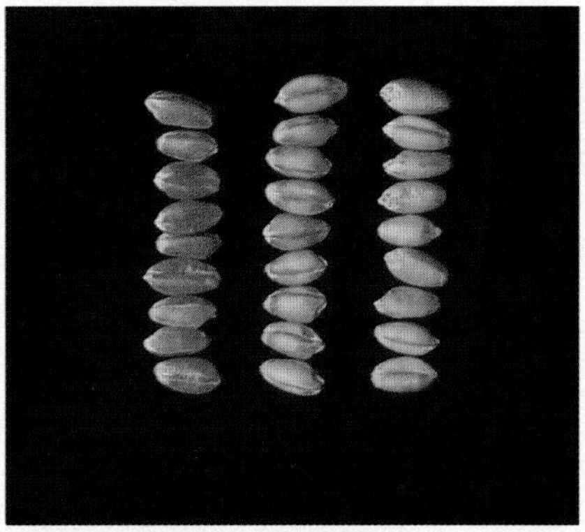

71. Normal kernels (left), yellow-spotted kernels (center), and yellow berries (right). (Courtesy A. L. Scharen)

72. White-tipping of plants deficient in copper. (British Crown Copyright. Used by permission of the Controller of Her Britannic Majesty's Stationery Office)

73. Leaves with interveinal iron chlorosis. (Courtesy T. G. Atkinson)

74. Wheat sown in soil with pH 4.5 (foreground) and pH 6.2 (background). (Courtesy M. V. Wiese)

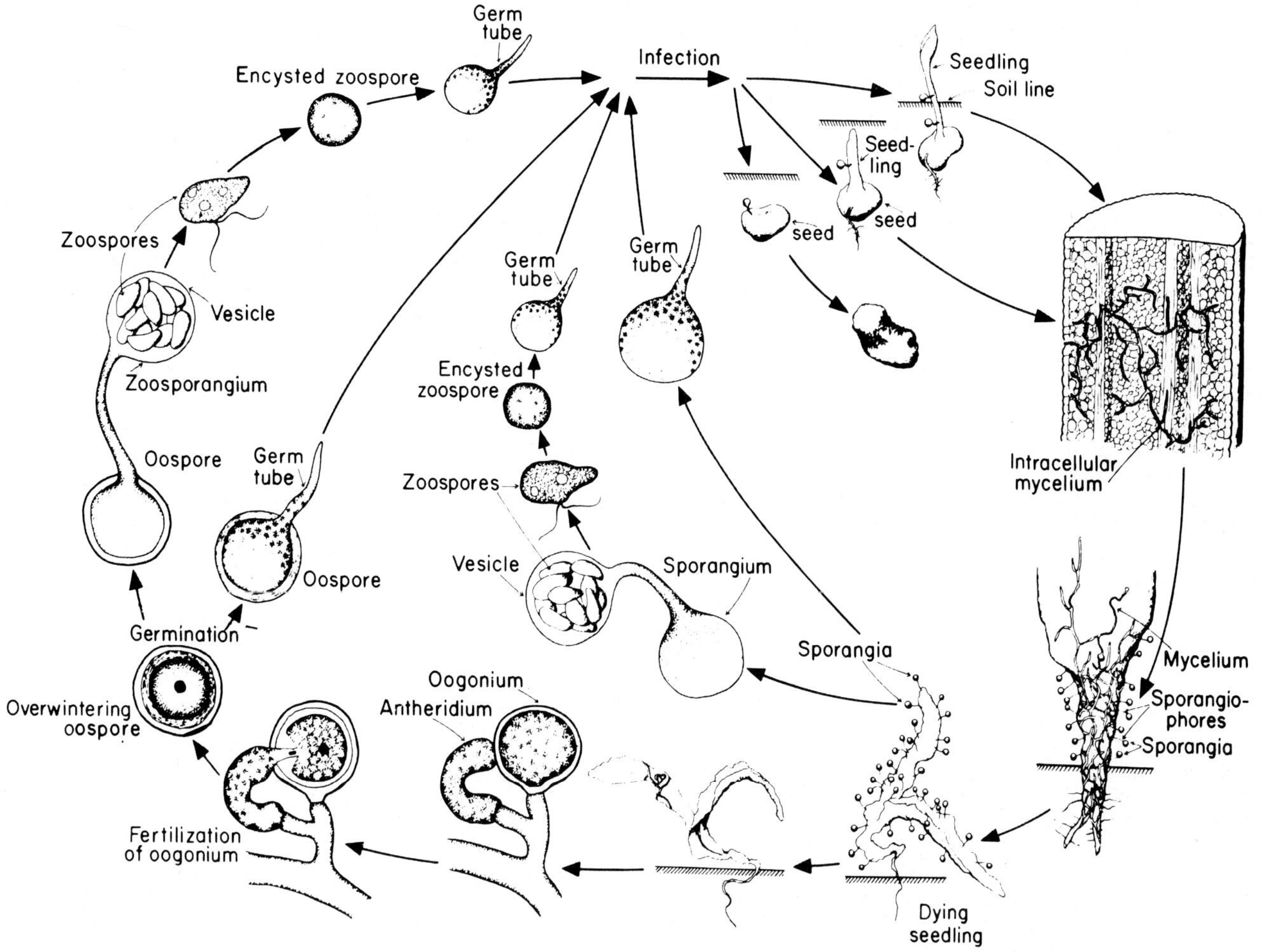

Fig. 65. Disease cycle of *Pythium* spp. on seed and seedlings. (Reprinted, by permission, from G. N. Agrios, Plant Pathology, Fig. 50, 2nd ed., copyright 1978, Academic Press, New York)

wheat growth in heavy crop residues infested with *Pythium* spp. is sometimes incorrectly diagnosed as nitrogen deficiency or stress from phytotoxins in rotting straw.

Control

Resistant wheat cultivars are not yet available, and oospores are too long-lived in soil and the host range of *Pythium* spp. is too broad for crop rotation to significantly deter Pythium root rot. Because *Pythium* is most destructive in wet soil, a seeding date that permits seedlings to emerge in a relatively dry or well-drained seedbed is beneficial. Ridge seeding, as opposed to furrow seedings, can also provide some control. High-quality seed and supplemental phosphorus should be used where this disease is a problem. Old seed that is fully germinable in the laboratory may not emerge in the field because of rapid invasion by *Pythium*. Systemic fungicides, such as metalaxyl, can provide important early seedling protection.

Selected References

Chamswarng, C., and Cook, R. J. 1985. Identification and comparative pathogenicity of *Pythium* species from wheat roots and wheat-field soils in the Pacific Northwest. Phytopathology 75:821–827.

Cook, R. J., and Haglund, W. A. 1982. Pythium root rot: A barrier to yield of Pacific Northwest wheat. Wash. State Univ. Coll. Agric. Res. Bull. No. XB0913. 20 pp.

Cook, R. J., Sitton, J. W., and Waldher, J. T. 1980. Evidence for *Pythium* as a pathogen of direct-drilled wheat in the Pacific Northwest. Plant Dis. 64:102–103.

Cook, R. J., and Zhang, B.-X. 1985. Degrees of sensitivity to metalaxyl within the *Pythium* spp. pathogenic to wheat in the Pacific Northwest. Plant Dis. 69:686–688.

Plaats-Niterink, A. J. van der. 1981. Monograph of the Genus *Pythium*. Centraalbureau voor Schimmelcultures, Baarn, The Netherlands. Inst. R. Neth. Acad. Sci. Lett. Stud. Mycol. 21. 242 pp.

Singleton, L. L., and Ziv, O. 1981. Effects of *Pythium arrhenomanes* and root-tip amputation on wheat seedling development. Phytopathology 71:316–319.

Vanterpool, T. C. 1962. Pythium root rot of wheat in Saskatchewan. Can. Plant Dis. Surv. 42:214–215.

Waterhouse, G. M. 1967. Key to *Pythium* Pringsheim. Mycological Paper 109. Commonwealth Mycological Institute, Kew, Surrey, England. 15 pp.

Waterhouse, G. M. 1968. The genus *Pythium* Pringsheim. Mycological Paper 110. Commonwealth Mycological Institute, Kew, Surrey, England. 50 pp.

Common (Dryland) Root and Foot Rot and Associated Leaf and Seedling Diseases

Infections by *Fusarium* and *Bipolaris* spp. are nonspecific and are manifested as root and foot rots, leaf spots, and seedling blights. The fungi also are seed and head pathogens (see Scab [Fig. 17 and Plate 10]). Their attacks on different plant parts and during different growth stages can be distinct but often are interrelated, and more than one pathogen may be involved. A severe phase of the disease, expressed as a decay and brown discoloration of the basal stem of drought-stressed plants, is called "dryland foot rot."

Root, crown, and foot rots are major diseases of wheat when moisture is deficient. The causal fungi are widely distributed as unspecialized pathogens on most small-grain cereals and numerous grasses. Root rot complexes on the Canadian Prairies are estimated to decrease wheat yields 5.7% annually. For most of North America, yield deficits of 3–4% are attributed to common root rot. Common root rot and associated leaf and head damage lower wheat yields by thinning stands, reducing head numbers, and causing premature ripening and shriveled grain.

Symptoms

Seedling blights. Death of seedlings before or soon after emergence is the most dramatic result of early fungus attack. Seedling blight caused by *Fusarium* spp. arises principally from seedborne inoculum (seeds from heads affected by scab). Soilborne inoculum rarely causes seedling blight but can damage wheat seedlings in warm, dry soils.

Brown lesions on coleoptiles, subcrown internodes, roots, and culms of seedlings are early symptoms of infection (Plate 38). Darkening of the subcrown internode is a usual symptom of *B. sorokiniana* infection. Soilborne *Fusarium* spp. tend to invade the base of the subcrown internode and secondary roots as they emerge from the crown (Plate 39). Infections that progress up the basal stem are most damaging and cause tiller abortion if they occur early or premature ripening (Plate 34) if they occur after heading.

Leaf spots. Wheat plants approaching maturity or supporting root infections by *B. sorokiniana* (syn. *Helminthosporium sativum*) are prone to attack aboveground by this same fungus. Leaf lesions frequently coincide with wet weather and with root infections that advance to the foot and basal stem. However, many independent foliar infections via airborne conidia also can occur. Leaf symptoms are most obvious after heading and most frequent on lower leaves. They appear as distinct, elongate, brown-black lesions (spot blotch) that rarely exceed 1 cm in diameter. The dark lesions contrast sharply with green or straw-colored leaf tissues (Plate 40). *Fusarium* spp. also cause light brown or ash-colored lesions on leaves in wet environments (Plate 41).

Common root rot. Common root rot, caused by *B. sorokiniana*, is difficult to diagnose and often goes unnoticed. The most diagnostic symptom is a dark brown or blackened subcrown internode (Plate 42). The infection and diagnostic discoloration may extend into the crown and a short distance up the culm. Diseased plants occur randomly or in irregular patches and appear stunted and chlorotic. Browning or blackening of the primary or secondary root systems may be observed on close examination of washed roots.

Fusarium foot rot. *Fusarium* spp. cause the roots, crown, and lower nodes and internodes of wheat plants to turn brown (Plate 43). The pathogen and discolored culm tissue may extend two or three internodes above the soil. Severely diseased plants mature early, produce mostly shriveled seed, appear bronzed or bleached, and have whiteheads (Plate 34). Usually whole plants, but occasionally one or more tillers on each plant, die prematurely from dryland foot rot. When *Fusarium* spp. progress up the culm, diagnostic cottony pink mycelium frequently develops within the culm or between the culm and lower leaf sheaths (see Scab).

Causal Organisms

B. sorokiniana (Sacc. in Sorok.) Shoem. (syns. *H. sativum* P.K. & B., *H. sorokinianum* Sacc. ex Sorok.) infects wheat roots, leaves, and seedlings. The fungus is easily isolated from rotted host tissues, especially subcrown internodes, using conventional media and is recovered directly from soil with selective media. Its mycelium is typically deep olive-brown. Conidiophores are single or clustered, simple, erect, and 6–8 × 110–150 μm, with septations. Conidia develop terminally and then remain laterally attached at pores beneath each newly produced conidiophore septum.

Conidia are olive-brown and ovate to oblong, with tapered ends and a prominent basal scar. They are 15–20 × 60–120 μm, three- to 10-septate, and slightly curved. Their walls are smooth and thickened noticeably at septa (Fig. 66).

The teleomorph is *Cochliobolus sativus* (Ito and Kurib) Drechs. ex Dastur, an ascomycete. Although not reported in nature, it develops in culture when opposite mating types are appropriately paired. Pseudothecia are black, 300–440 μm in diameter, and globose, with erect beaks 50–200 μm long. Asci are clavate and measure 20–45 × 120–250 μm. Ascospores are hyaline, uniformly filamentous, and spirally flexed within asci. They measure 5–10 × 200–450 μm and are four- to 14-septate.

F. culmorum and *F. graminearum* are the most common *Fusarium* spp. associated with water-stressed wheat. Less virulent and more localized as root and crown rot pathogens are *F. avenaceum, F. acuminatum, F. crookwellense*, and *F. poae*.

F. graminearum occurs in nature as two subpopulations referred to as Group I and Group II. Group I is associated with wheat, other small-grain cereals, and grasses and causes crown and foot rot. Group II is a stalk rot pathogen of corn and causes scab on wheat (see Fig. 17 and Plate 10) following corn.

All six *Fusarium* spp. survive in soil in host debris. In addition, *F. culmorum, F. crookwellense*, and *F. graminearum* Group I survive in soil as chlamydospores, which can be quantitatively isolated on selective media by soil dilution plating. Of these three species, *F. graminearum* Group I is the shortest-lived in soil, and its chlamydospores often are recovered only from host debris.

The teleomorphs of *F. graminearum* Group II and of *F. avenaceum* are *Gibberella zeae* and *G. avenacea* Cook, respectively. No teleomorphs are known for *F. graminearum* Group I, *F. culmorum*, or *F. crookwellense*. *Gibberella* spp. develop perithecia in clusters on basal nodes of mature plants (Fig. 67). Such perithecia range in color from buff to dark blue. They are globose, 125–265 μm in diameter, and rough-walled. Mature asci are clavate, measure 4–10 × 50–80 μm, and contain six to eight spores. Ascospores are hyaline, ellipsoid, 3.3–6.5 × 13–17 μm, and one- to three-septate.

Disease Cycle

The fungi responsible for dryland root, crown, and foot rot are ubiquitous. They occur on cereal and grass hosts, contaminate wheat seed, and persist in soil. Host debris, an important source of inoculum, is occupied primarily through previous parasitism rather than subsequent saprophytism. Conidia of *B. sorokiniana* and chlamydospores of *F. culmorum* can remain inactive in soil for months if hosts are unavailable. The mycelium, conidia, chlamydospores, and ascospores of *Fusarium* spp. all are infectious. The pathogens overwinter as perithecia and chlamydospores and in infested debris. *B. sorokiniana* persists principally as mycelium in crop refuse and as conidia in soil.

Primary infections occur on coleoptiles, subcrown internodes, and primary and secondary roots. Root rotting is

Fig. 66. Conidiophore and conidia of *Bipolaris sorokiniana* (syn. *Helminthosporium sativum*). (Courtesy W. G. Fields)

tolerated as long as supportive new roots are generated. Secondary conidia of *B. sorokiniana* develop when infections progress above the soil level. They are dispersed by wind and initiate lesions on leaves and culms later in the season. The seasonal progress of root rotting, however, is related only to the success of primary inoculum. Given favorable environmental conditions, the threshold population of *F. culmorum* necessary for a detectable effect on wheat yield is approximately 100 propagules (chlamydospores) per gram of soil.

Drought and warm temperatures predispose wheat to common root, crown, and foot rot fungi. Plants stressed by freezing or by Hessian flies also are subject to attack. Pathogen growth is optimal in culture media with low water potentials (−15 to −30 bars). For leaf infections (spot blotch) to occur, relative humidities near 100% and temperatures between 20 and 25° C are optimal.

F. graminearum (Groups I and II) is prevalent in warmer wheat-growing areas, *F. avenaceum* in cooler areas. *F. culmorum* is distributed widely in areas of intermediate temperature. *F. crookwellense* is uncommon but seems adapted to environments that favor *F. culmorum* and *F. graminearum*. *F. acuminatum* is reported in the United States in Wyoming and Colorado.

Control

Seedling infections are reduced by shallow seeding and by using clean or chemically disinfested seed. Also, late-autumn seeding of winter wheat is recommended to decrease seedling exposure to warm soil temperatures. Reduced seedling infection, however, is no guarantee against infection at later growth stages.

Soil fertility must be adequate and balanced to support vigorous root and shoot growth. However, excessive fertilization, especially with nitrogen, favors the diseases by promoting vegetative growth, which in turn increases transpiration and accelerates plant water stress.

Crop rotation is advised to limit alternative hosts and to control the more ephemeral *F. graminearum* and *F. avenaceum*. However, about half of the inoculum present after harvest is functional a year later, and about 10% can survive for nearly two years. Oats are especially supportive of *F. culmorum* and should not precede wheat in rotations where this pathogen is present.

Wheat cultivars differ greatly in resistance or tolerance to each of the major pathogens. Cultivars that tolerate or avoid plant water stress are less susceptible to dryland foot rot. Cultivars tolerant to *F. graminearum* are available in Australia. A few cultivars in Canada are resistant to *B. sorokiniana*.

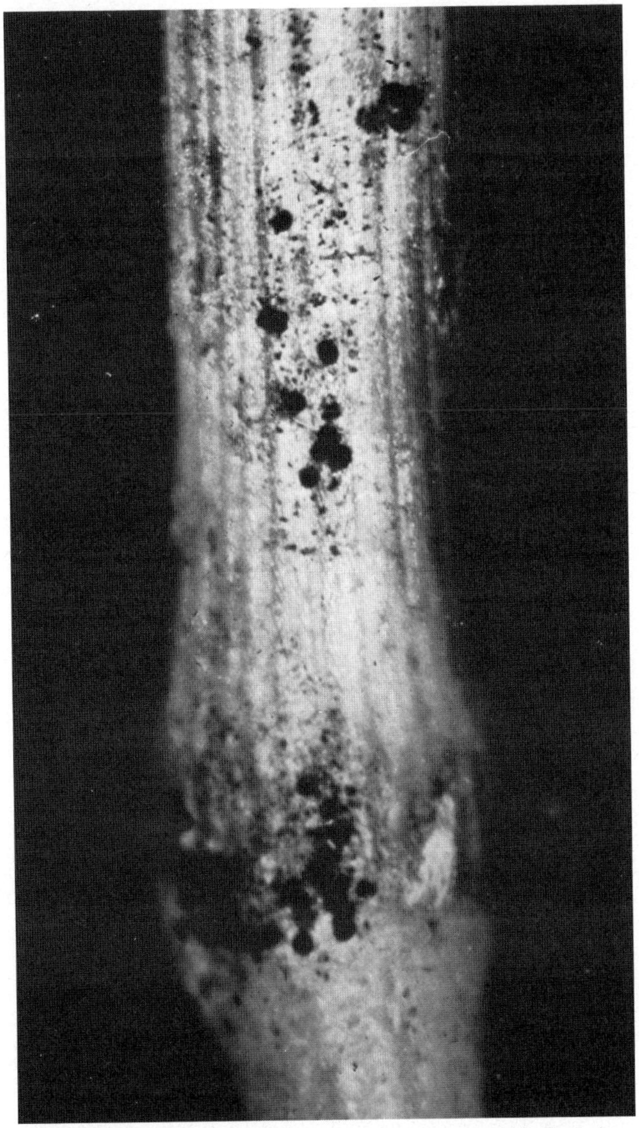

Fig. 67. Perithecia of *Gibberella avenacea* (anamorph *Fusarium avenaceum*) on a culm base. (Courtesy R. J. Cook, USDA)

Selected References

Cook, R. J. 1968. Fusarium root and foot rot of cereals in the Pacific Northwest. Phytopathology 58:127–131.

Cook, R. J. 1980. Fusarium foot rot of wheat and its control in the Pacific Northwest. Plant Dis. 64:1061–1066.

Fenster, C. R., Boosalis, M. G., and Weihing, J. L. 1972. Date of planting studies of winter wheat and winter barley in relation to root and crown rot, grain yields and quality. Nebr. Agric. Exp. Stn. Res. Bull. 250.

Francis, R. G., and Burgess, L. W. 1977. Characteristics of two populations of *Fusarium roseum* 'Graminearum' in eastern Australia. Trans. Br. Mycol. Soc. 68:421–427.

Hill, J. P., Fernandez, J. A., and McShane, M. S. 1983. Fungi associated with common root rot of winter wheat in Colorado and Wyoming. Plant Dis. 67:795–797.

Inglis, D. A., and Cook, R. J. 1986. Persistence of chlamydospores of *Fusarium culmorum* in wheat field soils of eastern Washington. Phytopathology 76:1205–1208.

Ledingham, R. J., Atkinson, T. G., Horricks, J. S., Mills, J. T., Piening, L. J., and Tinline, R. D. 1972. Wheat losses due to common root rot in the prairie provinces of Canada 1969–71. Can. Plant Dis. Surv. 53:113–122.

Liddell, C. M. 1985. The comparative pathogenicity of *Fusarium graminearum* Group 1, *Fusarium culmorum*, and *Fusarium crookwellense* as crown, foot and root rot pathogens of wheat. Australas. Plant Pathol. 14:29–31.

Papendick, R. I., and Cook, R. J. 1974. Plant water stress and development of Fusarium foot rot in wheat subjected to different cultural practices. Phytopathology 64:358–363.

Stack, R. W. 1977. A simple selective medium for isolation of *Cochliobolus sativus* from diseased cereal crowns and roots. Plant Dis. Rep. 61:521–522.

Verma, P. R., and Morrall, R. A. A. 1974. The epidemiology of common root rot in Manitou wheat: Disease progression during the growing season. Can. J. Bot. 52:1757–1764.

Verma, P. R., Morrall, R. A. A., and Tinline, R. D. 1976. The epidemiology of common root rot in Manitou wheat. IV. Appraisal of biomass and grain yield in naturally-infected crops. Can. J. Bot. 54:1656–1665.

Verma, P. R., Tinline, R. D., and Morrall, R. A. A. 1975. The epidemiology of common root rot in Manitou wheat. II. Effects of treatments, particularly phosphate fertilizer, on the incidence and intensity of disease. Can. J. Bot. 53:1230–1238.

Warren, H. L., and Kommedahl, T. 1973. Fertilization and wheat refuse effects on *Fusarium* species associated with wheat roots in Minnesota. Phytopathology 63:103–108.

Primitive (Zoosporic) Root Fungi

Several zoosporic fungi, whose morphology is simple and presumed to be evolutionarily primitive, parasitize a wide range of higher plants. They reside in soil and coexist in the root cells of wheat and other cereals. They often go unnoticed because, by themselves, they are not serious pathogens and to observe them requires close microscopic examination. Most can be found in plants that appear healthy as well as in plants with disease symptoms. All, however, are of interest as known or possible vectors of soilborne plant viruses. *Polymyxa graminis*, the vector of soilborne wheat mosaic virus and wheat yellow mosaic (wheat spindle streak mosaic) virus, is best known in this regard.

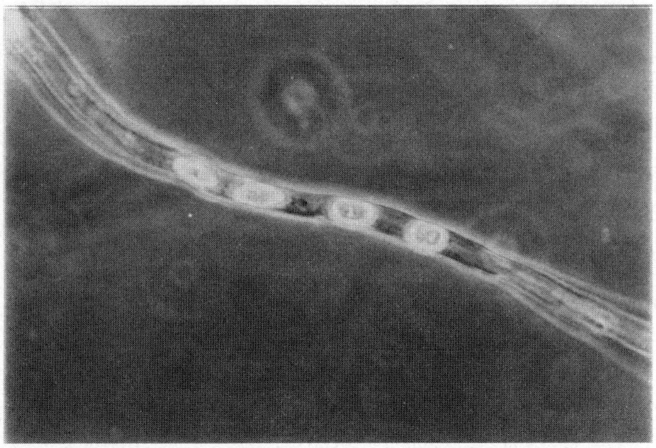

Fig. 68. Resting spores of *Lagena radicicola* in a root hair. (Courtesy D. J. S. Barr)

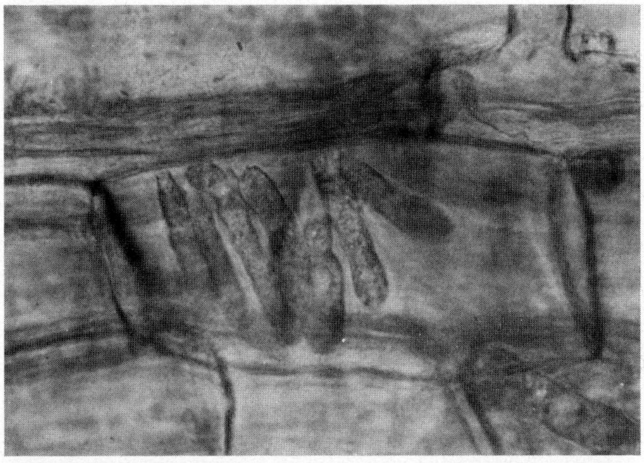

Fig. 69. Thalli of *Lagena radicicola* inside a root epidermal cell. (Courtesy D. J. S. Barr)

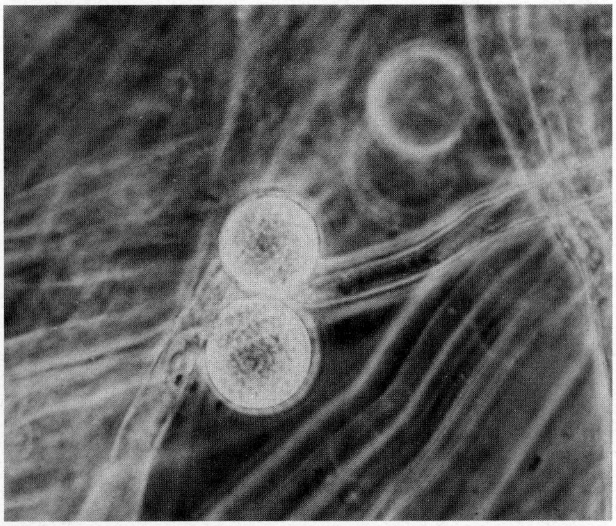

Fig. 70. Sporangia of *Rhizophydium graminis* on root hairs. (Courtesy D. J. S. Barr)

These primitive fungi lack mycelium and are identified from resting spores, sporangia, and zoospores within or on root cells. Although they thrive in wet soils, they require only sporadic rainfall to release zoospores, infect roots, and complete their life cycles. These parasites resemble and sometimes are confused with beneficial mycorrhizal fungi, primarily *Glomus, Acaulospora, Gigospora*, and *Sclerocystis* spp. (Endogonaceae), that also may be present in roots.

Lagena radicicola Vant. & Led., an oomycete, is an obligate parasite of cereals and grasses. It is best recognized by its resting spores in root hairs (Fig. 68). It is infrequently reported on wheat; however, it is not easily recognized.

Zoospores of *L. radicicola* gain access to root epidermal cells by means of a penetration peg. Once inside the penetrated cell, a thallus forms and remains attached to the host cell wall by a characteristic collar. The size and number of thalli depend on the number of infections and the size of the host cell. Thalli within host cells are 5–12 μm in diameter, ellipsoidal or tubular, and sometimes coiled (Fig. 69). When mature, thalli are multinucleate and develop into either sporangia or thick-walled resting spores. Zoospores form in sporangia or in vesicles outside the root. They are biflagellate, measure 6×11 μm, and are bean-shaped when actively swimming and pear-shaped when becoming immobile before encystment. The thick-walled resting spores are presumed asexual because no sexual reproduction has been documented.

The original description of *L. radicicola* mistakenly included oospores of a *Pythium* sp. with which it frequently coexists. Thus, whether *L. radicicola* causes root damage on its own is uncertain. In this complex, *L. radicicola* may be the primary invader, with *Pythium* spp. following as secondary invaders. In culture on wheat roots in the greenhouse, heavy infections may occur without apparent root necrosis.

Olpidium brassicae (Wor.) Dang. is a Chytridiomycete that invades a wide range of plant species within and outside the Gramineae. It is an obligate parasite and a primary invader of healthy roots. It is widely known as a vector of tobacco necrosis, tobacco stunt, and lettuce big vein viruses. It is easily recognized in wheat roots by its stellate resting spores, 8–30 μm in diameter, which appear to be asexual and are formed from sporangia.

Sporangia are thin-walled and generally globose and form one or more discharge tubes at maturity. Zoospores released through these tubes are fully formed, uniflagellate, and 3–6 μm in diameter.

Rhizophydium graminis Led. is a Chytridiomycete parasite on root hairs and epidermal cells of wheat and numerous other plant species. Sporangia develop on the outside of the host cell and are typically 6–36 μm in diameter (Fig. 70). The host cell is invaded by a fine, branched rhizoid that both anchors and nourishes the sporangium. When mature, the sporangia rupture and release uniflagellate zoospores that are globose and approximately 2–3 μm in diameter. In unfavorable environments, such as when moisture is insufficient, sporangia develop a thickened, slightly roughened, pale brown wall and survive as resting spores.

Ligniera pilorum Fron & Gaillat (Plasmodiophorales) is common on grasses and is found occasionally on wheat. It is closely related and morphologically similar to *P. graminis*. However, *L. pilorum* develops zoosporangia and resting spore clusters inside both root hairs and epidermal cells, whereas *P. graminis* produces these fruiting bodies almost exclusively in epidermal cells.

P. graminis Led. (Plasmodiophorales) is a soilborne fungus that persists as an obligate parasite in the root cells of cereals and grasses. It enters root hairs and epidermal cells of wheat during periods of high soil moisture via motile, biflagellate zoospores (Fig. 71). Zoospores average 4.2 μm in diameter and are important vectors of soilborne wheat mosaic virus and wheat yellow mosaic (wheat spindle streak mosaic) virus.

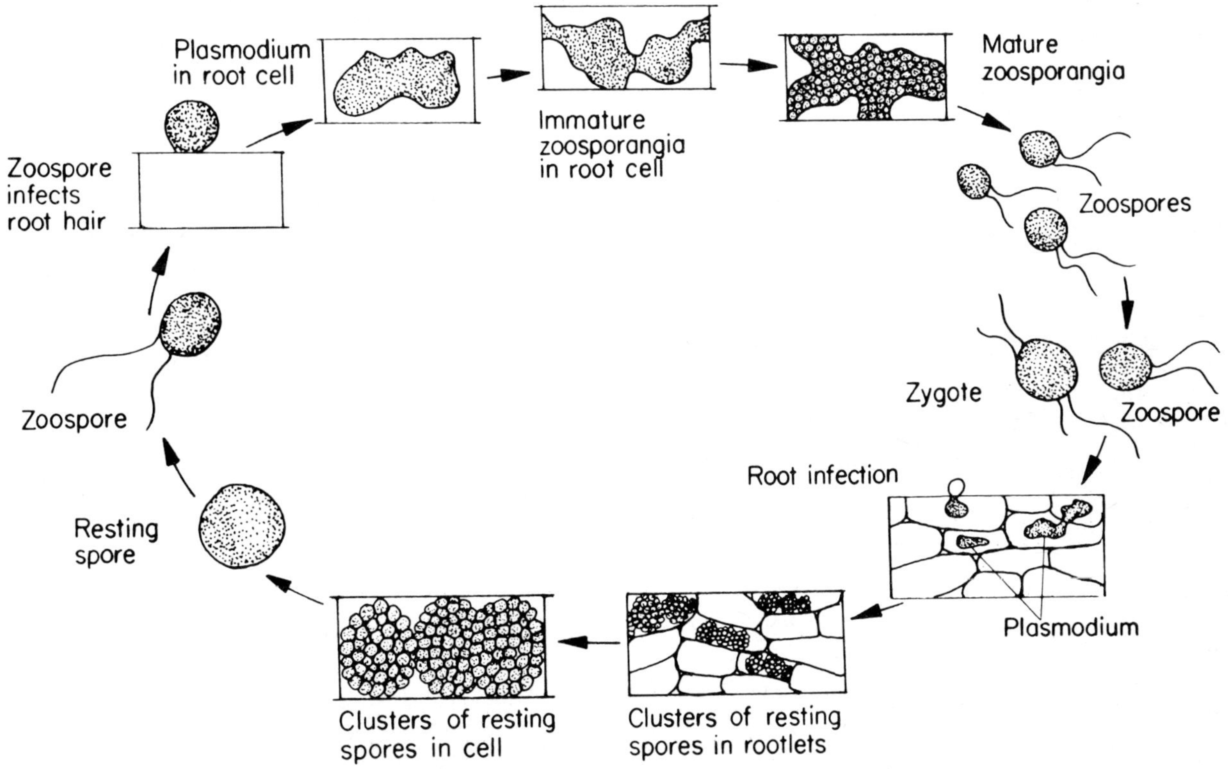

Fig. 71. Disease cycle of *Polymyxa graminis* in root cells. (Courtesy G. N. Agrios)

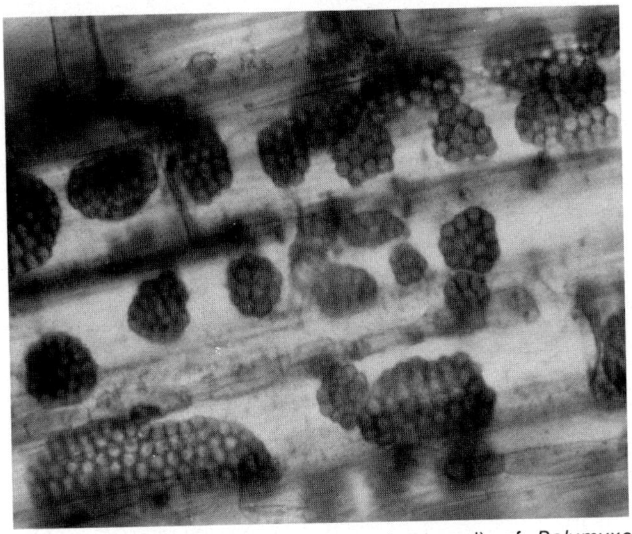

Fig. 72. Clustered resting spores (cystosori) of *Polymyxa graminis* inside root cells. (Courtesy D. J. S. Barr)

Inside host cells, *P. graminis* expands into plasmodia that replace the cell contents. The bodies either develop into zoosporangia or segment into tight clusters of smooth, thick-walled resting spores (cystosori) 5–7 μm in diameter (Fig. 72).

Selected References

Barr, D. J. S. 1973. *Rhizophydium graminis* (Chytridiales): Morphology, host range and temperature effect. Can. Plant Dis. Surv. 53:191–193.

Barr, D. J. S. 1979. Morphology and host range of *Polymyxa graminis, Polymyxa betae*, and *Ligniera pilorum* from Ontario and some other areas. Can. J. Plant Pathol. 1:85–94.

Barr, D. J. S., and Slykhuis, J. T. 1976. Further observations on zoosporic fungi associated with wheat spindle streak mosaic virus. Can. Plant Dis. Surv. 56:77–81.

Buczacki, S. T. 1983. Zoosporic Plant Pathogens, a Modern Perspective. Academic Press, London. 352 pp.

Gerdemann, J. W. 1968. Vesicular-arbuscular mycorrhiza and plant growth. Annu. Rev. Phytopathol. 6:397–418.

Macfarlane, I. 1970. *Lagena radicicola* and *Rhizophydium graminis*, two common and neglected fungi. Trans. Br. Mycol. Soc. 55:113–116.

Nolt, B. L., Romaine, C. P., Smith, S. H., and Cole, H., Jr. 1981. Further evidence for the association of *Polymyxa graminis* with the transmission of wheat spindle streak mosaic virus. Phytopathology 71:1269–1272.

Rao, A. S. 1968. Biology of *Polymyxa graminis* in relation to soilborne wheat mosaic virus. Phytopathology 58:1516–1521.

Temmink, J. H. M., and Campbell, R. N. 1969. The ultrastructure of *Olpidium brassicae*. III. Infection of host roots. Can. J. Bot. 47:421–424.

Diseases Caused by Nematodes

Nematodes (eelworms) are unsegmented roundworms that inhabit soil and water in great numbers. Most species are nonparasitic and live in compatible association with plants, animals, and other organisms. Some, however, are parasitic on plants or animals. Visible under low magnification, plant-parasitic nematodes typically are transparent and vermiform (eel-shaped). Exceptions are the females of some genera, which become swollen and saclike. Nematodes have bitubular bodies that are anatomically differentiated for feeding, digestion, locomotion, and reproduction. Important anatomic features used for species identification include phasmids, cuticular patterns, the structure of digestive and reproductive organs, and the size and shape of the stylet, head, and tail.

All nematodes develop from eggs and pass through a succession of juvenile (larval) stages (usually four, each ending in a molt) before adulthood (Fig. 73). Reproduction may be sexual or parthenogenetic, with generation times as short as two to four weeks or as long as nine months.

Most nematodes are mobile within water films, and motility is greatest in soils near field capacity. Nematodes are dispersed in soil, in running water, and on plant parts and packing material. Wind, higher animals, machinery, and other agents that move soil also disseminate nematodes.

Nematodes feed on roots and occasionally on aboveground plant parts. They mechanically penetrate host cells by means of a sharp, hollow stylet or a grooved, dorsal tooth. Their feeding reduces plant vigor and induces lesions, rots, deformations, galls, and root knots. Affected crops appear uneven, usually with patches of stunted, yellow plants (see, for example, Fig. 83).

Nematodes feed from the plant surface (ectoparasitic) or from within and between cells (endoparasitic). They puncture cells (Fig. 74) and feed by injection and extraction mechanisms. They damage plants mechanically and chemically (by introducing toxins or enzymes) and predispose plants to other diseases by providing points of entry for other pathogens. The stress of nematode infections lowers the general disease resistance of host plants.

Nematodes can be mechanically recovered directly from host plants and from soil. Collecting nematodes from soil samples involves incubation, elutriation, sieving, flotation, and/or centrifugation techniques. Such collections normally yield mixtures of pathogenic and nonpathogenic species. A number of parasitic and saprophagous species are likely to persist on or about wheat roots. Inferences about the pathologic importance of each species are drawn from their feeding habits and the relationship of plant symptoms to species frequency. It may be necessary to inoculate plants with populations of each different species to demonstrate virulence.

Next to fungi, nematodes are the oldest known parasites of wheat. The seed-gall nematode, *Anguina tritici*, was first characterized in 1743. Earlier, in 1594, Shakespeare referred to it in "Sowed cockle, reap'd no corn." Today, nematodes are well recognized as wheat pathogens, but their significance in limiting wheat yields remains poorly understood.

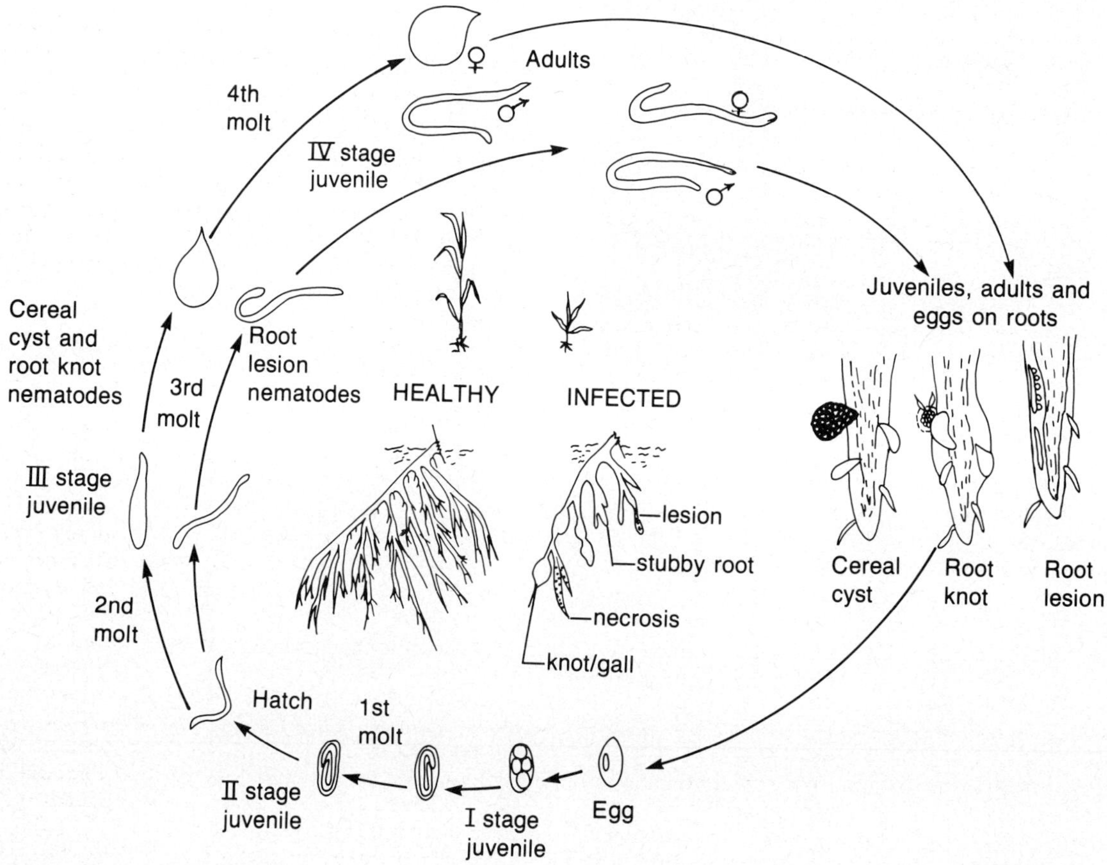

Fig. 73. Generalized life cycles of cereal cyst, root-knot, and root-lesion nematodes on wheat. (Modified from original by R. B. Malek and D. J. Royce)

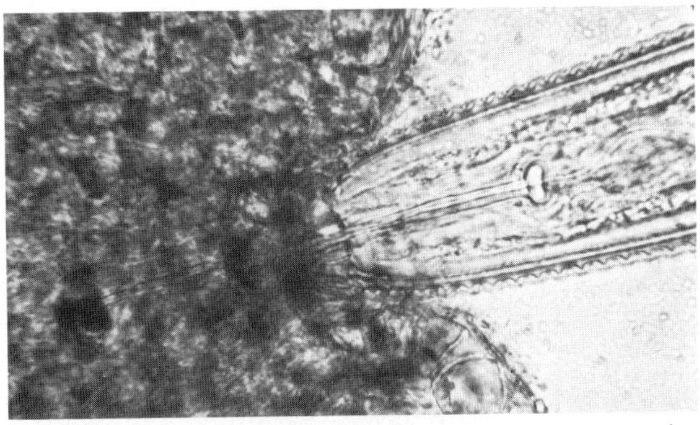

Fig. 74. Head and stylet of ectoparasitic nematode penetrating and feeding on root tissue. (Courtesy K. Kuiper, Netherlands Plant Protection Service)

Selected References

Bird, A. F. 1971. The Structure of Nematodes. Academic Press, New York. 318 pp.

Brown, R. H. 1985. The selection of management strategies for controlling nematodes in cereal. Agric. Ecosyst. Environ. 12:371–388.

Brown, R. H., and Kerry, B. R. 1987. Principles and Practice of Nematode Control in Crops. Academic Press, Sydney. (In press)

Kort, J. 1972. Nematode diseases of cereals of temperate climates. Pages 97–126 in: Economic Nematology. J. M. Webster, ed. Academic Press, New York.

Mai, W. F., and Lyon, H. H. 1975. Pictorial Key to the Genera of Plant Parasitic Nematodes. Comstock Publishing Assoc. Cornell Press, Ithaca, NY. 219 pp.

Nickle, W. R. 1984. Plant and Insect Nematodes. Marcel Dekker, New York. 925 pp.

Wallace, H. R. 1973. Nematode Ecology and Plant Disease. Crane, Russak and Co., New York. 228 pp.

Zuckerman, B. M., Mai, W. F., and Rohde, R. A., eds. 1971. Plant Parasitic Nematodes. 2 vols. Academic Press, New York.

Zuckerman, B. M., and Rohde, R. A., eds. 1981. Plant Parasitic Nematodes. Vol. 3. Academic Press, New York. 508 pp.

Cereal Cyst Nematode

The cereal cyst nematode, *Heterodera avenae* (also called cereal root eelworm and oat cyst nematode), was first described on oats in Germany in 1874. The pest was first reported on wheat in England in 1908 and was identified in the United States (Oregon) in 1972. Today the nematode is widely known as an important pathogen of wheat and other cereals and grasses. Cereal crops grown on newly tilled grasslands have been most subject to attack. The nematode occurs in New Zealand, Peru, and the United States, throughout Europe and the Mediterranean basin, and in southeastern Canada. It damages wheat in Australia, Africa, Japan, Israel, and the Soviet Union. In India it causes "molya" disease of wheat and barley. In the United States, it has recently caused economic losses in the Pacific Northwest.

H. avenae parasitizes only the Gramineae. Oats are most susceptible, and rye, triticale, and corn also serve as hosts. Damage is greatest in, but not restricted to, light-textured, well-drained soils. Spring wheat is more prone to damage than winter wheat. Yield losses are predictable based on nematode populations present at seeding.

Symptoms

Symptoms are observed more readily on seedlings than on older plants. Seedling roots have multiple short branches and moderate swellings (knots). Root growth is shallow and noticeably impaired (Fig. 75). Cysts (swollen and transformed bodies of female nematodes) may be seen loosely attached to infested roots (Plate 44). The cysts are glistening white-gray initially and dark brown when mature. Attached loosely at their necks, many cysts are dislodged when roots are harvested for examination. Cysts are best seen on harvested roots that are gently floated in water.

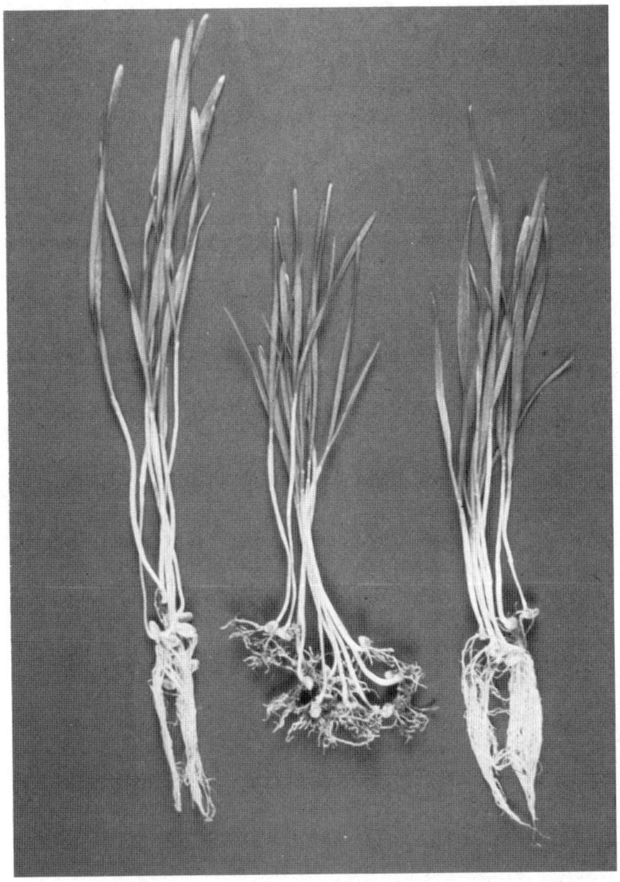

Fig. 75. Healthy seedlings (left and right) and seedlings parasitized by the cereal cyst nematode, *Heterodera avenae* (center). (Courtesy R. H. Brown)

Seedlings weakened by nematodes often are invaded by other pathogens, especially soil fungi that cause secondary root lesions and rots. *Rhizoctonia solani*, for example, can interact with cereal cyst nematodes and increase plant damage. With their root systems impaired, infected plants often appear stunted, as if from moisture or nutritional stress.

Causal Organism

The cereal cyst nematode, *H. avenae* Woll. (syn. *H. major* (Schmidt) Frank.) (Fig. 76), is one of several cyst-forming nematodes that parasitize cereals. Others, such as *H. latipons* Franklin, *H. hordecalis* Andersson, *H. bifenestra* Cooper, and *H. zeae* Koshy are normally less prevalent and less damaging but also may exist in association with wheat roots.

Heterodera spp. are distinguished by the egg-carrying habit of the female. Several hundred eggs are borne in her swollen, saclike body and remain in place even after her death and transformation to a cyst. Cysts of *H. avenae* are lemon-shaped and measure 0.3–0.5 × 0.55–0.77 mm. They have a rugose cuticle with large, irregularly spaced punctations. Their head region is annular, with six rounded lips. Stylets are approximately 26 μm long. A light-colored, subcrystalline layer is sloughed off as females transform into cysts.

Males, normally 1–1.6 mm long, have stylets 26–29 μm long with rounded basal knobs. Juveniles resemble adult males but are smaller (0.52–0.61 mm long), lack reproductive organs, and have sharp rather than rounded tails.

Currently 10 pathotypes of *H. avenae* have been distinguished by their aggressiveness on different cereal hosts. The Australian pathotype appears to be the most damaging.

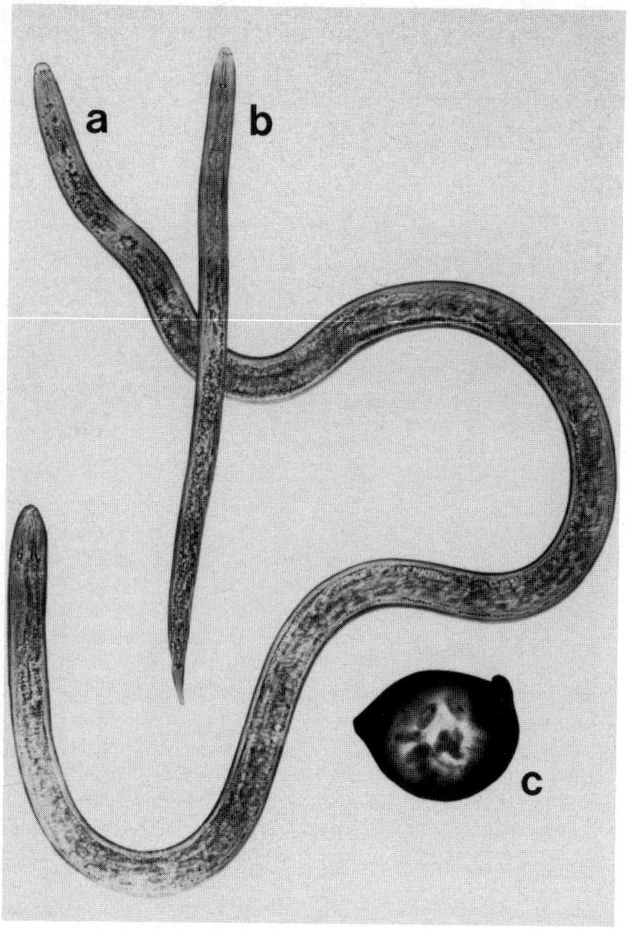

Fig. 76. The cereal cyst nematode, *Heterodera avenae*: adult male (a), juvenile (b), and adult female (c). (Adapted from W. F. Mai and H. H. Lyon, Pictorial Key to the Genera of Plant Parasitic Nematodes, 1975, Comstock Publishing Assoc. Cornell Press, Ithaca, NY)

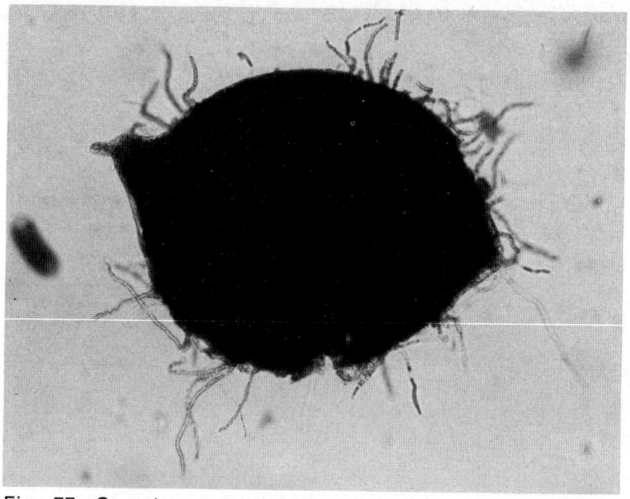

Fig. 77. Cereal cyst nematode (*Heterodera avenae*) female parasitized by the nematophagous fungus *Nematophthora gynophila*. The fungal sporangia penetrating the cyst eventually release infective zoospores into the soil. (Reprinted, by permission, from B. R. Kerry and D. H. Crump. 1980. Two fungi parasitic on females of cyst nematodes (*Heterodera* spp.). Trans. Br. Mycol. Soc. 74:119–125)

Disease Cycle

In addition to cereals, numerous perennial and annual grasses sustain *H. avenae*. Eggs in cysts may lie dormant in soil for several years, but free juveniles persist only a few weeks if hosts are unavailable. Juveniles emerge in spring from eggs within cysts that have overwintered. Moist soils and temperatures near 10° C are optimal for hatching and emergence, but hatching can continue through autumn if not interrupted by periods of soil drying. Hatching does not appear to be influenced by root exudates.

Juveniles penetrate roots just behind the growing point and develop to adulthood near the vascular cylinder. Males retain their vermiform shape and eventually leave the roots and return to the soil. Females, however, are sedentary and mechanically erupt through root tissues as they swell and mature. In the process they induce root branching, cell enlargements (syncytia or giant cells), and root swellings. The females continue to grow, begin egg production, and then are transformed into cysts. Most cysts appear within 10 weeks after infection; however, in the Northern Hemisphere, autumn infections by juveniles do not produce cysts until the next growing season.

Only one generation is completed each year. Eggs in soilborne cysts and nematodes in various stages of juvenile development lie dormant in roots over the winter. Each year about half of the eggs present in cysts hatch. Dispersal of cysts in windblown soil is common.

Control

Cereal cyst nematode populations are influenced by the frequency of host crops. Wheat and other small grains should be grown as infrequently as possible. Grassy weeds like wild oats (*Avena fatua* L.) and ryegrass (*Lolium* spp.) maintain nematode populations and should be controlled.

Larvae invade all wheat cultivars. Some cultivars, however, do not support cyst development, and their use in rotations is encouraged, even though certain pathotypes of the nematode may limit their effectiveness. The cultivar Katyil is resistant in Australia.

Chemical nematicides are effective but usually not economical. Crop performance may improve after their use, especially when additional pests and diseases are controlled. In Australia, seed treatments and low dosages of nematicides applied in the drill row at seeding increase yields as much as 50% (approximately 200,000 ha were treated in 1985). A bioassay to determine nematode incidence at seeding is used to prescribe nematicide treatments.

Providing the crop with adequate nutrition and moisture increases its tolerance to nematodes. Nitrogen should be applied cautiously because nematode populations increase when nitrogen is excessive. Practices that build soil fertility, humus levels, and water-holding capacity are encouraged.

In northern Europe, populations of nematophagous fungi such as *Nematophthora gynophila* Kerry & Crump and *Verticillium chlamydosporium* Goddard increase under cereal monocultures and provide effective biological control of *H. avenae* (Fig. 77).

Selected References

Brown, R. H. 1984. Ecology and control of cereal cyst nematode (*Heterodera avenae*) in southern Australia. J. Nematol. 16:216–222.

Brown, R. H., and Young, R. M. 1982. Katyil, a wheat resistant to cereal cyst nematode. Agnote. Department of Agriculture, Victoria, Australia. 2 pp.

Kerry, B. 1981. Fungal parasites: A weapon against cyst nematodes. Plant Dis. 65:390–393.

Kerry, B. R. 1982. The decline of *Heterodera avenae* populations. EPPO Bull. 12:491–496.

Meagher, J. W. 1974. The morphology of the cereal cyst nematode (*Heterodera avenae*) in Australia. Nematologica 20:1–8.

Meagher, J. W. 1977. World dissemination of the cereal cyst nematode (*Heterodera avenae*) and its potential as a pathogen of wheat. J. Nematol. 9:9–15.

Simon, A. 1980. A plant assay of soil to assess potential damage to wheat by *Heterodera avenae*. Plant Dis. 64:917–919.

Swarup, G., and Singh, D. 1961. Molya disease of wheat and barley in Rajasthan. Indian Phytopathol. 14:127–133.

Williams, T. D., and Siddiqi, M. R. 1972. *Heterodera avenae.* Descriptions of Plant-Parasitic Nematodes, No. 2. Commonwealth Institute of Parasitology, St. Albans, Herts., England.

Grass Cyst Nematode

The grass cyst nematode, *Punctodera punctata*, is named for its association with grasses and for its conspicuously punctate cuticle. Its life cycle, habits, and effects on wheat are similar to those described for *Heterodera avenae. P. punctata* Mul. & Stone (syn. *H. punctata* Thorne) was first described on wheat in Canada in 1926. Since then, it has been reported in Europe, Mexico, and the Soviet Union. It is more prevalent on cereals than on grasses, but its economic importance is poorly understood. Its importance in the Great Plains of the United States may be underestimated.

Selected References

Spears, J. F. 1956. Occurrence of the grass cyst nematode, *Heterodera punctata*, and *Heterodera cacti* group cysts in North Dakota. Plant Dis. Rep. 40:583–584.
Thorne, G. 1928. *Heterodera punctata* n. sp., a nematode parasitic on wheat roots from Saskatchewan. Sci. Agric. 8:707–711.
Webley, D. P., and Lewis, S. 1977. *Punctodera punctata.* Descriptions of Plant-Parasitic Nematodes, No. 102. Commonwealth Institute of Parasitology, St. Albans, Herts., England.

Root-Gall Nematode

Swellings (galls) on roots of grasses and cereals have been known for nearly a century. Many result from and contain larvae of the root-gall nematode, *Subanguina radicicola.* The pest is sustained on wheat in Europe, North America, and the Soviet Union. Barley, oats, and rye also are hosts, but damage to cereals is minor relative to damage to grasses. Root-gall nematodes and root galls were recently confirmed on unthrifty wheat in Oregon.

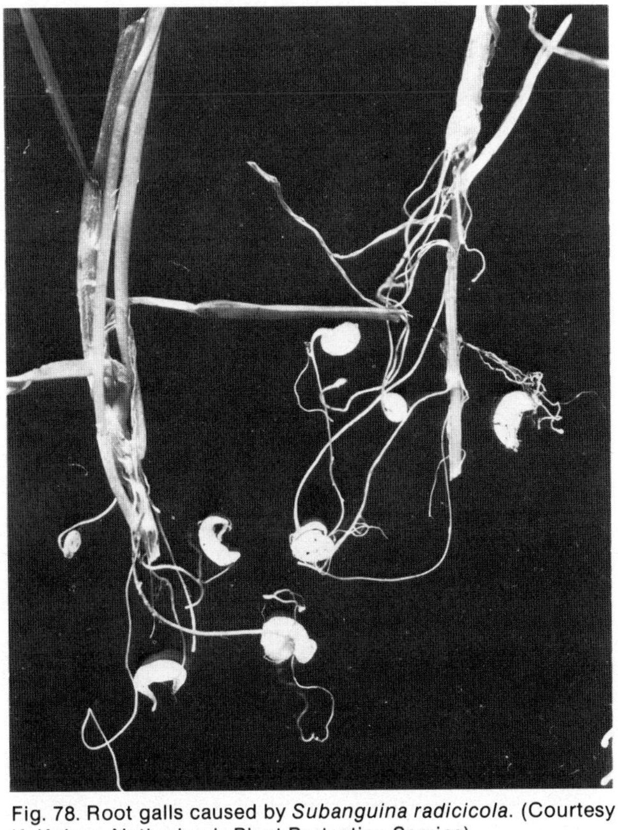

Fig. 78. Root galls caused by *Subanguina radicicola.* (Courtesy K. Kuiper, Netherlands Plant Protection Service)

Symptoms

Washed wheat roots must be examined closely to confirm the presence of root-gall nematodes. Galls are usually inconspicuous, varying in diameter from 0.5 to 6 mm. They are firm, with a smooth-textured surface. Young galls may resemble root knots caused by *Meloidogyne* spp. Roots frequently are bent at gall sites (Fig. 78).

Hypertrophied cortical and epidermal root cells make up the bulk of each swelling. At the center of larger galls is a cavity with clusters of nematode larvae. Juveniles also are found among cortical cells of horizontal roots within 10 cm of the soil surface.

Affected seedlings sometimes show leaf chlorosis and reduced top growth.

Causal Organism

Subanguina radicicola (Grf.) Param. (syns. *Ditylenchus radicicola* (Grf.) Filip., *Anguillulina radicicola* (Grf.) Gdy.) is an endoparasite. All specimens have a finely striated cuticle. The lip region of adults is set off by a constriction, and adult stylets are 12–16 μm long.

Females are 1.2–3.2 mm long. They produce numerous eggs (40×70–150 μm) and have prominent labia. Males are 1.2–2.0 mm long. Populations of the nematode may differ in virulence on wheat and grass hosts.

Disease Cycle

S. radicicola is presumed to survive by continuous habitation in host roots. Juveniles penetrate roots and develop rapidly through successive molts in cortical tissues. Root galls develop within two weeks, and newly mature females begin egg production. The galls eventually weaken and release hatched juveniles that establish secondary infections on the same root system. A generation is normally completed within 60 days.

Control

Rotating wheat, cereal, or grass crops with legumes and root crops limits root-gall nematode populations. Unless additional diseases or pathogens are controlled, the use of nematicides for *S. radicicola* in wheat is not profitable.

Selected References

Jatala, P., Jensen, H. J., and Shimabukuro, R. A. 1973. Host range of the "grass root-gall nematode," *Ditylenchus radicicola*, and its distribution in Willamette Valley, Oregon. Plant Dis. Rep. 57:1021–1023.
Jenkins, W. A. 1948. A root rot disease-complex of small grains in Virginia. Phytopathology 38:519–527.
Lewis, S., and Webley, D. 1966. Observations on two nematodes infesting grasses. Plant Pathol. 15:184–186.
Vanterpool, T. C. 1948. *Ditylenchus radicicola* (Greeff) Filipjev, a root-gall nematode new to Canada, found on wheat and other Gramineae. Sci. Agric. 28:200–205.

Root-Knot Nematodes

Root-knot nematodes, *Meloidogyne* spp., are the best-known plant-parasitic nematodes because they induce conspicuous root swellings or galls on over 1,000 plant species. They are widely distributed and are often involved in disease complexes. *M. hapla* Chit., for example, is a common parasite on dicotyledonous plants in the United States and Europe. Others, like *M. naasi*, are better known on wheat, barley, and grasses.

Symptoms

Root-knot nematodes cause swellings or clublike thickenings (galls or knots) on wheat roots (Fig. 79). The knots, most visible

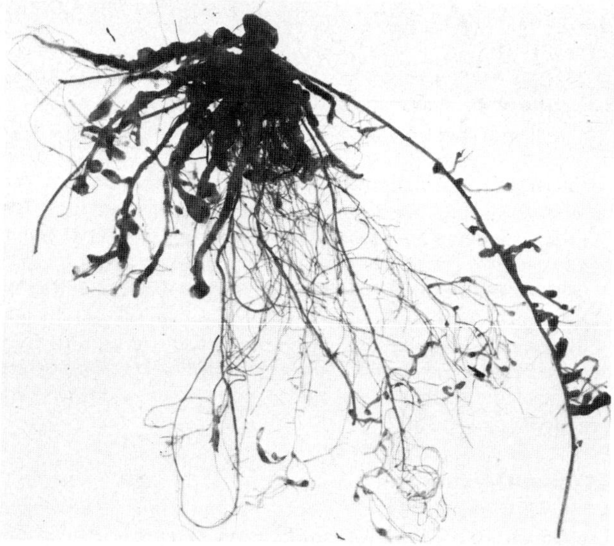

Fig. 79. Roots knotted by *Meloidogyne naasi*. (Courtesy K. Kuiper, Netherlands Plant Protection Service)

in spring and summer, are composed of expanded cortical cells and the swollen bodies of nematodes. Each knot houses one or more female nematodes with large, gelatinous egg masses. The egg masses, normally transparent, darken when exposed to air to resemble cysts of *Heterodera avenae*. Root knots may also contain vermiform juveniles at feeding sites near the vascular cylinder.

Heavily infested wheat plants may appear stunted and chlorotic.

Causal Organisms

Four principal species of *Meloidogyne* cause root knots as sedentary endoparasites in wheat. *M. naasi* Frank. is most important. *M. incognita* (Kofoid & White) Chit. (syn. *M. incognita* var. *acrita* Chit.), *M. javanica* (Treub) Chit., and *M. chitwoodi* Golden et al are encountered less frequently.

Meloidogyne females are spherical to saclike and 0.2–0.7 mm in diameter. Necks protruding from their swollen bodies are 0.1–0.2 mm long. Phasmids are conspicuous, and their cuticle is coarsely ridged in the dorsal region.

Males measure $0.02–0.04 \times 0.8–1.3$ mm and have stylets 16–19 μm long with prominent rounded knobs. Spicules are 20–30 μm long, and tails are short and blunt. Three annules surround the head. Eggs are oval and 70–100 μm long. Emerging juveniles measure $0.01–0.02 \times 0.4–0.5$ mm.

Disease Cycle

Most infections occur in early spring, but juveniles may enter roots at any time. By midsummer, swollen females extrude their egg masses but otherwise remain embedded in root tissues. The nematodes overwinter principally as eggs, and *M. naasi*, for example, completes only one generation each year. Damage to wheat is directly related to the population of egg masses in soil.

Control

M. naasi is controlled by rotating wheat with nonhosts such as oats and certain root crops. However, crops like potatoes and sugar beets may be attacked by other *Meloidogyne* spp. Some wheat cultivars and grasses support nematode feeding but not reproduction. Root stunting and gall formation may occur, but the nematodes either die or migrate to other feeding sites. Autumn-sown wheat tolerates large numbers of spring-hatched juveniles better than spring-sown grain.

Selected References

Franklin, M. T. 1973. *Meloidogyne naasi*. Descriptions of Plant-Parasitic Nematodes, No. 19. Commonwealth Institute of Parasitology, St. Albans, Herts., England.

Lewis, S., and Webley, D. 1966. Observations on two nematodes infesting grasses. Plant Pathol. 15:184–186.

Mitchell, R. E., Malek, R. B., Taylor, D. P., and Edwards, D. I. 1973. Races of the barley root knot nematode, *Meloidogyne naasi*. I. Characteristics by host preference. J. Nematol. 5:41–44.

Nyczepir, A. P., Inserra, R. N., O'Bannon, J. H., and Santo, G. S. 1984. Influence of *Meloidogyne chitwoodi* and *M. hapla* on wheat growth. J. Nematol. 16:162–165.

Orton Williams, K. J. 1973. *Meloidogyne incognita*. Descriptions of Plant-Parasitic Nematodes, No. 18. Commonwealth Institute of Parasitology, St. Albans, Herts., England.

Roberts, P. A., and van Gundy, S. D. 1981. The development and influence of *Meloidogyne incognita* and *M. javanica* on wheat. J. Nematol. 13:345–352.

Roberts, P. A., van Gundy, S. D., and McKinney, H. E. 1981. Effects of soil temperature and planting date of wheat on *Meloidogyne incognita* reproduction, soil populations and grain yield. J. Nematol. 13:338–345.

Santo, G. S., and O'Bannon, J. H. 1981. Pathogenicity of the Columbia root-knot nematode (*Meloidogyne chitwoodi*) on wheat, corn, oats and barley. J. Nematol. 13:548–550.

Thomason, I. J. 1962. Reaction of cereals and sudan grass to *Meloidogyne* spp. and the relation of soil temperature to *M. javanica* populations. Phytopathology 52:787–791.

Root-Lesion Nematodes

Root-lesion nematodes, *Pratylenchus* spp., live freely in soil as migratory endoparasites. They feed on a variety of cultivated and wild hosts in most regions of the world. Often, more than one species can be found in soil samples and in association with tobacco, corn, rye, peanuts, red clover, fruit trees, and vegetables. Those that feed on wheat predispose plants to attack by other soilborne pathogens. Their feeding induces root lesions, which, when invaded by secondary organisms, can lead to root rots.

The economic importance of root-lesion nematodes on cereals depends on the species and populations that are present. In the United Kingdom, *P. fallax* can stunt the growth of cereals, whereas *P. neglectus* is relatively unimportant. In the United States, Mexico, and Queensland, Australia, *P. neglectus* and especially *P. thornei* are the principle species that limit wheat yields.

Symptoms

Young roots of wheat, oats, and occasionally corn may be darkened and stunted. Such plants with restricted or rotted root systems set little grain. Root lesions and rots are usual symptoms that typically result from secondary infections. The mechanical injuries to host cells and tissues that result from nematode feeding serve as infection courts for fungi and other soilborne organisms. *P. neglectus* in combination with *Rhizoctonia solani*, for example, causes a serious root and crown rot of wheat in Canada. Affected areas of fields appear yellow and drought-stricken, but they can be restored to normal productivity by soil fumigation.

Causal Organisms

Adult females of *P. neglectus* (Rensch) Filip. & S. Stekho. are 0.3–0.6 mm long and have two annules in their lip region. Their stylets are 16–18 μm long, with anteriorly cupped knobs. Their tails are bluntly rounded. Males are about 0.4 mm long but are rare and appear sexually nonfunctional.

Females of *P. thornei* Sher & Allen are 0.4–0.8 mm long, with three annules in their lip region. Their stylets are 17–19 μm long and have prominent, anteriorly truncate basal knobs. Their tails are blunt and dorsally convex. Males are extremely rare, approximately 0.6 mm long, and nonfunctional.

Disease Cycle

Root-lesion nematodes are motile within soil and host tissues. Juveniles and adults penetrate roots but do not induce root swellings or form giant cells, galls, or cysts. They move through cortical cells, and females deposit eggs as they migrate. Older, dying roots are vacated in a constant search for young cells. Eggs and most other growth stages are quiescent in soil during winter or periods when hosts are unavailable.

Control

Approved soil fumigants reduce nematode populations and increase wheat growth, but their use generally is not economical. Seeding winter wheat in late autumn when soil temperatures are below 13°C decreases disease risk. Plants given adequate nutrients better tolerate nematode feeding. Crop rotations are of limited value because *Pratylenchus* spp. parasitize a wide range of hosts. However, in Queensland, Australia, wheat grown in rotation with barley and sorghum is less damaged.

Selected References

Benedict, W. G., and Mountain, W. B. 1956. Studies on the etiology of root rot of winter wheat in southwestern Ontario. Can. J. Bot. 34:159–174.

Corbett, D. C. M. 1972. The effects of *Pratylenchus fallax* on wheat, barley and sugarbeet roots. Nematologica 18:303–308.

Fortuner, R. 1977. *Pratylenchus thornei*. Descriptions of Plant Parasitic Nematodes, No. 93. Commonwealth Institute of Parasitology, St. Albans, Herts., England.

Inserra, R. N., Vovlas, N., and Bradonisio, A. 1978. Nematodi endoparassiti associati a colture di cereali deperimento nell' Italia meridionale. Nematol. Mediterr. 6:163–174.

O'Brien, P. C. 1983. A further study on the host range of *Pratylenchus thornei*. Aust. Plant Pathol. 12:1–3.

Orion, D., Amir, J., and Krikum, J. 1984. Field observations on *Pratylenchus thornei* and its effects on wheat under arid conditions. Rev. Nematol. 7:341–345.

Townshend, J. L., and Anderson, R. V. 1976. *Pratylenchus neglectus*. Descriptions of Plant Parasitic Nematodes, No. 82. Commonwealth Institute of Parasitology, St. Albans, Herts., England.

van Gundy, S. D., Perez, B. J. G., Stolzy, L. H., and Thomason, I. J. 1974. A pest management approach to the control of *Pratylenchus thornei* on wheat in Mexico. J. Nematol. 6:107–116.

Seed-Gall Nematode

"Seed gall" was one of the earliest recognized wheat diseases, and the seed-gall nematode, *Anguina tritici*, was the first nematode described as a plant pathogen. Seed gall (also called cockles, purples, peppercorns, eelworm disease, and hard smut) was characterized in England in 1743. Through its association with wheat seed, *A. tritici* once occurred worldwide and caused significant crop losses. Today, it parasitizes rye and the wheat ancestors emmer, spelt, and *Aegilops* but is consequential only on wheat. Currently, seed gall is best known in eastern Asia, India, Yugoslavia, and southeastern Europe. It also appears occasionally in the southeastern United States. The disease has largely disappeared because of modern seed-cleaning techniques.

Symptoms

Seed gall symptoms occur on living plants and among harvested seed. Before heading, wheat plants are swollen near ground level, and leaves appear wrinkled, twisted, or rolled (Fig. 80). At heading, leaf distortions are less apparent, but plants are stunted and slow to mature. Diseased heads are small and have broadly open glumes (see Common Bunt and Dwarf Bunt) because most or all kernels are displaced by darkened, seedlike nematode galls averaging 2–3 × 3–5 mm in size (Fig. 81).

Galls are brown-black and lack the brush and embryo markings of normal kernels. They are less easily crushed and without odor compared to bunt balls. Galls are distinguished from weed seeds and ergot sclerotia (Fig. 11) in grain samples by the release of motile larvae when wetted (Fig. 81).

Causal Organism

A. tritici (Steinb.) Chit. (syns. *Tylenchus tritici* (Steinb.) Bast., *Anguillulina tritici* Gerv. & Bened.) is an obligate parasite. Its body cuticle is finely striated, and its lip region is narrow, flattened, and annulate. Stylets are 8–10 μm long with rounded basal knobs. Adult females average 3.8 mm in length, are stouter and less motile than males, and are more coiled ventrally. Males are approximately 2.5 mm long.

Fig. 80. Seedlings distorted by the seed-gall nematode, *Anguina tritici*. (Courtesy K. Kuiper, Netherlands Plant Protection Service)

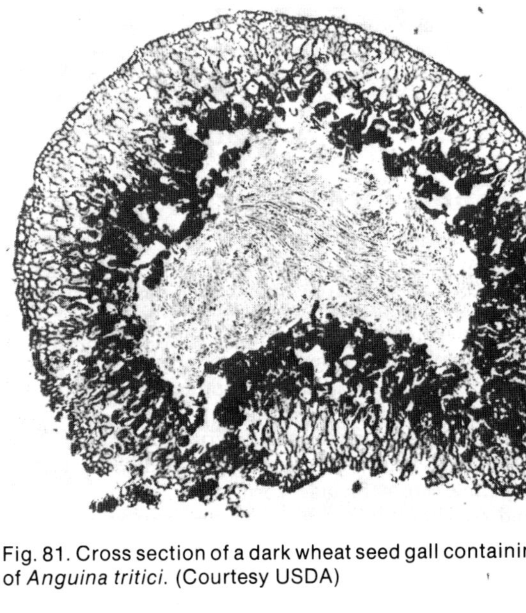

Fig. 81. Cross section of a dark wheat seed gall containing larvae of *Anguina tritici*. (Courtesy USDA)

Trichodorus and *Xiphinema*. Queensland J. Agric. Sci. 21:77–123.
Kerr, E. D. 1967. Population dynamics and fumigation studies of certain nematodes associated with roots of wheat in Nebraska. Plant Dis. Rep. 51:637–641.
Norton, D. C. 1959. Relationship of nematodes to small grains and native grasses in north and central Texas. Plant Dis. Rep. 43:227–235.
Sechler, D. T., Tappan, W. B., and Luke, H. H. 1967. Effect of variety and fumigation on nematode populations in oats, wheat and rye. Plant Dis. Rep. 51:915–919.
Taylor, D. P., and Schleder, E. G. 1959. Nematodes associated with corn, barley, oats, rye and wheat. Plant Dis. Rep. 43:329–333.

Diseases Caused by Viruses and Viruslike Agents

Viruses are submicroscopic agents composed only of a DNA or RNA genome usually surrounded by a protective protein coat. By themselves, virus particles (virions) are inert, but within cells of microorganisms, plants, animals, and humans, they can modify the host genome and thereby become infectious, replicative, and metabolically influential.

Plants acquire viruses through grafts, sap, seed, and pollen; from animal, plant, and fungal vectors; and through deliberate or accidental manipulation by humans. Most viruses in wheat are spread by insects. Others depend on mites, soilborne fungi, or nematodes for transmission. A few are transmitted through seed.

Plant viruses are identified by their specific effects on host plants (symptomatology); their morphology, mode of transmission, and host specificity; and their biochemical, physical, serologic, and genomic properties. Most plant viruses are named for the original or major host attacked and for its response. However, the inherently high mutation rates among viruses make their identification and nomenclature a challenge. Many wheat viruses exist as a collection of related mutants (strains).

Virus infections in wheat range from latent (symptomless) to lethal. They induce stunting, rosetting, mosaics, yellowing, and necrosis. The action of two or more viruses in a given plant (multiple infections) can be additive, synergistic, or cross-protective.

Viruses in wheat cause patterned foliar chlorosis in the form of mottles, dashes, blotches, or streaks (Plate 45). Many wheat viruses are mechanically transmissible (through sap), and nearly all wheat cells are susceptible to infection. Some wheat viruses, however, are restricted to phloem cells, where they interfere with nutrient transport and induce general chlorosis. Such phloem-limited viruses are transmitted principally by insects.

Virions in wheat are filamentous (rigid or flexuous rods), roughly spherical (isometric), or bacilliform (see Figs. 86, 87, 89, 90, 92, and 99). Most contain RNA, and a few geminiviruses, like maize streak virus and wheat dwarf virus, contain DNA genomes. Virions cannot be observed without an electron microscope. They range in diameter from 10 to 70 nm, and rod lengths can range up to 5 μm. Numerous particles are sometimes organized into crystals in infected cells.

Until 1950, except for the pioneering study of soilborne wheat mosaic, reports of viruses in wheat were rare even though viruses in other plants were well known. Today, approximately 30 different viruses have been associated with the crop (Table 1), and a viroid (small circular molecule of RNA) has recently been associated with a wheat disorder in China (see Seedborne Wheat Yellows).

Cereal virology is a dynamic and challenging science. More wheat viruses and viroids may be awaiting identification because infections tend to be overlooked or dismissed as a nutritional or other nonparasitic disorder, which they sometimes greatly resemble. On the other hand, some suspected viral diseases, like aster yellows, have proved to have nonviral causes. Some disorders, like European wheat striate mosaic and wheat spot mosaic, resemble virus diseases but still have unknown causes. Others, like soilborne wheat mosaic and wheat streak mosaic, have a well-known and well-characterized viral etiology. Barley yellow dwarf, now known to be caused by a complex of related virus strains, is one of the most serious diseases of small-grain cereals in the world. Some viruses once regarded as distinct, like wheat spindle streak mosaic virus and wheat yellow mosaic virus, now appear to be strains of the same virus.

Selected References

Ball, E. M., and Brakke, M. K. 1968. Leaf-dip serology for electron microscopic identification of plant viruses. Virology 36:152–155.
Brakke, M. K. 1987. Virus diseases of wheat. Chapter 8 in: Wheat and Wheat Improvement, 2nd ed. E. G. Heyne, ed. Agronomy Monograph 13. American Society of Agronomy, Crop Science Society of America, and Soil Science Society of America, Madison, WI. (In press)
Francki, R. I. B., Milne, R. G., and Hatta, T. 1985. Atlas of Plant Viruses. 2 vols. CRC Press, Boca Raton, FL.
Holland, J., Spindler, K., Horodyski, F., Grabau, E., Nichol, S., and Vandepol, S. 1982. Rapid evolution of RNA genomes. Science 215:1577–1585.
Jackson, A. O., Milbrath, G. M., and Jedlinski, H. 1981. Rhabdovirus diseases of the Gramineae. Pages 51–76 in: Virus and Virus-like Diseases of Maize in the United States. D. T. Gordon, J. K. Knoke, and G. E. Scott, eds. Southern Cooperative Series Bulletin 247. Ohio Agricultural Research and Development Center, Wooster, OH. 218 pp.
Matthews, R. E. F. 1981. Plant Virology. Academic Press, New York. 897 pp.
Noordam, D. 1973. Identification of Plant Viruses; Methods and Experiments. Centre for Agricultural Publishing and Documentation, Wageningen, The Netherlands. 207 pp.
Slykhuis, J. T. 1976. Virus and virus-like diseases of cereal crops. Annu. Rev. Phytopathol. 14:189–210.

Agropyron Mosaic

Agropyron mosaic (also called Agropyron green mosaic, yellow mosaic, streak mosaic, and couch grass streak mosaic) was first identified on quackgrass (*Agropyron repens* L.) in 1934 in Virginia. The disease is most common among grasses in the northern United States, Canada, and northern Europe. It occurs infrequently on wheat, rye, and barley and is of little economic importance.

Symptoms

Agropyron mosaic virus (AgMV) induces pale green or yellow mosaics, streaks, or dashes on leaf blades of quackgrass and wheat (Fig. 84). These symptoms become less conspicuous as plants approach maturity, but moderate stunting persists. Diseased plants often appear in patches or along grassy field borders.

TABLE 1. Viruses that Occur Naturally in Wheat

Virus	Particle	Distribution of Natural Wheat Infections	Transmission	Major Symptoms	Control	Additional Hosts
Agropyron mosaic (potyvirus)	Flexuous rod, 15 × 710–717 nm	Northeastern USA, Canada, N. Europe	Sap, mite (*Abacarus hystrix*)	Yellow-green mosaic, streaks, dashes	Tolerant cultivars, late-autumn seeding, clean cultivation	Rye, barley, quackgrass
Barley stripe mosaic (hordeivirus)	Rigid rod, 20–25 × 100–150 nm	China, Israel, (rare) USA	Seed, sap, pollen	Yellow-brown stripes, chlorotic spots, mottle	Virusfree seed, clean cultivation	Barley, rye, wild oats, millet, corn, grasses, some dicots
Barley yellow dwarf (luteovirus)	Isometric, 21–26 nm	Worldwide	Aphids (*Macrosiphum, Rhopalosiphum, Schizaphis,* and other species)	General chlorosis, reddening, purpling, stunting	Resistant cultivars, vector control, late-autumn or early-spring seeding	Barley, oats, corn, sorghum, millet, grasses
Barley yellow striate mosaic (rhabdovirus)	Bacilliform, 45–60 × 260–330 nm	Italy	Planthopper (*Laodelphax striatellus*)	Chlorosis, striations	None prescribed	Oats, barley, grasses
Barley yellow stripe	Unknown	Turkey	Leafhopper (*Euscelis plebejus*)	Leaf stripes, yellowing	None prescribed	Barley, oats, grasses
Brome mosaic (bromovirus)	Isometric, 25 nm	USA, Europe, S. Africa	Sap, cereal leaf beetles, nematodes (*Xiphinema* spp.)	Mild yellow-green mosaic, streaks, dashes, stunting	Clean cultivation	Brome, oats, corn, barley, rye, grasses, some dicots
Northern cereal mosaic (rhabdovirus)	Bacilliform rod, 40–60 × 300–400 nm	Japan	Planthoppers (*Laodelphax, Unkanodes,* and *Delphacodes* spp.)	Yellow mosaic, stunting, streaks	Resistant cultivars	Rice, barley, oats, rye, grasses
African cereal streak	Isometric, 24 nm	Kenya	Planthopper (*Toya catilina*)	Streaks, yellowing, head distortion	None prescribed	Barley, oats, rye, triticale, rice, grasses
Cereal tillering (reovirus)	Shelled, isometric, 65–73 nm	Sweden, Italy	Planthoppers (*Laodelphax* and *Dicranotropis* spp.)	Stunting, rosetting, sterility	None prescribed	Barley, oats, corn, rye, grasses
Cocksfoot mottle (sobemovirus)	Isometric, 30 nm	United Kingdom	Sap, beetles (*Oulema* spp.)	Leaf mottle, chlorosis, necrosis	None prescribed	Cocksfoot, oats, barley
Enanismo	Unknown	Colombia, Ecuador	Leafhopper (*Cicadulina pastusae*)	Dwarfing, blotches, stripes, galls	Resistant cultivars, delayed planting	Barley, oats
Maize streak (geminivirus)	Geminate, 18–20 nm	Africa, Southeast Asia	Leafhoppers (*Cicadulina* spp.)	Stunting, fine chlorotic streaks	Resistant cultivars; isolated, clean cultivation; late seeding	Corn, oats, sugarcane, barley, millet, grasses
Oat sterile dwarf (reovirus)	Shelled, isometric, 65–70 nm	Sweden, Czechoslovakia	Planthoppers (*Javesella* and *Dicranotropis* spp.)	Stunting, leaf twisting, enations, tillering, sterility	Clean cultivation, isolation from oats	Oats, corn, barley, millet, grasses
Rice black-streaked dwarf (reovirus)	Shelled, isometric, 75–80 nm	Japan	Planthoppers (*Laodelphax* and *Unkanodes* spp.)	Leaf swelling, twisting, stunting	Clean, isolated cultivation	Rice, corn, oats, barley
Rice hoja blanca	Flexuous rod, 10 × 1,000–3,000 nm	Central and South America, southern USA	Planthoppers (*Sogata* spp.)	Chlorosis, yellow-white discoloration, sterility	Resistant cultivars, isolated cultivation	Rice, rye, oats, barley, grasses
Tobacco mosaic (tobamovirus)	Rigid rod, 18 × 300 nm	Kansas	Sap	Local lesions, mild mosaic	None prescribed	Tobacco, tomato, many dicots, rye, barley

Continued on next page

TABLE 1—*Continued*

Virus	Particle	Distribution of Natural Wheat Infections	Transmission	Major Symptoms	Control	Additional Hosts
Wheat chlorotic streak (rhabdovirus) (see Barley yellow striate mosaic)	Bacilliform, enveloped, 62 × 345–365 nm	France	Planthopper (*Laodelphax striatellus*)	Fine, chlorotic stripes, stunting	Resistant cultivars	*Agropyron repens*, grasses
Wheat dwarf (geminivirus)	Geminate, 18–20 nm	Soviet Union, Sweden, E. Europe	Leafhoppers (*Psammotettix alienus*)	Stunting, yellow-brown blotches, chlorosis	Resistant cultivars	Barley, rye
Soilborne wheat mosaic (furovirus)	Hollow, rigid rod, 20 × 110–160 and 280–300 nm	USA, Japan, Canada, Egypt, Brazil, Italy, Argentina	Sap, fungus (*Polymyxa graminis*)	Yellow-green mosaic, stunting, rosetting	Resistant cultivars, crop rotation, late-autumn seeding	Rye, barley, grasses, sorghum
Wheat (cardamom) mosaic streak	Isometric, 40 nm	India	Sap, aphids (*Brachycaudus* and *Rhopalosiphum* spp.)	Yellow-green mosaic	Resistant cultivars, isolated cultivation	Cardamom, arrowroot
Wheat rosette stunt (rhabdovirus) (see Northern cereal mosaic)		China				
Wheat spindle streak mosaic (potyvirus) (see Wheat yellow mosaic)						
Wheat spot mosaic	Unknown	Alberta, Ohio	Mite (*Aceria tulipae*)	Distinct chlorotic spots	Same as for wheat streak mosaic	Corn, barley, rye, grasses
Wheat streak mosaic (potyvirus)	Flexuous rod, 15 × 700 nm	N. America, Europe, Soviet Union	Sap, mite (*Aceria tulipae*)	Mosaic, yellow-green streaks, stunting	Destruction of volunteer wheat, isolated cultivation, late-autumn seeding	Barley, corn, rye, millet, oats, grasses
American wheat striate mosaic (rhabdovirus) (see Russian winter wheat mosaic)	Bacilliform, enveloped, 65–75 × 250–270 and 415 nm	U.S.-Canadian border	Leafhoppers (*Endria* and *Elymana* spp.)	Delicate streaks, necrotic blotches	Tolerant cultivars, late-autumn and early-spring seeding	Barley, oats, corn, grasses
Australian wheat striate mosaic (geminivirus) (see Chloris striate mosaic)						
Chloris striate mosaic (geminivirus)	Geminate, 18 × 30 nm	Australia	Leafhopper (*Nesoclutha obscura*)	Fine yellow-gray streaks, stunting	None prescribed	Oats, barley, corn, grasses
Eastern wheat striate	Isometric, 40 nm	India	Leafhopper (*Cicadulina mbila*)	Leaf stripes, chlorosis, dwarfing	None prescribed	Unknown
European wheat striate mosaic	Unknown	Europe	Planthoppers (*Javesella* spp.)	Yellow-white streaks, necrosis, stunting	Late seeding	Oats, barley, rye, corn, grasses
Wheat yellow leaf (closterovirus)	Flexuous rod, 10 × 1,600–1,850 nm	Japan	Aphid (*Rhopalosiphum* spp.)	Leaf yellowing, blight	None prescribed	Barley
Wheat yellow mosaic (potyvirus)	Flexuous rod, 14–18 × 200–2,000 nm	Japan, southeastern Canada, USA, India, France, Germany	Sap, fungus (*Polymyxa graminis*)	Green-yellow mosaic, dashes, spindles	Resistant cultivars, crop rotation, late-autumn seeding	Rye, barley
Russian winter wheat mosaic (rhabdovirus) (see American wheat striate mosaic)	Bacilliform, enveloped, 60 × 260 nm	Soviet Union, E. Europe	Leafhoppers (*Psammotettix* and *Macrosteles* spp.)	Mosaic, streaks, stunting, rosetting	Resistant cultivars	Oats, barley, rye, grasses

Causal Agent

Particles of AgMV, a potyvirus, are flexuous rods (15 × 710–717 nm) similar to particles of hordeum mosaic virus, ryegrass mosaic virus, and wheat streak mosaic virus. AgMV most resembles wheat streak mosaic virus but is a weak pathogen in oats and has a different mite vector, *Abacarus hystrix* (Nal.) (Fig. 85). AgMV is sap-transmissible between gramineous hosts. It is inactivated after 10 min in sap at 50° C. It forms pinwheels (see Fig. 100) and crystalline inclusions in cells of tolerant and sensitive plants. Severe and mild strains of the virus are differentiated by symptom intensity.

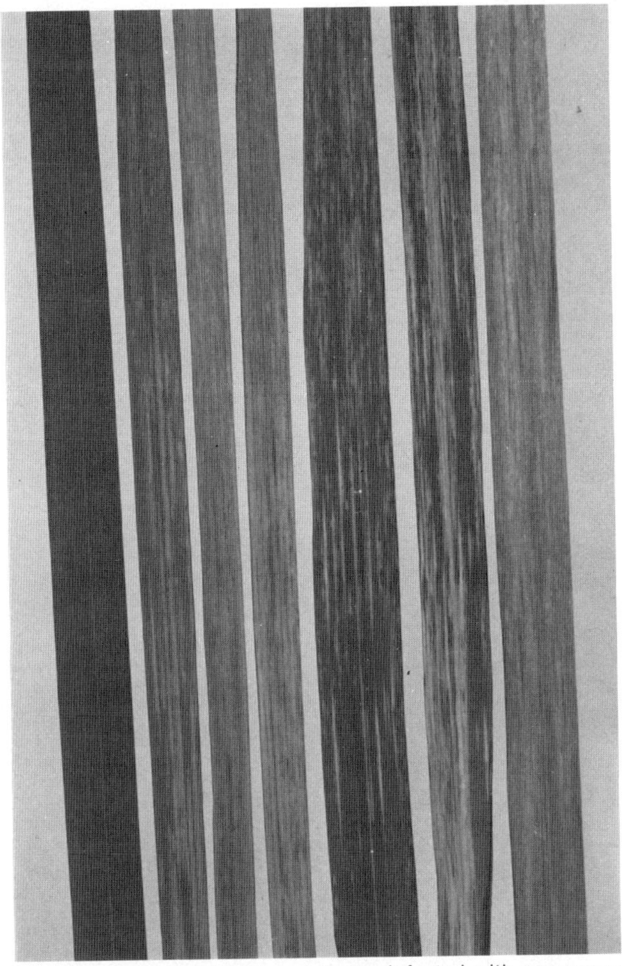

Fig. 84. Healthy leaf (left) and leaves infected with agropyron mosaic virus. (Courtesy J. T. Slykhuis)

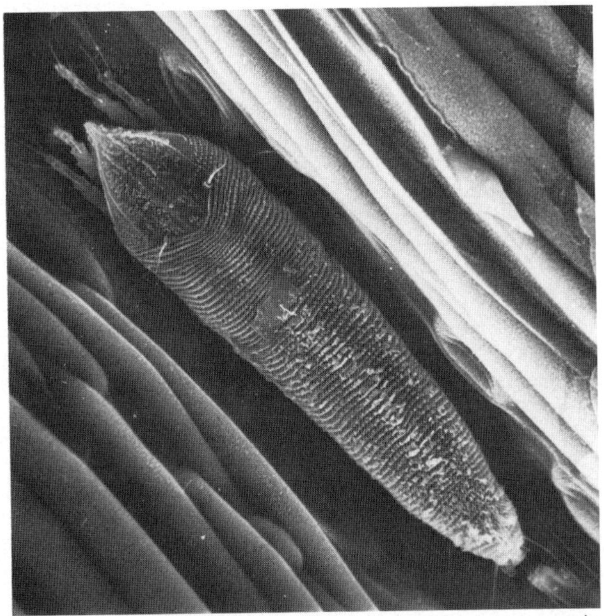

Fig. 85. *Abacarus hystrix*, the mite vector of agropyron mosaic virus. (Courtesy R. W. Gibson)

Disease Cycle

Wheat infections occur from spring through autumn, especially near quackgrass and other grass reservoirs of mites and AgMV. Like *Aceria tulipae*, mite vector of wheat streak mosaic virus, *Abacarus hystrix* is dispersed primarily by wind. At temperatures between 15 and 25° C, wheat develops symptoms within two weeks after mite feeding (within six to 10 days when inoculated with sap from diseased plants). Symptoms develop optimally at 15° C and rarely occur below 10 or above 35° C. Autumn and early-spring infections in winter wheat are most damaging.

Control

In winter wheat, Agropyron mosaic is reduced by late-autumn seeding. Clean cultivation to eliminate volunteer wheat and alternative grassy hosts is beneficial. Tolerant wheat cultivars are available.

Selected References

Bremer, K. 1964. Agropyron mosaic virus in Finland. Ann. Agric. Fenn. 3:324–333.

Catherall, P. L., and Chamberlain, J. A. 1975. Occurrence of agropyron mosaic virus in Britain. Plant Pathol. 24:155–157.

Plumb, R. T., and Lennon, E. A. 1981. Agropyron mosaic virus in wheat. Pages 55–60 in: Proceedings of the 3rd Conference on Virus Diseases of Gramineae in Europe. R. T. Plumb, ed. Rothamsted Experiment Station, Harpenden, Herts., England.

Shepard, J. F. 1968. Occurrence of agropyron mosaic virus in Montana. Plant Dis. Rep. 52:139–141.

Slykhuis, J. T. 1969. Transmission of Agropyron mosaic virus by the eriophyid mite, *Abacarus hystrix*. Phytopathology 59:29–32.

Slykhuis, J. T. 1973. Agropyron mosaic virus. Descriptions of Plant Viruses, No. 118. Commonwealth Mycological Institute, Association of Applied Biologists, Kew, Surrey, England.

Barley Stripe Mosaic

Barley stripe mosaic was described in Wisconsin as "barley false stripe" in 1910, making it perhaps the first cereal virus disease described in the United States. Today, the disease is known to occur on barley in North America, Europe, Japan, Australia, the Soviet Union, China, and Korea.

The causal virus, barley stripe mosaic virus (BSMV), was characterized in 1951 and may be the only virus in the Gramineae efficiently transmitted through seed. In nature, BSMV is rare in wheat except perhaps in China and Israel. It sometimes occurs in wild oats, bromegrass (*Bromus* spp.), perennial ryegrass, corn, and millet, and some dicotyledonous plants are experimental hosts. In hosts other than barley, seed transmission is relatively inefficient. Seed transmission in wheat occurs through both gametes but is limited by severe symptoms and poor seed set on diseased plants.

Symptoms

Wheat infections usually progress through severe mosaics, yellow-white blotches, dwarfing, rosetting, excessive tillering, and necrosis. Normally, chlorotic stripes develop on leaf blades and become increasingly yellow or brown. Additional symptoms and their severity depend on the strain of BSMV and the mode and time of infection. Plants that acquire BSMV through sap develop locally acute chlorosis and necrosis on inoculated leaves. With time, a systemic mosaic appears and then subsides on new leaves.

Causal Agent

BSMV is a multicomponent RNA virus and the type member of the hordeivirus group. Its particles are straight, hollow rods normally 25 nm wide and from 100 to 150 nm long, depending on the strain (Fig. 86). Apart from a weak cross-reaction with poa semilatent virus and lychnis ringspot virus, none of the more than 20 known strains of BSMV appear serologically related to other cereal viruses. BSMV in sap is thermally inactivated after 10 min at 70° C but can persist for years in dried leaves at −18° C.

BSMV is verified in suspect plants by using an electron microscope to identify the virions in leaf-dip preparations and by specific antibody decoration. Its presence in seed is determined serologically and by monitoring barley stripe mosaic symptoms on seedlings grown from seed samples.

Disease Cycle

BSMV is transmitted through seed, sap, and pollen. A recent report from South Africa suggests that the Russian aphid (*Diuraphis noxia*) and the oat bird-cherry aphid (*Rhopalosiphum padi*) are additional vectors. The virus can persist in seed embryos for several years. Natural wheat infections apparently result primarily from plant-to-plant contact and sap transfer. Sap is easily transmitted manually and may be transmitted between adjacent plants in the field when leaves are damaged by wind, hail, or animals.

Wheat plants are susceptible in all premature growth stages and develop symptoms within two weeks of inoculation. Symptoms are best expressed between 22 and 30° C. Below 20° C, symptoms may be restricted or masked because of slowed systemic spread of the virus.

Control

No economic losses have been reported from barley stripe mosaic, and no controls are prescribed for the disease on wheat. Because no latent or mild strains of BSMV are known in wheat, seed from healthy-appearing fields is assumed to be virusfree. Clean cultivation or rotation of wheat apart from barley or other hosts of the virus avoids infection through sap. No wheat cultivars are known to be resistant.

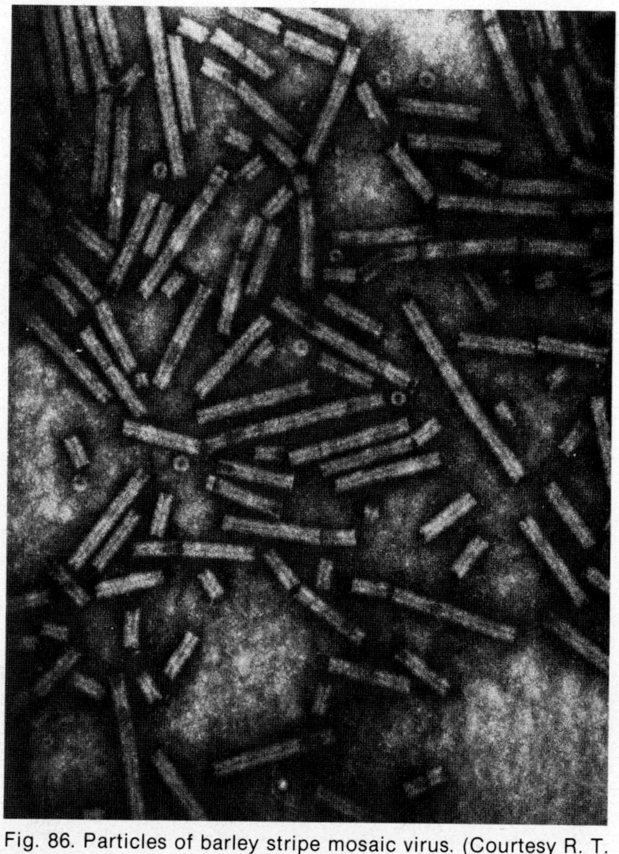

Fig. 86. Particles of barley stripe mosaic virus. (Courtesy R. T. Plumb)

Selected References

Atabekov, J. G. 1971. Barley stripe mosaic virus. Descriptions of Plant Viruses, No. 68. Commonwealth Mycological Institute, Association of Applied Biologists, Kew, Surrey, England.

Carroll, T. W. 1980. Barley stripe mosaic virus: Its economic importance and control in Montana. Plant Dis. 64:136–140.

Carroll, T. W., and Mayhew, D. E. 1976. Anther and pollen infection in relation to the pollen and seed transmissibility of two strains of barley stripe mosaic virus in barley. Can. J. Bot. 54:1604–1621.

Gardner, W. S. 1967. Electron microscopy of barley stripe mosaic virus: Comparative cytology of tissues infected during different stages of maturity. Phytopathology 57:1315–1326.

Hagborg, W. A. F. 1954. Dwarfing of wheat and barley by the barley stripe mosaic (false stripe) virus. Can. J. Bot. 32:24–37.

Langenberg, W. G. 1986. Deterioration of several rod-shaped wheat viruses following antibody decoration. Phytopathology 76:339–341.

Lister, R. M., Carroll, T. W., and Zaske, S. K. 1981. Sensitive serologic detection of barley stripe mosaic virus in barley seed. Plant Dis. 65:809–814.

McFarland, J. E., Brakke, M. K., and Jackson, A. O. 1983. Complexity of the Argentina mild strain of barley stripe mosaic virus. Virology 130:397–400.

McNeal, F. H., and Afanasiev, M. M. 1955. Transmission of barley stripe mosaic through seed in 11 varieties of spring wheat. Plant Dis. Rep. 39:460–462.

Ohmann-Kreutzberg, G. 1962. A contribution to the analysis of the virus diseases of graminaceous plants in central Germany. I. Barley stripe mosaic. Phytopathol. Z. 45:260–288.

Von Wechmar, M. B. 1982. Russian aphid spreads Gramineae viruses. Pages 38–41 in: Proc. 4th Int. Conf. Comparative Virology, Banff, Alberta, Canada.

Barley Yellow Dwarf

Barley yellow dwarf (also called cereal yellow dwarf, yellow dwarf, and red leaf) is probably the most widely distributed and the most destructive virus disease of cereals. It occurs throughout North America, Europe, Australia, Asia, New Zealand, and Africa and in parts of South America. In many regions of the world, such as in the United Kingdom and parts of the United States, barley yellow dwarf is the most economically important virus disease of wheat. It undoubtedly damaged cereal crops long before 1951, when it was first described on barley.

The disease occurs on most cereals and numerous grasses but is not known to affect dicotyledonous plants. Many of its hosts remain symptomless. Damage to wheat varies with the cultivar, virus strain, time of infection, and environmental conditions. Severely infected crops often yield no grain. Over large production areas, barley yellow dwarf is frequently estimated to reduce yields 5–25%.

Symptoms

Symptoms caused by barley yellow dwarf are ambiguous and are often overlooked or associated with nutritional or other nonparasitic disorders. They somewhat resemble symptoms of aster yellows. Barley yellow dwarf is tentatively diagnosed from the presence of aphid vectors and the occurrence of yellowed, stunted plants singly or in small groups among normal plants (Plate 46). Definitive diagnosis requires serologic tests such as enzyme-linked immunosorbent assay (ELISA) and/or recovery and transmission of the virus by aphids.

Leaf discoloration in shades of yellow, red, or purple, especially from tip to base and from margin to midrib, is typical (Plate 47). Seedling infections, if not lethal, slow plant growth and cause prominent or brilliant yellowing of older leaves.

Yellowed or reddened flag leaves on otherwise normal plants are indicative of postseedling infections. Some cultivars show stunting but no leaf discoloration.

Diseased plants have less flexible leaves and underdeveloped root systems. Phloem tissues in all diseased organs often are darkened and necrotic. Symptom expression is enhanced in plants grown in ample light (>12,000 lux) and cool temperatures (16–20° C).

Barley yellow dwarf symptoms may include a response of wheat cultivars to aphid feeding apart from any influence of the virus. Toxic secretions introduced by some vectors, such as *Schizaphis graminum*, produce a pattern of tiny spots on leaves or culms (see Plate 59). The spots, normally less than 1 mm in diameter, become brown-black, and adjacent tissues turn yellow, tan-brown, or necrotic (see Melanism).

Causal Agent

Barley yellow dwarf virus (BYDV), a luteovirus, is a group of closely related virus strains. Strains of BYDV differ serologically and in virulence, host range, and vector specificity. Many strains can coexist in wheat and aphids, but others are mutually exclusive (see Vectors, below). BYDV is not transmissible through seed, soil, or sap or by insects other than aphids.

Virions associated with barley yellow dwarf occur in phloem cells and are polyhedrons 21–26 nm in diameter. The particles are infrequent and are difficult to observe in leaf-dip or thin-sectioned preparations for electron microscopy. Unless tissue grinding is highly efficient, purification procedures normally recover less than 50 μg of virus per liter of plant sap. Roots of infected plants often yield more virus per gram than leaves.

Vectors

More than 20 aphid species transmit BYDV. The most important are *Rhopalosiphum padi* L., the oat bird-cherry aphid; *R. maidis* (Fitch), the corn leaf aphid; *Macrosiphum avenae* Fabr., the English grain aphid; and *S. graminum* (Rondani), the greenbug. In some areas, *Metopolophium (Acyrthosiphon) dirhodum* (Walker) is an important vector. In South Africa, the Russian aphid (*Diuraphis noxia*) is a reported vector.

Aphids acquire BYDV by feeding on infected plants for periods as short as 30 min. However, feeding periods of 12–30 hr are most efficient. After a latent period of one to four days, the virus is transmitted in subsequent feedings of 4 hr or more. Viruliferous aphids remain infectious through successive molts, but the virus is not passed through eggs.

Although some strains of BYDV are transmitted equally well by various aphids, others show high degrees of vector specificity. This specificity is related to the quality of the BYDV protein coat and barriers for virus passage from the aphid gut to its saliva. Specificity also is influenced by the physiologic and behavioral characteristics of aphids, i.e., their capacity to overwinter and multiply, their feeding and flight habits, and their host preferences. Five prominent strains of BYDV are differentiated as follows: RMV, transmitted efficiently by *R. maidis* and weakly virulent in Coast Black oats; RPV, transmitted efficiently by *R. padi* and weakly virulent in Coast Black oats; MAV, transmitted efficiently by *M. avenae* and moderately virulent on Coast Black oats; PAV, vector-nonspecific and highly virulent on Coast Black oats; and SGV, transmitted efficiently by *S. graminum*.

Disease Cycle

BYDV persists in small-grain cereals, in corn, in perennial and annual grasses, and in its vectors. Over 80 species of Gramineae are susceptible. The spread of BYDV depends entirely on vector movement (see Insects [aphids]). Aphids, overwintering as adults on grasses or winter cereals, are immediately active as vectors in the spring. Others developed from eggs acquire BYDV in the spring by feeding on infected host plants during migration.

Barley yellow dwarf epidemics are most apt to occur in cool (10–18° C), moist seasons that favor grass and cereal growth as well as aphid multiplication and migration. Aphid flights can be localized or, when assisted by wind, can cover hundreds of miles. The development of barley yellow dwarf on wheat and grasses from south to north in North America coincides with aphid movement. The greenbug, for example, overwinters in Mexico and moves northward in the jet stream in spring. Other aphids move into wheat from trees and shrubs.

Infections occur throughout the growing season. They are most numerous in spring where aphids overwinter. In winter wheat, autumn infections are most damaging. Adjacent fields of corn or spring small-grain cereals increase the incidence and potential of BYDV and aphids in autumn-sown wheat.

Inoculated plants become systemically infected and develop symptoms within two weeks at 20° C and within four weeks at 25° C. No symptoms develop when temperatures are above 30° C.

Control

The lack of definitive symptoms makes it difficult to assess barley yellow dwarf damage and to prescribe controls. Cultivar selection is important, although no high-level resistance is available in wheat. Wheats that are tolerant to the virus and are poor hosts for the vectors offer the most promise.

Controlling aphids in wheat fields with insecticides can reduce the incidence of barley yellow dwarf and increase yields, especially when autumn infections are reduced. However, the economics of this practice must be carefully weighed because BYDV may be spread by insects that escape treatment or that immigrate from untreated areas.

Barley yellow dwarf is best controlled by late-autumn or early-spring seeding so that the susceptible seedling stage avoids periods of high aphid activity. Late-summer or early-autumn seeding should be avoided.

Selected References

Brown, J. K., Wyatt, S. D., and Hazelwood, D. 1984. Irrigated corn as a source of barley yellow dwarf virus and vectors in eastern Washington. Phytopathology 74:46–49.

Bruehl, G. W. 1961. Barley Yellow Dwarf, A Virus Disease of Cereals and Grasses. Monograph 1. American Phytopathological Society, St. Paul, MN. 52 pp.

Carrigan, L. L., Ohm, H. W., Foster, J. E., and Patterson, F. L. 1981. Response of winter wheat cultivars to barley yellow dwarf virus infection. Crop Sci. 21:377–380.

Cisar, G., Brown, G. M., and Jedlinski, H. 1982. Diallel analysis for tolerance in winter wheat to barley yellow dwarf virus. Crop Sci. 22:328–333.

Clement, D. L., Lister, R. M., and Foster, J. E. 1986. ELISA-based studies on the ecology and epidemiology of barley yellow dwarf virus in Indiana. Phytopathology 76:86–92.

Fargette, D., Lister, R. M., and Hood, E. L. 1982. Grasses as a reservoir of barley yellow dwarf virus in Indiana. Plant Dis. 66:1041–1045.

Gill, C. C. 1980. Assessment of losses on spring wheat naturally infected with barley yellow dwarf virus. Plant Dis. 64:197–203.

Jensen, S. G. 1973. Systemic movement of barley yellow dwarf virus in small grains. Phytopathology 63:854–856.

Lister, R. M., and Rochow, W. F. 1979. Detection of barley yellow dwarf virus by enzyme-linked immunosorbent assay. Phytopathology 69:649–654.

Paliwal, Y. C. 1978. Purification and some properties of barley yellow dwarf virus. Phytopathol. Z. 92:240–246.

Plumb, R. T. 1983. Barley yellow dwarf virus—A global problem. Pages 185–198 in: Plant Virus Epidemiology. R. T. Plumb and J. M. Thresh, eds. Blackwell Scientific Publications, Oxford, England.

Rochow, W. F. 1970. Barley yellow dwarf virus. Descriptions of Plant Viruses, No. 32. Commonwealth Mycological Institute, Association of Applied Biologists, Kew, Surrey, England.

Smith, H. C. 1963. Control of barley yellow dwarf virus in cereals. N.Z. J. Agric. Res. 6:229–244.

Von Wechmar, M. B. 1982. Russian aphid spreads Gramineae viruses. Pages 38–41 in: Proc. 4th Int. Conf. Comparative Virology, Banff, Alberta, Canada.
Wallin, J. R., and Loonan, D. V. 1971. Low-level jet winds, aphid vectors, local weather, and barley yellow dwarf virus outbreaks. Phytopathology 61:1068–1070.

Barley Yellow Striate Mosaic

For years the virus causing barley yellow striate mosaic was known only by its circulative transmissibility from the planthopper *Laodelphax striatellus* (Fallén) to barley, wheat, oats, and grasses in the greenhouse. An additional planthopper, *Toya propinqua* (Fieber), is now implicated as a vector. The disease was first reported in 1972 on volunteer wheat plants in Italy and is now reported to occur also in France, China, Japan, and Morocco on wheat, barley, and oats. Today barley yellow striate mosaic, wheat chlorotic streak, Cynodon chlorotic streak, maize sterile stunt, and northern cereal mosaic all appear to be caused by related rhabdoviruses.

Barley yellow striate mosaic virus (BYSMV) is bacilliform and measures 45–60 × 260–330 nm. In wheat it causes mild yellowing on basal leaves and striations on upper leaves. After acquisition feeding, BYSMV is transmitted following an incubation period of two to three weeks. Planthoppers injected with purified virus also are efficient vectors. Sap transmission does not occur.

No economic importance is attached to barley yellow striate mosaic. Its incidence is extremely low, perhaps because *L. striatellus* is most active in early summer when maturing wheat is a nonpreferred host. No controls are prescribed.

Selected References

Conti, M. 1972. Barley yellow striate mosaic isolated from plants in the field. Phytopathol. Z. 73:39–45.
Conti, M. 1980. Vector relationships and other characteristics of barley yellow striate mosaic virus. Ann. Appl. Biol. 95:83–92.
Conti, M., and Appiano, A. 1973. Barley yellow striate mosaic virus and associated viroplasms in barley cells. J. Gen. Virol. 21:315–322.
Lockhart, B. E. L., Khaless, N., El Maataoui, M., and Lastra, R. 1985. Cynodon chlorotic streak virus, a previously undescribed plant rhabdovirus infecting Bermuda grass and maize in the Mediterranean area. Phytopathology 75:1094–1098.
Milne, R. G., Masenga, V., and Conti, M. 1986. Serological relationships between the nucleocapsids of some planthopper-borne rhabdoviruses of cereals. Intervirology 25:83–87.

Barley Yellow Stripe

Barley yellow stripe occurs in Turkey on wheat, oats, barley, and bromegrass. The causal agent has not been identified but is transmitted by the leafhopper *Euscelis plebejus* (Fallén), which, by itself, causes enations on host plants (see Enanismo [Plate 48]). Symptoms of barley yellow stripe on wheat include fine continuous stripes on leaves, sometimes followed by yellowing and death.

Barley yellow stripe is not sap-transmissible. It occurs especially along field borders and near grassy reservoirs of *E. plebejus*.

Selected Reference

Bremer, K., and Raatikainen, M. 1975. Cereal diseases transmitted or caused by aphids and leafhoppers in Turkey. Ann. Acad. Sci. Fenn. Ser. A4 203:6–11.

Brome Mosaic

Brome mosaic was initially described on bromegrass (*Bromus inermis* Leyss.) in 1941. It occurs in the central United States, the Soviet Union, South Africa, Yugoslavia, and central-northern Europe. Ryegrass streak virus (*Weidelgrasmosaikvirus*), described in the 1960s, is a strain of brome mosaic virus (BMV).

BMV infects wheat, oats, corn, barley, and rye, but natural wheat infections are infrequent and of little economic importance. However, yield losses usually exceed 50% when wheat seedlings are manually inoculated. The virus also infects certain dicotyledonous plants and one or more species of *Bromus, Agropyron, Agrostis, Lolium, Festuca, Phleum*, and *Poa*, an unusually wide host range.

Symptoms

Some wheat cultivars are symptomless carriers of BMV, whereas others develop a prominent, streaklike mottle. Diseased leaves develop a yellow-green mosaic that subsides as plants age. Mild stunting and head deformations also may occur. BMV sometimes occurs in complexes with other viruses that produce symptoms resembling those of barley yellow dwarf.

Causal Agent

BMV (a bromovirus) is a sap-transmissible polyhedron with an average diameter of 25 nm (Fig. 87) and contains four RNA components. It appears in parenchyma tissues, from which it can be extracted and purified in large quantities (1–5 g/kg of tissue). A German strain of BMV has been identified with five RNA components.

BMV causes a diagnostic lethal reaction in sap-inoculated seedlings of sweet corn cultivars Golden Giant and North Star. It has a high thermal inactivation point (10 min in sap at 72–78° C).

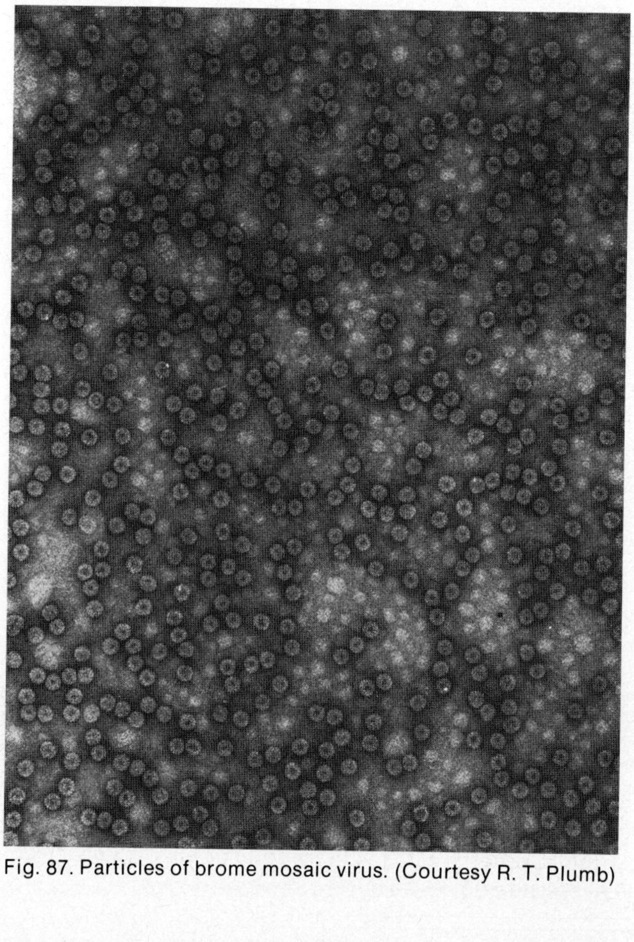

Fig. 87. Particles of brome mosaic virus. (Courtesy R. T. Plumb)

Disease Cycle

BMV persists in cereal and perennial grass hosts. Wheat appears to be infected via mechanical (sap) transmission from grasses and cultivated cereals, although the Russian aphid (*Diuraphis noxia*) was recently reported as a vector in South Africa. In Europe the dagger nematodes *Xiphinema coxi* Tarjan and *X. paraelongatum* Alt. are inefficient vectors. In North America the cereal leaf beetle transmits BMV in controlled environments.

BMV is probably spread locally through plant contact and by incidental transfer of sap during grazing, mowing, cultivation, and windy periods. It can survive for months in dry leaf tissues, perhaps increasing the likelihood of wind dissemination and winter survival.

Control

Control of BMV in wheat has not been seriously explored and at present appears unnecessary. Clean cultivation to eliminate alternative grassy hosts and the use of tolerant cultivars should prove beneficial.

Selected References

Bremer, K. 1973. The brome grass mosaic virus as a cause of cereal disease in Finland. Ann. Agric. Fenn. 12:207–214.

Chiu, R.-J., and Sill, W. H., Jr. 1963. Purification and properties of bromegrass mosaic virus. Phytopathology 53:1285–1291.

Lane, L. C. 1979. Bromovirus group. Descriptions of Plant Viruses, No. 215. Commonwealth Mycological Institute, Association of Applied Biologists, Kew, Surrey, England.

Lane, L. C. 1981. Bromoviruses. Pages 334–376 in: Handbook of Plant Virus Infection and Comparative Diagnosis. E. Keerstak, ed. Elsevier North-Holland Biomedical Press, Amsterdam.

Paliwal, Y. C. 1970. Electron microscopy of bromegrass mosaic virus in infected leaves. J. Ultrastruct. Res. 30:491–502.

Schmidt, H. B., Fritsche, R., and Lehmann, W. 1963. Die Ubertragung des Weidelgrasmosaikvirus durch Nematoden. Naturwissenschaften 50:386.

Slykhuis, J. T. 1967. *Agropyron repens* and other perennial grasses as hosts for bromegrass mosaic virus from the USSR and the United States. FAO Plant Prot. Bull. 15:65–66.

Von Wechmar, M. B. 1982. Russian aphid spreads Gramineae viruses. Pages 38–41 in: Proc. 4th Int. Conf. Comparative Virology, Banff, Alberta, Canada.

Von Wechmar, M. B., and Rybicki, E. P. 1985. Brome mosaic virus infection mimics barley yellow dwarf disease symptoms in small grains. Phytopathol. Z. 114:332–337.

Northern Cereal Mosaic

Cereal mosaic occurs in China, Japan, Korea, and Siberia. Known since 1910, the disease causes stunting and a yellow mosaic and chlorotic streaks on wheat leaves. Cereal mosaic virus is transmitted by the planthopper *Laodelphax striatellus* (Fallén) to some 15 grass species in addition to wheat, barley, oats, rice, and rye (corn and sorghum are nonhosts). More recently, *Unkanodes sapporona* Mats. and *Delphacodes* spp. also have been described as vectors. Some resistant wheat cultivars have been reported.

Particles of cereal mosaic virus, a rhabdovirus, are bacilliform, measure 40–60 × 300–400 nm, and are phloem-limited. The virus is thermally inactivated after 10 min in sap at 55° C. It is similar to oat pseudorosette virus and barley yellow striate mosaic virus.

In China, wheat rosette stunt virus, a strain of northern cereal mosaic virus, invades winter wheat and is controlled by late seeding and cultivation to eliminate alternate grassy weed hosts.

Selected References

Hyung, L. S., and Shikata, E. 1977. Occurrence of northern cereal mosaic virus in Korea. Plant Prot. 16:87–92.

Ito, S., and Fukushi, T. 1944. Studies on northern cereal mosaic. J. Sapporo Soc. Agric. For. 36(3):62–89, 36(4):65–88.

Milne, R. G., Masenga, V., and Conti, M. 1986. Serological relationships between the nucleocapsids of some planthopper-borne rhabdoviruses of cereals. Intervirology 25:83–87.

Murayama, D., and Lu, Y. T. 1967. Some physical properties of northern cereal mosaic virus. J. Fac. Agric. Hokkaido Univ. 55:182–190.

Ruan, Y. L., Lin, D. D., and Xu, R. Y. 1983. Transmission characteristics of the northern cereal mosaic virus by the small brown planthopper (*Laodelphax striatellus* Fallen). Acta Phytopathol. Sin. 13:20–24.

Shikata, E., and Lu, Y. T. 1967. Electron microscopy of northern cereal mosaic virus in Japan. Proc. Jpn. Acad. 43:918–923.

Shirako, Y., and Ehara, Y. 1985. Composition of northern cereal mosaic virus and its detection by enzyme-linked immunosorbent assay with antinucleocapsid serum. Phytopathology 75:453–457.

African Cereal Streak

African cereal streak reached epidemic levels at lower elevations in Kenya during 1969–1972. Today infections are sporadic, but wheat fields with up to 40% diseased plants have been reported. The disease, not known elsewhere except perhaps in Ethiopia, also occurs on barley, oats, rye, triticale, rice, and several grasses.

Symptoms on all cultivated hosts include faint, broken, chlorotic leaf streaks that originate near the base of the blade (Fig. 88). Streaking eventually encompasses the entire blade, and leaves may become totally yellowed. Infected seedlings are stunted, develop severe chlorosis, and eventually die. Plants infected later develop twisted or variously distorted heads. As with barley yellow dwarf, diseased plants exhibit phloem necrosis and contain isometric virions, 24 nm in diameter, in phloem cells. The particles are serologically related to cynosurus mottle virus.

Fig. 88. Leaf symptoms induced by African cereal streak virus. (Courtesy D. E. Harder)

The virus is transmitted by the planthopper *Toya catilina* (Fennah) but not through sap, seed, or soil or by aphids. Native grasses are presumed to be natural reservoirs for both the virus and vectors. Temperatures above 20°C favor disease development.

Selected References

Harder, D. E. 1975. Electron microscopy of African cereal streak diseased plants. Can. J. Bot. 53:565–581.

Harder, D. E., and Bakker, W. 1973. African cereal streak, a new disease of cereals in East Africa. Phytopathology 63:1407–1411.

Cereal Tillering

Reported in Italy and Sweden on barley, cereal tillering is similar to oat sterile dwarf. Wheat infections produce excessive tillering, dwarfing, dark green coloration, and poor grain yields. Occasionally, leaves appear malformed, with serrated margins.

The causal virus, cereal tillering virus (CTV) (a reovirus), is a phloem-limited, two-shelled sphere 65–73 nm in diameter. Morphologically, the virions of CTV resemble those of oat sterile dwarf virus (OSDV), rice black-streaked dwarf virus (RBSDV), and maize rough dwarf virus (MRDV). However, CTV is distinguished from OSDV by its more severe reaction in corn and from RBSDV by its negligible reaction in rice; MRDV, unlike CTV, has *Javesella pellucida* (Fabr.) as a vector. All four viruses persist in planthopper vectors and are acquired and transmitted within similar feeding intervals.

CTV is circulative within *Laodelphax striatellus* (Fallén) and *Dicranotropis hamata* Boh. but is not passed through eggs. It is transmitted to barley, rye, oats, corn, and grasses in addition to wheat. Of these hosts, barley is most susceptible. Symptoms of cereal tillering appear within one to 15 days after inoculation feedings.

Selected References

Lindsten, K. 1974. Planthopper transmitted virus diseases of cereals in Sweden. Acta Biol. Iugosl. 11:55–66.

Lindsten, K. 1979. Planthopper vectors and plant disease agents in Fennoscandia. Pages 155–178 in: Leafhopper Vectors and Plant Disease Agents. K. Maramorosch and K. F. Harris, eds. Academic Press, New York.

Lindsten, K., Gerhardson, B., and Pettersson, J. 1973. Cereal tillering disease in Sweden and some comparisons with oat sterile dwarf and maize rough dwarf. Natl. Swed. Inst. Plant Prot. Contrib. 15:375–397.

Milne, R. G., Lindsten, K., and Conti, M. 1975. Electron microscopy of the particles of cereal tillering disease virus and oat sterile dwarf virus. Ann. Appl. Biol. 79:371–373.

Cocksfoot Mottle

Cocksfoot mottle virus (CoMV), a sobemovirus, infects cocksfoot (orchardgrass [*Dactylis glomerata* L.]) in the United Kingdom. Occasionally, wheat is a natural host. The virus is isometric and 30 nm in diameter and is readily transmitted through sap to wheat and oat seedlings. Inoculated plants show leaf mottling within 14 days, followed by yellowing and premature death.

The cereal leaf beetles *Oulema melanopa* L. and *O. lichenis* L. are efficient vectors, but sap transmission accounts for most plant-to-plant spread of the virus in grass crops. CoMV appears to be the only cereal virus efficiently transmitted by a chewing insect.

Selected References

Benigno, D. A., and A'Brook, J. 1972. Infection of cereals by cocksfoot mottle and Phleum mottle virus. Ann. Appl. Biol. 72:43–52.

Catherall, P. L. 1970. Cocksfoot mottle virus. Descriptions of Plant Viruses, No. 23. Commonwealth Mycological Institute, Association of Applied Biologists, Kew, Surrey, England.

Serjeant, E. P. 1967. Some properties of cocksfoot mottle virus. Ann. Appl. Biol. 59:31–38.

Toriyama, S. 1983. Cocksfoot mottle virus in Japan. Ann. Phytopathol. Soc. Jpn. 48:514–520.

Enanismo

Enanismo (also called cereal dwarf) was described in the 1950s as an important disease of wheat in Colombia and Ecuador. It also occurs on barley and oats, seriously limiting yields in localized highland areas. In 1960, the leafhopper *Cicadulina pastusae* Rup. & DeLg. was found to cause part of the disease syndrome. No additional causal agent has been described, but one or more viruses and a toxin from *C. pastusae* appear to be involved.

Symptoms

Infected seedlings normally are killed or severely stunted. Infections at later growth stages cause less stunting but induce prominent leaf blotches and earlike enations (galls) (Plate 48). Plants infected near heading develop distorted heads with incompletely filled kernels. Galls develop one to three weeks after insect feeding and appear on new leaves rather than on the leaves that were fed upon. A translocatable toxin from *C. pastusae* appears to induce gall formation.

Disease Cycle and Control

Leafhopper adults and nymphs of both sexes transmit the causal agent in circulative fashion. Females pass the pathogen transovarially and are more efficient vectors than males. The minimum inoculation feeding time is 24 hr. *C. pastusae* apparently regulates the incidence and distribution of enanismo.

The disease is controlled by using tolerant and resistant wheat cultivars and is avoided by seeding in late spring after vector activity declines.

Selected References

Calvache, H., Guerrero, O., and Martínez-López, G. 1975. Relationship studies between the leafhopper *Cicadulina pastusae* (Ruppel & DeLong) and the transmission of two disease agents to barley and wheat in southwestern Colombia. (Abstr.) Proc. Am. Phytopathol. Soc. 2:74.

Gálvez-E., G. E. 1965. Toxin from *Cicadulina pastusae*, vector of enanismo virus, causes galls on leaves of small grains. (Abstr.) Phytopathology 55:1059.

Gálvez, G. E., Thurston, H. D., and Bravo, G. 1963. Leafhopper transmission of enanismo of small grains. Phytopathology 53:106–108.

Maize Streak

Maize streak virus (MSV), a geminivirus, causes a serious disease of corn in Africa and southeast Asia. A *Pennisetum* strain of MSV infects wheat in India, and other strains infect sugarcane, guinea grass (*Panicum maximum* Jacq.), oats, barley, and certain wild grasses. No hosts outside the Gramineae are known.

Symptoms

MSV incites fine, linear, chlorotic leaf streaks. Shortened, curled leaves, excessive tillering, sterility, and shortened culms

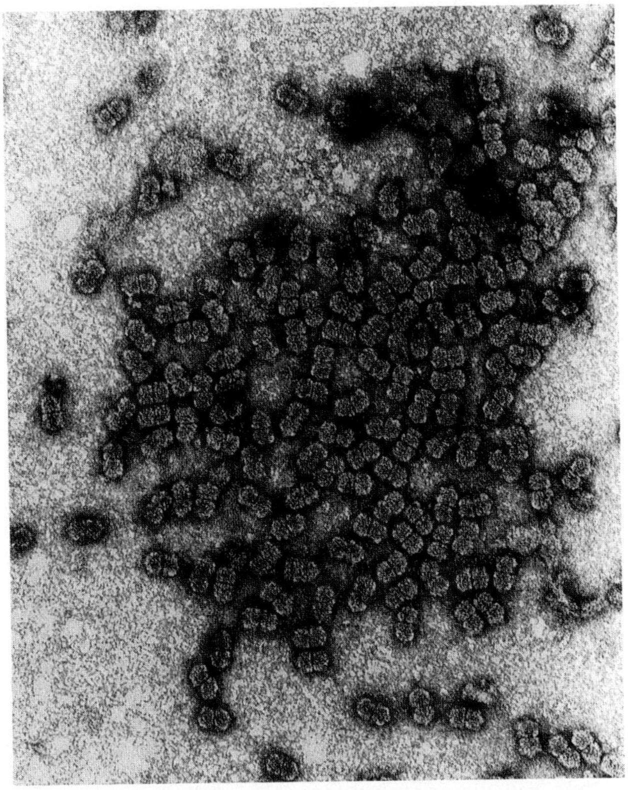

Fig. 89. Paired particles of maize streak virus. (Copyright Plant Pathology Department, Rothamsted Experimental Station. Used by permission. Photo by R. D. Woods)

and spikes may also occur. Wheat is most apt to develop symptoms when grown in association with corn.

Causal Agent

Particles of MSV contain single-stranded DNA. They are paired (Fig. 89), measure 18–20 nm in diameter, and occur in phloem and mesophyll cells. The virus is transmitted circulatively by six species of *Cicadulina* leafhoppers, especially *C. mbila* Naudé. At optimal temperatures (20–25° C), the vectors acquire MSV within 1-hr feedings and transmit in 5-min feedings after an incubation period of 6–12 hr. As with most other leafhopper-transmitted viruses, latent periods in wheat exceed three days. All stages of leafhoppers transmit MSV, but transovarial passage has not been demonstrated. MSV is thermally inactivated after 10 min in sap at 60° C but can be stored for years in dried leaf tissue at −125° C. Strains of MSV are distinguished by host range and serology.

Control

Maize streak is avoided by delayed autumn seeding of winter wheat and by seeding in areas removed from corn or grasses that harbor MSV. Many wheat cultivars are resistant to the more common strains of the virus.

Selected References

Bock, K. R. 1974. Maize streak virus. Descriptions of Plant Viruses, No. 133. Commonwealth Mycological Institute, Association of Applied Biologists, Kew, Surrey, England.

Bock, K. R., Guthrie, E. J., and Woods, R. D. 1974. Purification of maize streak virus and its relationship to viruses associated with streak diseases of sugarcane and *Panicum maximum*. Ann. Appl. Biol. 77:289–296.

Damsteegt, V. D. 1983. Maize streak virus: I. Host range and vulnerability of maize germ plasm. Plant Dis. 67:734–737.

Damsteegt, V. D. 1984. Maize streak virus: Effect of temperature on vector and virus. Phytopathology 74:1317–1319.

Oat Sterile Dwarf

Oat sterile dwarf occurs in eastern and northern Europe. The disease especially affects oats, but wheat, barley, millet, grasses, and corn also are susceptible. Symptoms, better expressed on oats than on wheat, include stunting, excessive tillering, helical leaf twisting, floret sterility, and enations on leaf veins (see Enanismo [Plate 48]).

Oat sterile dwarf virus (OSDV) (Reoviridae) is transmitted persistently by planthoppers, especially *Javesella pellucida* (Fabr.). Other *Javesella* spp. and *Dicranotropis hamata* Boh. are other possible vectors. The virus is transmitted after three to four weeks' incubation in its vectors and is rarely passed through eggs. Symptoms develop on host plants after 18–26 days. OSDV is a phloem-limited, double-shelled sphere 65–70 nm in diameter and is serologically related to lolium enation virus.

Controlling grassy weeds and avoiding oat cover crops limits vector populations and the disease. No resistant wheat cultivars are available.

Selected References

Boccardo, G., and Milne, R. G. 1980. Oat sterile dwarf virus. Descriptions of Plant Viruses, No. 217. Commonwealth Mycological Institute, Association of Applied Biologists, Kew, Surrey, England.

Brcak, J., Kralik, O., and Vacke, J. 1972. Virus origin of oat sterile dwarf disease. Biol. Plant. 14:302–304.

Ikaheimo, K. 1961. A virus disease of oats in Finland similar to oat sterile dwarf disease. Maataloustiet. Aikak. 33:81–87.

Lindsten, K. 1970. Investigations on the spread and control of oat sterile dwarf. Natl. Swed. Inst. Plant Prot. Contrib. 14:407–446.

Lindsten, K. 1979. Planthopper vectors and plant disease agents in Fennoscandia. Pages 155–178 in: Leafhopper Vectors and Plant Disease Agents. K. Maramorosch and K. F. Harris, eds. Academic Press, New York.

Milne, R. G., Lindsten, K., and Conti, M. 1975. Electron microscopy of the particles of cereal tillering disease virus and oat sterile dwarf virus. Ann. Appl. Biol. 79:371–373.

Vacke, J. 1966. Study of transovarial passage of oat sterile dwarf virus. Biol. Plant. 8:127–130.

Rice Black-Streaked Dwarf

Rice black-streaked dwarf was first described in 1952 and is currently known only in Japan. The disease occurs principally on rice, but wheat, corn, and barley are also damaged. Oats and other Gramineae are less important hosts.

The spherical particles of rice black-streaked dwarf virus (RBSDV) (Reoviridae) are 75–80 nm in diameter. Their transmission and host range are closely related to the feeding habits of the planthoppers *Laodelphax striatellus* (Fallén), *Unkanodes sapporona* Mats., and *U. albifascia* Mats. The virus is not transmitted through sap or seed. RBSDV in sap is thermally inactivated after 10 min at 60° C. It is similar to cereal tillering virus, rice dwarf virus, and maize rough dwarf virus.

In wheat, RBSDV causes severe stunting and characteristically twisted leaves. Waxy veinal swellings (hypertrophied phloem tissues) may develop on the undersurface of leaves and on culms. Both wheat and barley maintain high concentrations of RBSDV in infected tissues and are desirable host species for virus maintenance and multiplication. Corn, which develops obvious white streaks, is a useful assay host.

Winter wheat harbors viruliferous insects that move to rice in the spring and back to wheat in the autumn. Vectors acquire the virus within 30-min feedings and transmit within 5-min feedings after an incubation period of seven to 35 days. The virus is retained from season to season in all stages of planthoppers but is not transmitted through eggs. Normally, nymphs are more efficient vectors than adults.

The use of resistant rice cultivars decreases the chance of virus transmission to wheat. Where possible, wheat and rice crops should be grown in isolation.

Selected References

Ou, S. H. 1972. Black-streaked dwarf. Pages 25–28 in: Rice Diseases. Commonwealth Mycological Institute, Association of Applied Biologists, Kew, Surrey, England.

Shikata, E. 1969. Electron microscope studies of rice viruses. Pages 223–240 in: The Virus Diseases of the Rice Plant. Johns Hopkins Press, Baltimore, MD.

Shikata, E. 1974. Rice black-streaked dwarf virus. Descriptions of Plant Viruses, No. 135. Commonwealth Mycological Institute, Association of Applied Biologists, Kew, Surrey, England.

Rice Hoja Blanca

Rice hoja blanca (also called white leaf) occurs in South and Central America, the Caribbean, and the south-central United States. Since its recognition in 1935, it has occurred sporadically. Rice hoja blanca virus (RHBV) causes mottling, striping, and chlorosis on leaves of rice, wheat, rye, barley, oats, and some grasses. Sterility and death of the plant normally follow. Wheat becomes infected especially when grown adjacent to rice and shows a characteristic and prominent gray-white discoloration of the spike and uppermost leaves sometimes referred to as "white tip" or "white spike."

RHBV is transmitted persistently by tropical planthoppers of the genus *Sogata*, especially *S. cubana* (Crawford) and *S. orizicola* (Muir). In these insects RHBV is circulative and is transmitted through eggs. It is maintained through as many as 10 generations without further acquisition feedings. RHBV is acquired during feedings of 15 min to 1 hr and is transmitted after an incubation period of six to 37 days.

Particles of RHBV are incompletely characterized. Particles 10 nm wide and up to 3 μm long (similar to those of wheat yellow mosaic virus) and spiral, filamentous particles 3 nm in diameter have been reported.

Maize hoja blanca, a similar "white tip" disease in corn, is vectored by *Peregrinus maidis* (Ashm.). White-tipped corn plants contain both threadlike particles and isometric particles 40–45 nm in diameter.

Wheat grown apart from rice or in association with resistant rice cultivars is rarely infected.

Selected References

Galvez, G. 1969. Hoja blanca disease of rice. Pages 35–49 in: The Virus Diseases of the Rice Plant. Johns Hopkins Press, Baltimore, MD.

Gibler, J. W., Jennings, P. R., and Krull, C. F. 1961. Natural occurrence of hoja blanca on wheat and oats. Plant Dis. Rep. 45:334.

Kitajima, E. W., Caetano, V. R., and Costa, A. S. 1971. Intracellular inclusions associated with 'white spike' of wheat. Bragantia 30:101–108. (English summary)

Lamey, H. A., McMillian, W. W., and Hendrick, R. D. 1964. Host ranges of the hoja blanca virus and its insect vector. Phytopathology 54:536–541.

Morales, F. J., and Niessen, A. I. 1983. Association of spiral filamentous viruslike particles with rice hoja blanca. Phytopathology 73:971–974.

Ou, S. H. 1972. Hoja blanca. Pages 28–33 in: Rice Diseases. Commonwealth Mycological Institute, Association of Applied Biologists, Kew, Surrey, England.

Varón de Agudelo, F., and Martinez-López, G. 1983. Maize hoja blanca: A complex of viruses transmitted by *Peregrinus maidis*. (Abstr.) Phytopathology 73:125.

Tobacco Mosaic

Tobacco mosaic virus (TMV), a tobamovirus, is perhaps the best known and most investigated plant virus. Its ready availability and high stability have made it a model for numerous studies of basic and practical plant virology.

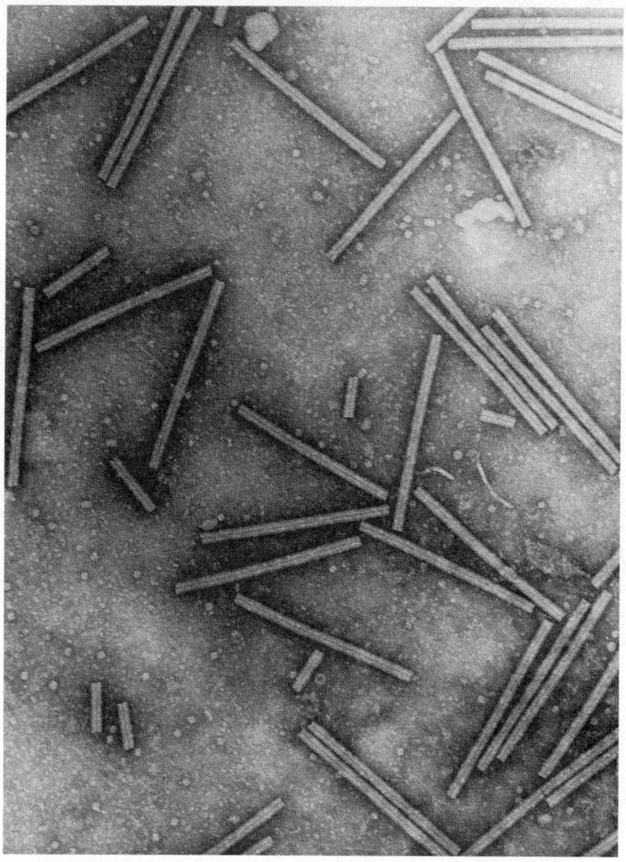

Fig. 90. Rigid, hollow particles of tobacco mosaic virus. (Copyright Plant Pathology Department, Rothamsted Experimental Station. Used by permission. Photo by R. D. Woods)

TMV occurs worldwide and has a large host range among dicotyledonous species. Tobacco and tomato are perhaps the most economically important hosts. TMV was not known among the Gramineae until 1970, when manually inoculated barley was found to be susceptible at high temperatures. Later, wheat and rye were identified as hosts. In 1975 a strain of TMV was found in wheat plants in Kansas that bore symptoms of and were infected with soilborne wheat mosaic virus (SBWMV). Neither the origin nor the practical importance of the natural double infection is known.

Wheat inoculated with TMV-infected sap may remain symptomless or may develop local lesions or a mild systemic mosaic. Infections are reproducible and are most successful at temperatures between 25 and 30°C.

TMV particles are rigid rods measuring 18 × 300 nm (Fig. 90) and are similar to particles of SBWMV. They are abundant in host plants and are readily purified. Tobacco plants, *Nicotiana tabacum* L. 'Xanthi-nc', rubbed with sap from TMV-infected plants develop characteristic local lesions and are widely used in TMV assays.

Selected References

Oxefelt, P. 1974. Multiplication of tobacco mosaic virus in oats, rye and wheat. Phytopathol. Z. 79:281–284.

Paulsen, A., Niblett, C. L., and Willis, W. G. 1975. Natural occurrence of tobacco mosaic virus in wheat. Plant Dis. Rep. 59:747–750.

Zaitlin, M., and Israel, H. W. 1975. Tobacco mosaic virus (type strain). Descriptions of Plant Viruses, No. 151. Commonwealth Mycological Institute, Association of Applied Biologists, Kew, Surrey, England.

Wheat Chlorotic Streak

In 1971, fine chlorotic striations were observed on the uppermost leaves of durum wheat and wheatgrass (*Agropyron repens*) in France (Fig. 91). Affected plants were more numerous near field borders, and often their heads were confined within the leaf sheath. These symptoms are now known to be associated with wheat chlorotic streak virus (WCSV).

Particles of WCSV, a rhabdovirus, measure 62 × 345–365 nm, are tubular and bacilliform, and are sheathed by an electron-transparent membrane (Fig. 92). WCSV appears to be closely related to barley yellow striate mosaic virus. It is transmitted by the planthopper *Laodelphax striatellus* (Fallén) in a persistent manner. Transmission is accomplished within four days after a minimum acquisition feeding period of 48 hr. Wheat plants develop symptoms within 10–14 days after being fed upon by viruliferous insects.

Wheat chlorotic streak occurs infrequently. It limits the growth and yield of individual plants but is not economically important on a larger scale. Grasses, especially *A. repens* and *Phalaris arundinacea*, appear to maintain both the virus and vectors and are presumed to be the source of wheat infections.

Selected References

Signoret, P. A. 1974. Les maladies à virus des graminées dans le midi de la France. (The virus diseases of grasses in southern France.) Mikrobiologija 11:115–120.

Signoret, P. A., Conti, M., Leclant, F., Alliot, B., and Giannotti, J. 1977. Données nouvelles sur la maladie des striés chlorotiques du blé (wheat chlorotic streak virus). Ann. Phytopathol. 9:381–385.

Signoret, P. A., Giannotti, J., and Alliot, B. 1972. Particules de type viral chez *Triticum durum* Desf. présentant des symptomes de striés chlorotiques. (Viruslike particles in *Triticum durum* Desf. with chlorotic streak symptoms.) Ann. Phytopathol. 4:45–53.

Wheat Dwarf

Wheat dwarf was first described in Czechoslovakia in 1960. In Europe the disease is economically important on spring and winter wheat and also affects rye and barley. In Sweden it occurs in association with heavy infestations of leafhoppers, which were originally thought to be its cause. Wheat dwarf somewhat resembles, but apparently is distinct from, Russian winter wheat mosaic and occurs in approximately the same locations.

Symptoms

Infected plants develop vein distortions on the underside of leaves and fine light green to yellow-brown spots and blotches on both leaf surfaces. The spots may coalesce eventually to cause prominent yellowing and necrosis. Eventually, dwarfing becomes the prominent symptom. It results from the combined effects of virus infection and intense leafhopper feeding. Plants infected as seedlings attain about 15% of their normal height and do not head; some die. Stunting is less pronounced in plants infected at later growth stages. If heads emerge, seeds are sparse and shriveled.

Causal Agent

Particles of wheat dwarf virus (WDV), a geminivirus, are isometric, 18–20 nm in diameter, and usually paired (see Maize Streak [Fig. 89]). They have a genome of single-stranded DNA. Leafhoppers, primarily *Psammotettix alienus* (Dahlbom), transmit WDV efficiently. The virus is not transmitted through sap, seed, or soil. As with Russian winter wheat mosaic (also transmitted by *P. alienus*), nymphs are more efficient vectors than adults but must acquire the disease agent from infected plants because it is not passed through eggs. Transmission is accomplished within one to two days after acquisition feeding. *Lolium* spp. and other grasses serve as reservoirs of WDV.

Control

Resistant cultivars provide the best defense against wheat dwarf. Soft wheat cultivars appear less susceptible than hard wheats.

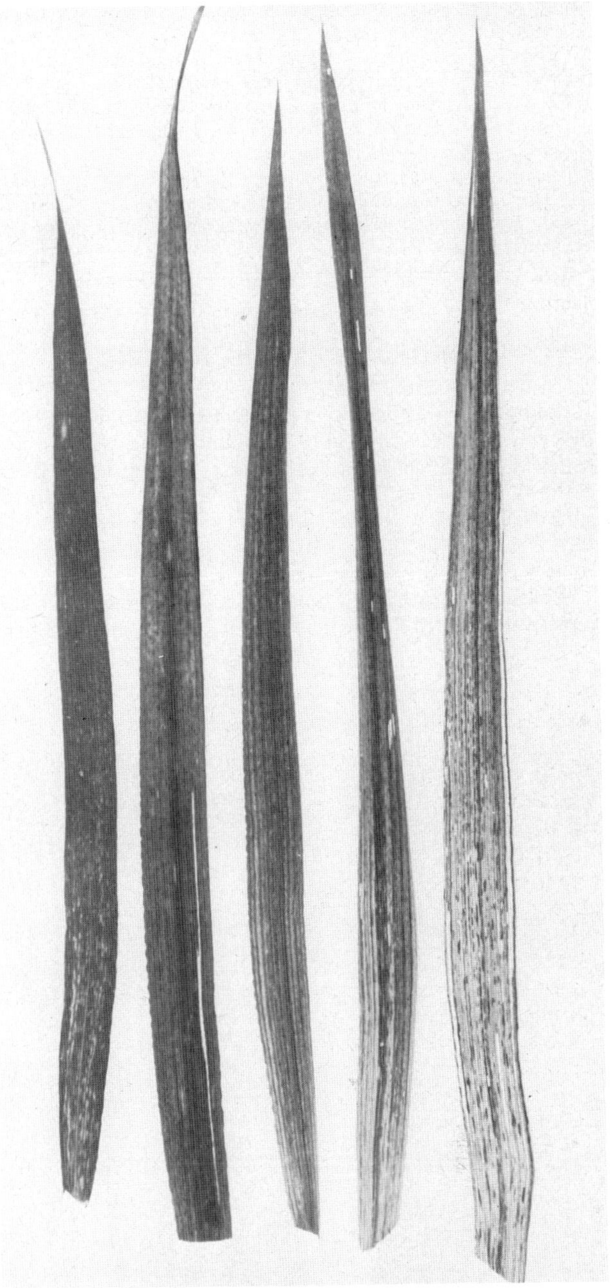

Fig. 91. Leaf symptoms caused by wheat chlorotic streak virus. (Courtesy P. A. Signoret)

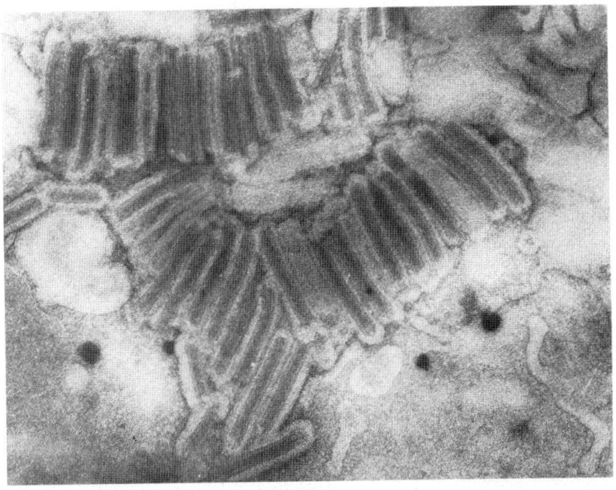

Fig. 92. Membrane-bound, bacilliform particles of wheat chlorotic streak virus. (Courtesy P. A. Signoret)

Selected References

Lindsten, K., Lindsten, B., Abdelmoeti, M., and Juntti, N. 1981. Purification and some properties of wheat dwarf virus. Pages 27–34 in: Proceedings of the 3rd Conference on Virus Diseases of Gramineae in Europe. R. T. Plumb, ed. Rothamsted Experiment Station, Harpenden, Herts., England.

Lindsten, K., Vacke, J., and Gerhardson, B. 1970. A preliminary report of three cereal virus diseases new to Sweden spread by *Macrosteles* and *Psammotettix* leafhoppers. Natl. Swed. Inst. Plant Prot. Contrib. 14:128.

Vacke, J. 1961. Wheat dwarf virus disease. Biol. Plant. 3:228–233.

Soilborne Wheat Mosaic

Soilborne wheat mosaic first attracted attention in 1919 in the central United States. In 1923, soilborne wheat mosaic virus (SBWMV) was identified as the cause, making it one of the earliest known wheat viruses and the first to be characterized as soilborne. In early reports, forms of soilborne wheat mosaic were described as "green mosaic," "yellow mosaic," "mosaic rosette," and "eastern wheat mosaic."

Soilborne wheat mosaic now occurs throughout the eastern and central United States and in China, Japan, France, Egypt, Italy, Argentina, and Brazil. Normally only autumn-sown wheats develop symptoms, although spring wheats also are susceptible. Symptoms occasionally develop on rye, barley, and hairy bromegrass (*Bromus* spp.).

Among winter wheats, losses to soilborne wheat mosaic vary with the cultivar, virus strain, and environment. Entire fields or areas of fields can be damaged, and in some fields, harvest is abandoned. In the central grain belt of the United States, the disease reduces yields annually and rivals barley yellow dwarf in economic importance.

Symptoms

Symptoms of soilborne wheat mosaic range from mild green to prominent yellow leaf mosaics (Fig. 93). Stunting, too, can be moderate or severe. Certain strains of the virus cause rosetting in highly susceptible cultivars (Fig. 94). Fields may be uniformly diseased but more often show patterned chlorosis corresponding to the distribution of the fungus vector, *Polymyxa graminis*, which preferentially inhabits low-lying wet areas (Fig. 95). Symptoms are most prominent on early-spring growth and rarely appear in autumn. As new leaves unfold, they appear mottled and develop parallel dashes and streaks. Leaf sheaths are also distinctly mottled. Warming spring temperatures slow and eventually stop disease development, normally confining symptoms to lower leaves.

Causal Agent

A new group, furovirus, has been proposed for SBWMV and other fungusborne, rod-shaped viruses. Particles of wild-type

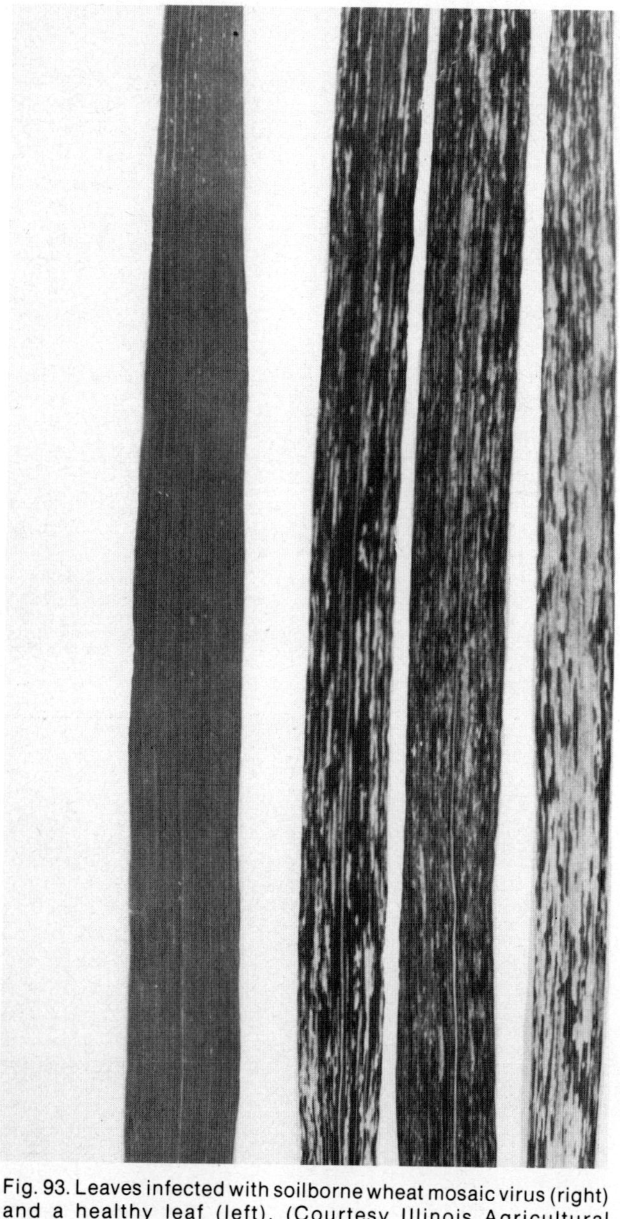

Fig. 93. Leaves infected with soilborne wheat mosaic virus (right) and a healthy leaf (left). (Courtesy Illinois Agricultural Experiment Station)

Fig. 94. Severe stunting (rosetting) caused by soilborne wheat mosaic virus. (Courtesy Illinois Agricultural Experiment Station)

Fig. 95. Typical distribution of soilborne wheat mosaic (chlorotic area) in low-lying portion of field. (Courtesy C. L. Niblett)

SBWMV are hollow, rigid rods 20 nm wide and of two principal lengths, 140 and 280 nm. Particles of both sizes are necessary for infection. The longer rods (component I) account for approximately 5% of all rods and resemble particles of tobacco mosaic virus (Fig. 90). The RNA of 140-nm rods (component II) mutates frequently by deletion to yield rods 90–140 nm long, which predominate in infected plants. All SBWMV rods contain single-stranded RNA.

Compared to other mosaic viruses, SBWMV is difficult to transmit through sap, but it causes diagnostic local lesions on mechanically inoculated leaves of *Chenopodium* spp. It is best transmitted through root washings that contain zoospores of *P. graminis*.

SBWMV is thermally inactivated in sap after 10 min at 65°C but is stable for years in dried leaf tissues. Infected host cells often contain amorphous and crystalline inclusion bodies that contain virions in an orderly (paracrystalline) array.

Vector

The natural mode of wheat infection with SBWMV is through *P. graminis* Led., a soilborne plasmodiophoraceous fungus (see Primitive Root Fungi). *P. graminis* is an obligate parasite in roots of many higher plants. It enters wheat root hairs and epidermal cells during periods of high soil moisture via motile, biflagellate zoospores (see Fig. 71). The zoospores average 4.2 μm in diameter and appear to carry SBWMV internally or tightly bound externally. After penetrating host cells, *P. graminis* expands into plasmodial bodies that replace the cell contents. The plasmodia eventually segment into additional zoospores or, within two to four weeks after infection, develop into smooth, thick-walled resting spores 5–7 μm in diameter. Clusters of resting spores are visible in cortical and epidermal cells under low-power magnification (see Fig. 72).

Disease Cycle

SBWMV survives in soil in close association with *P. graminis*. Soils may remain infested for years. *P. graminis* infects wheat roots during cool, wet periods in autumn and possibly in early spring. Normally, spring infections occur too late to cause damage before warmer temperatures and crop maturation prevent their progress. Autumn infections permit the virus to multiply to damaging proportions and predispose plants to other diseases and winter injury. Temperatures from 10 to 20°C (optimum near 16°C) promote soilborne wheat mosaic. Disease progress is prevented above 20°C (see Wheat Yellow Mosaic).

SBWMV is spread by cultivation, wind, water, and other factors that disperse infested soil. However, for reasons not understood, the virus sometimes spreads more rapidly and over longer distances than can be explained by soil or water movement.

Control

Resistant or tolerant cultivars offer the best control of soilborne wheat mosaic. The hard red cultivars Homestead and Newton and many soft red wheats sustain little damage. The nature of resistance in wheat to SBWMV is unknown. Some cultivars that are resistant in the field are susceptible to the vector and to mechanical inoculation.

Less effective controls include crop rotation and late-autumn seeding. Continuous wheat cultivation should be avoided. Soil fumigation destroys most soil microflora, including *P. graminis*, but cost limits its use to special circumstances.

Selected References

Brakke, M. K. 1971. Soil-borne wheat mosaic virus. Descriptions of Plant Viruses, No. 77. Commonwealth Mycological Institute, Association of Applied Biologists, Kew, Surrey, England.

Brakke, M. K., Estes, A. P., and Schuster, M. L. 1965. Transmission of soil-borne wheat mosaic virus. Phytopathology 55:79–86.

Campbell, L. G., Heyne, E. G., Gronau, D. N., and Niblett, C. 1975. Effect of soilborne wheat mosaic virus on wheat yield. Plant Dis. Rep. 59:472–476.

Larsen, H. J., Brakke, M. K., and Langenberg, W. G. 1985. Relationships between wheat streak mosaic virus and soilborne wheat mosaic virus infection, disease resistance, and early growth of winter wheat. Plant Dis. 69:857–862.

Merkle, O. G., and Smith, E. L. 1983. Inheritance of resistance to soilborne mosaic in wheat. Crop Sci. 23:1075–1076.

Peterson, J. F. 1970. Electron microscopy of soil-borne wheat mosaic virus in host cells. Virology 42:304–310.

Rao, A. S. 1968. Biology of *Polymyxa graminis* in relation to soil-borne wheat mosaic virus. Phytopathology 58:1516–1521.

Rao, A. S., and Brakke, M. K. 1969. Relation of soil-borne wheat mosaic virus and its fungal vector, *Polymyxa graminis*. Phytopathology 59:581–587.

Shirako, Y., and Brakke, M. K. 1984. Two purified RNAs of soil-borne wheat mosaic virus are needed for infection. J. Gen. Virol. 65:119–127.

Signoret, P. A. 1974. Gramineae virus diseases in southern France. Mikrobiologija 11:115–120.

Sill, W. H., Jr. 1958. A comparison of some characteristics of soil-borne wheat mosaic viruses in the Great Plains and elsewhere. Plant Dis. Rep. 42:912–924.

Tsuchizaki, T., Hibino, H., and Saito, Y. 1973. Comparisons of soil-borne wheat mosaic virus isolates from Japan and the United States. Phytopathology 63:634–639.

Wheat (Cardamom) Mosaic Streak

In India, a virus originally described in large cardamom (*Amomum subulatum* Roxb.) as the cause of "chirke" or "cardamom mosaic" also occurs in wheat and arrowroot. The 40-nm isometric virus is transmissible through sap and by the aphids *Brachycaudus helichrysi* (Kalt.) and *Rhopalosiphum maidis* (Fitch). Wheat infections occur near reservoirs of the virus in cardamom.

Most Indian wheat cultivars are susceptible and show chronic yellow-green mosaic symptoms. Infected plants are mildly stunted and are predisposed to infection by *Bipolaris sorokiniana* (syn. *Helminthosporium sativum*). The cultivar Ridley is reported to be resistant, and wheats grown apart from cardamom avoid the disease.

Selected References

Adlakha, K. L., and Raychaudhuri, S. P. 1975. Interaction between *Helminthosporium sativum* and wheat mosaic streak. Z. Pflanzenkr. Pflanzenschutz 82:201–206.

Raychaudhuri, S. P., and Chatterjee, S. N. 1965. Transmission of chirke disease of large cardamom by aphid species. Indian J. Entomol. 27:272–276.

Raychaudhuri, S. P., and Ganguly, B. 1968. A mosaic streak of wheat. Phytopathol. Z. 62:61–65.

Wheat Spot Mosaic

During investigations of wheat streak mosaic virus (WSMV) and its transmission by the wheat curl mite, *Aceria tulipae* Keifer, a second disease agent was uncovered. The second agent is also miteborne but is not sap-transmissible. It causes spotting rather than mosaic symptoms, followed by a chlorotic mottle.

Wheat spot mosaic (also called wheat spot and wheat spot chlorosis) has been reported in Alberta in association with WSMV and in Ohio on wheat and corn. There is increasing evidence that the disease may be widely distributed across the northern United States. Barley, rye, and some grasses also are hosts, but wheat is most susceptible.

Symptoms

Within three to eight days after infectious mites feed on wheat seedlings, light green spots 0.5–1 mm in diameter develop on the youngest leaves. The spots intensify, become necrotic, enlarge,

Fig. 96. Leaf symptoms of wheat spot mosaic. (Courtesy J. T. Slykhuis)

and coalesce to form generally golden yellow leaves and leaf areas (Fig. 96). In addition to the diagnostic spotting, mottling, chlorosis, leaf tip necrosis, stunting, and plant death may occur, especially when WSMV occurs in combination with the wheat spot mosaic pathogen.

Causal Agent

The wheat spot mosaic pathogen has not been identified. It is acquired by immature mites within 1-hr feedings on diseased plants and is retained during molts. Adult mites apparently cannot acquire the virus nor retain it for more than 10 days.

Infected wheat tissues contain double-membrane-bound ovoid bodies, 100–200 nm in diameter, in the cytoplasm of parenchyma, phloem, and epidermal cells. The bodies are not clearly mycoplasmal or viral in nature. Many have an outer membrane that is continuous with membranes of the host cell. Infected cells also contain small vesicles 2–20 nm in diameter.

Control

The incidence of wheat spot mosaic, like that of wheat streak mosaic, depends on the activity of *A. tulipae*. Thus, the controls described for wheat streak mosaic are also effective against wheat spot mosaic. Wheat spot mosaic could become a problem in areas where mite controls are relaxed in favor of cultivars resistant to wheat streak mosaic.

Selected References

Atkinson, T. G., and Larson, R. I. 1975. Virus-vector considerations in utilizing sources of resistance to wheat streak mosaic. (Abstr.) Proc. Can. Phytopathol. Soc. 42:26.

Bradfute, O. E., Whitmoyer, R. E., and Nault, L. R. 1970. Ultrastructure of plant leaf tissue infected with mite-borne viral-like pathogens. Proc. Electron Microsc. Soc. Am. 28:178–179.

Nault, L. R., and Styer, W. E. 1970. Transmission of an eriophyid-borne wheat pathogen by *Aceria tulipae*. Phytopathology 60:1616–1618.

Nault, L. R., Styer, W. E., Gordon, D. T., Bradfute, O. E., LaFever, H. N., and Williams, L. E. 1970. An eriophyid-borne pathogen from Ohio and its relation to wheat spot mosaic virus. Plant Dis. Rep. 54:156–160.

Slykhuis, J. T. 1956. Wheat spot mosaic, caused by a mite-transmitted virus associated with wheat streak mosaic. Phytopathology 46:682–687.

Whitmoyer, R. E., Nault, L. R., and Bradfute, O. E. 1972. Fine structure of *Aceria tulipae* (Acarina: Eriophyidae). Ann. Entomol. Soc. Am. 65:201–215.

Wheat Streak Mosaic

Wheat streak mosaic is an important and widely distributed wheat disease. First recognized in Nebraska as "yellow mosaic" in 1922, it occurs throughout central and western North America, eastern Europe, and parts of the Soviet Union. It is most prevalent in the central Great Plains of the United States, where it destroys a significant percentage of the winter wheat crop annually. In some years, losses in Kansas alone exceed $30 million. Diseased areas can cover hundreds of hectares or be localized in a few fields. Yield losses range from insignificant to complete. Wheat streak mosaic also occurs on spring wheat, barley, corn, rye, oats, and a number of annual and perennial grasses.

Symptoms

The distribution of wheat streak mosaic is closely related to the dispersal of its mite vector. Margins of fields are often the first and at times the only areas affected. Symptoms vary with the cultivar, strain of the virus, time of infection, and environmental conditions. Infected plants normally are stunted, with mottled and streaked leaves. Leaf streaks are green-yellow, parallel, and discontinuous (Fig. 97). Mite-infested leaves tend to remain erect, with their lateral margins rolled toward the upper midrib.

Autumn infections are frequent, but symptoms rarely develop before spring. As spring temperatures rise, stunting and yellowing become more obvious. Heads, if formed, are totally or partially sterile. Many leaves, and sometimes entire plants, may become yellow and necrotic.

The mite vector feeds preferentially on the upper surface and near the margin of leaves, causing them to curl upward and inward toward the midvein (see Fig. 104). Mites frequently are enclosed protectively within the rolled leaves.

Causal Agent

Particles of wheat streak mosaic virus (WSMV), a potyvirus, occur in most leaf cells as flexuous rods measuring 15 × 700 nm. Membranous pinwheels and aggregates of virions also appear in infected cells (see Wheat Yellow Mosaic [Fig. 100]).

WSMV is difficult to purify in quantity. It is thermally inactivated at 54° C (10 min in sap) and has several characteristics in common with agropyron mosaic virus, ryegrass mosaic virus, and hordeum mosaic virus. It is easily transmissible through sap.

Vector

The wheat curl mite, *Aceria tulipae* Keifer (syn. *Eriophyes tulipae* Keifer), and in Yugoslavia, *A. tosichella* Keifer, are the natural means of WSMV dissemination. The vectors thrive on the lush, young growth of wheat and many grasses. Where temperatures and predators are not limiting, mites develop from eggs to adults within eight to 10 days. Their numbers can increase markedly during relatively short periods when the environment is favorable.

Mites are approximately 0.3 mm long and are best observed under magnification. They are normally white and cylindrical (Plate 49). Their four diminutive legs next to their head provide limited movement. Like *Abacarus hystrix*, they are dispersed from plant to plant and from field to field by wind.

WSMV is carried in the midgut and hindgut of all larval and adult mite stages but is not passed through eggs. Acquisition feeding periods can be as short as 15 min, and most mites remain infective for seven to nine days without additional acquisitions.

Disease Cycle

WSMV and its mite vectors persist on wheat, corn, millet, and susceptible grasses like buffalograss (*Buchloë dactyloides* (Nutt) Engelm.) and foxtail (*Alopecurus* spp.). The host ranges of WSMV and its mite vectors are not identical but overlap sufficiently for their cosurvival. From spring through autumn, winds distribute mites to new cereal and grass hosts. Mites and WSMV persist on living susceptible plants, but neither can survive on ripened grain or grasses. In mature or recently harvested wheat fields, mites persist on green shoots or on volunteer plants. Wheat streak mosaic is severe where a host continuum exists for both mite and virus between spring and autumn-sown crops. Such conditions are found in areas where both spring and winter wheats are grown and where wheat is seeded in late summer or early autumn as forage for livestock. Late-maturing summer crops and volunteer stands from grain shattered by hailstorms or harvest operations can also play an important role in completing the disease cycle. Mites and WSMV are perpetuated in the south-central United States, where crops of wheat and irrigated corn overlap in spring and autumn.

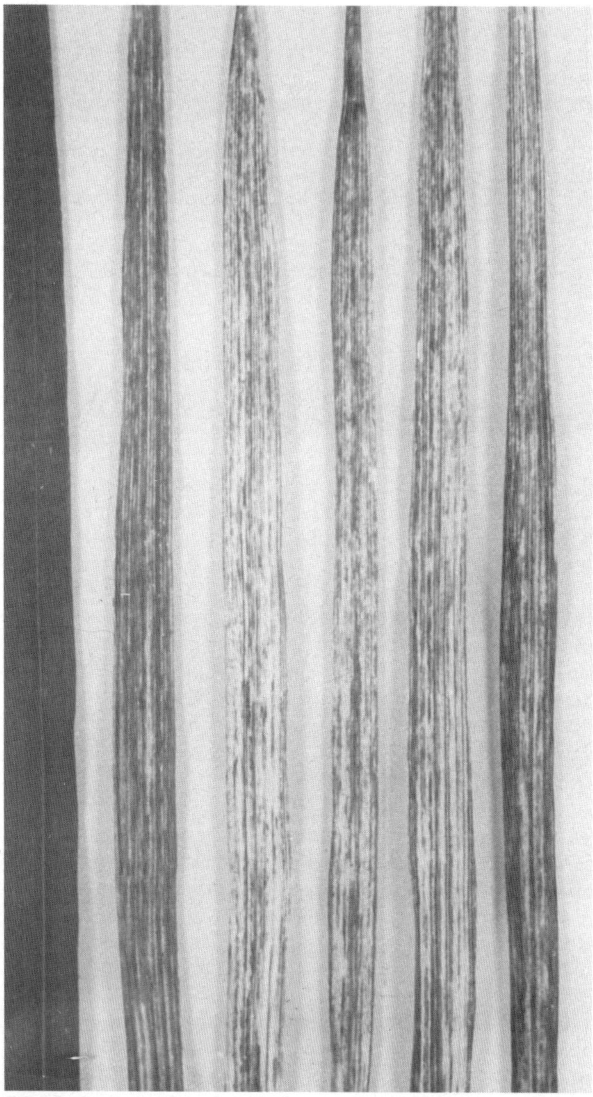

Fig. 97. Leaves infected with wheat streak mosaic virus (right) and a healthy leaf (left). (Courtesy Illinois Agricultural Experiment Station)

Control

Wheat streak mosaic is controlled by cultural practices that minimize sources of WSMV and mites when new wheat crops emerge in autumn. Thus, the destruction of volunteer wheat, coupled with late seeding, controls the disease. Late-autumn seeding can ensure that new crops develop after summer crops have matured and mite populations have declined. Late seeding postpones infections and gives mites and wheat streak mosaic less time to reach damaging proportions.

Wheats tolerant to WSMV or to mite feeding have been identified and are being adapted for commercial use. Chromosome substitutions from *Agropyron* spp. are principal sources of resistance but are difficult to incorporate into acceptable wheat cultivars. Currently grown wheat cultivars differ in tolerance; Oslo, PR2369, and Butte are tolerant spring wheats along the central U.S.–Canadian border.

Because corn and certain grasses may harbor WSMV and mites, clean cultivation and avoidance of susceptible alternative crops may be beneficial. Wheat, however, appears to be the principal host involved in wheat streak mosaic epidemiology.

Selected References

Ashworth, L. J., Jr., and Futrell, M. C. 1961. Sources, transmission, symptomatology and distribution of wheat streak mosaic virus in Texas. Plant Dis. Rep. 45:220–224.

Atkinson, T. G., and Grant, M. N. 1967. An evaluation of streak mosaic losses in winter wheat. Phytopathology 57:188–192.

Brakke, M. K. 1971. Wheat streak mosaic virus. Descriptions of Plant Viruses, No. 48. Commonwealth Mycological Institute, Association of Applied Biologists, Kew, Surrey, England.

Bremer, K. 1971. Wheat streak mosaic virus in Turkey. Phytopathol. Mediterr. 10:280–282.

del Rosario, M. S. E., and Sill, W. H., Jr. 1965. Physiological strains of *Aceria tulipae* and their relationships to the transmission of wheat streak mosaic virus. Phytopathology 55:1168–1175.

Martin, T. J., Harvey, T. L., Bender, C. G., and Seifers, D. L. 1984. Control of wheat streak mosaic virus with vector resistance in wheat. Phytopathology 74:963–964.

McMullen, C. R., and Gardner, W. S. 1980. Cytoplasmic inclusions induced by wheat streak mosaic virus. J. Ultrastruct. Res. 72:65–75.

Orlob, G. B. 1966. Feeding and transmission characteristics of *Aceria tulipae* Keifer as vector of wheat streak mosaic virus. Phytopathol. Z. 55:218–238.

Pfannenstiel, M. A., and Niblett, C. L. 1978. The nature of the resistance of agrotricums to wheat streak mosaic virus. Phytopathology 68:1204–1209.

Shepard, J. F., and Carroll, T. W. 1967. Electron microscopy of wheat streak mosaic virus particles in infected plant cells. J. Ultrastruct. Res. 21:145–150.

Slykhuis, J. T. 1955. *Aceria tulipae* Keifer (Acarina: Eriophyidae) in relation to the spread of wheat streak mosaic. Phytopathology 45:116–128.

Tosic, M. 1973. Transmission of wheat streak mosaic virus to different host plants by *Aceria tosichella*. Plant Prot. (Belgrade) 126:317–321.

Whitmoyer, R. F., Nault, L. R., and Bradfute, O. E. 1972. Fine structure of *Aceria tulipae* (Acarina: Eriophyidae). Ann. Entomol. Soc. Am. 65:210–215.

American Wheat Striate Mosaic

American wheat striate mosaic occurs near the central U.S.-Canadian border. Since its detection in South Dakota in 1953, it has been observed sporadically on winter and spring wheats, barley, oats, and corn. Durum wheats appear most susceptible. *Eragrostis, Bromus, Panicum*, and other grass genera also are susceptible. The disease is of economic importance only in isolated areas.

Symptoms

Early symptoms include faint chlorotic streaks along veins of the leaf undersurface (Fig. 98). Later, fine streaks appear over the entire leaf. Leaves often become wholly chlorotic, with necrotic brown stripes, and sometimes die back from tip to base. Many diseased plants are stunted and do not head. They may be generally distributed but more often occur in patches or along field borders. Symptoms of American wheat striate mosaic markedly resemble those of European wheat striate mosaic, Chloris striate mosaic (Australian wheat striate mosaic), and Russian winter wheat mosaic.

Causal Agent

Virions of American wheat striate mosaic virus (AWStMV) (a rhabdovirus) in infected plants are short-bacilliform (65–75 × 250–270 nm), bullet-shaped (75 × 200 nm), and long-bacilliform (75 × 415 nm). The bacilliform particles are most abundant and are often associated with undulating filaments 5 nm in diameter and of indeterminate length. Virions are dispersed or loosely aggregated in the cytoplasm and nuclei of parenchyma and phloem cells, where they often appear membrane-bound. The virus is thermally inactivated at 55°C (10 min) in sap.

Vectors

AWStMV is transmitted principally by the painted leafhopper, *Endria inimica* Say, but *Elymana virescens* F. is also a vector. Acquisition occurs within 15-min feedings, and an incubation period of six to 21 days precedes transmission. Both nymphs and adults can remain infectious for 71 days without renewed acquisition feeding. The virus is not passed through eggs nor is it transmissible through sap.

Disease Cycle

AWStMV overwinters in wheat and grasses in association with *E. inimica*, which overwinters as eggs on similar hosts. The leafhoppers, in various stages of development, are dispersed during spring and summer by wind. Winged adults also are migratory. Wheat infections occur from spring through autumn; however, autumn infections may not produce symptoms until the following spring. Warm temperatures (25–33°C) are optimal for symptom development and are prerequisite during early summer for epidemics of American wheat striate mosaic. In general, occurrences of American wheat striate mosaic in the United States are associated with heavy leafhopper infestations.

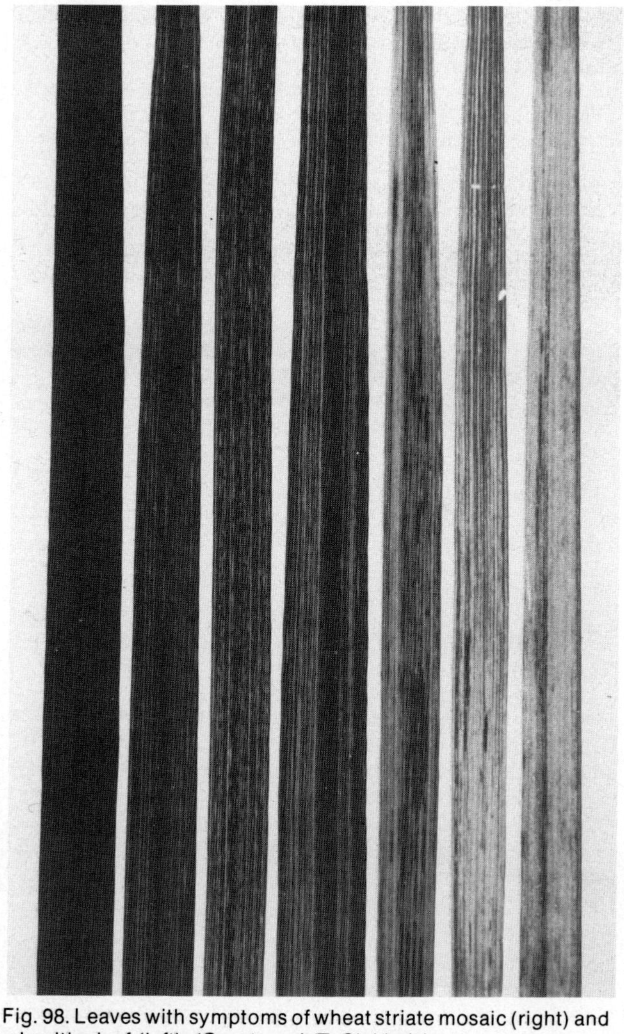

Fig. 98. Leaves with symptoms of wheat striate mosaic (right) and a healthy leaf (left). (Courtesy J. T. Slykhuis)

Control

Some spring and winter wheat cultivars are tolerant to American wheat striate mosaic, but none are highly resistant. Wheat sown late in autumn or early in spring is more likely to avoid high vector populations and escape infection than wheat seeded at more standard times.

Selected References

Ahmed, M. E., Sinha, R. C., and Hochster, R. M. 1970. Purification and some morphological characteristics of wheat striate mosaic virus. Virology 41:768–771.

Jons, V. L., Timian, R. G., Gardner, W. S., Stromberg, E. L., and Berger, P. 1981. Wheat striate mosaic virus in the Dakotas and Minnesota. Plant Dis. 65:447–448.

Lee, P. F., and Bell, W. 1963. Some properties of wheat striate mosaic. Can. J. Bot. 41:767–771.

Sinha, R. C. 1971. Distribution of wheat striate mosaic virus in infected plants and some morphological characteristics of the virus in situ and in vitro. Virology 44:342–351.

Sinha, R. C., and Behki, R. M. 1972. American wheat striate mosaic virus. Descriptions of Plant Viruses, No. 99. Commonwealth Mycological Institute, Association of Applied Biologists, Kew, Surrey, England.

Slykhuis, J. T. 1962. Wheat striate mosaic, a virus disease to watch on the prairies. Can. Plant Dis. Surv. 42:135–141.

Slykhuis, J. T. 1963. Vector and host relations of North American wheat striate mosaic virus. Can. J. Bot. 41:1171–1185.

Thottappilly, G., and Sinha, R. C. 1973. Serological analysis of wheat striate mosaic virus and its soluble antigen. Virology 53:312–318.

Chloris Striate Mosaic (Australian Wheat Striate Mosaic)

In 1962 a viruslike disease of grasses and small grains was recognized in Australia. The disease appears primarily associated with windmill grass, *Chloris gayana* Kunth, in eastern Australia, where it also has been identified on *Ixophorus, Pennisetum, Phalaris,* and *Dactylis* spp. Subsequent investigation uncovered a geminivirus associated with the chlorotic leaf symptoms on grasses and small grains. The pathogen has a DNA genome and paired particles measuring 18 × 30 nm (see Maize Streak and Fig. 89). It is transmitted to wheat by the leafhoppers *Nesoclutha obscura* Evans and *N. pallida* (Evans). It is not transmissible through sap, nor does it pass through the eggs of its vectors.

N. obscura is common in Australia, where it overwinters as eggs on the leaves of corn and grasses that also serve as reservoirs of the virus. It can complete three generations in one season. Populations are greatest in late summer. The vector has the unusual habit of feeding, acquiring the pathogen, and introducing the pathogen in mesophyll rather than phloem tissues.

Many cultivars of wheat, oats, barley, and corn are susceptible and develop symptoms as early as eight days after leafhopper feeding. Symptoms on wheat include yellow, broken streaks, fine grayish striping, and dwarfing, much like symptoms caused by American wheat striate mosaic, European wheat striate mosaic, and Russian winter wheat mosaic. The disease is of little economic importance, perhaps because

Lovisolo, O. 1971. Maize rough dwarf virus. Descriptions of Plant Viruses, No. 72. Commonwealth Mycological Institute, Association of Applied Biologists, Kew, Surrey, England.
Niblett, C. L., and Paulsen, A. Q. 1975. Purification and further characterization of panicum mosaic virus. Phytopathology 65:1157–1160.
Ou, S. H. 1972a. Dwarf. Pages 6–14 in: Rice Diseases. Commonwealth Mycological Institute, Kew, Surrey, England.
Ou, S. H. 1972b. Stripe. Pages 14–22 in: Rice Diseases. Commonwealth Mycological Institute, Kew, Surrey, England.
Paul, H. L., Huth, W., and Querfurth, G. 1973–1974. Cocksfoot mild mosaic virus—phleum mottle virus: A comparison. Intervirology 2:253–260.
Pirone, P. P. 1972. Sugarcane mosaic virus. Descriptions of Plant Viruses, No. 88. Commonwealth Mycological Institute, Association of Applied Biologists, Kew, Surrey, England.
Polak, Z., and Slykhuis, J. T. 1972. Comparisons of poa semilatent and barley stripe mosaic viruses. Can. J. Bot. 50:263–267.
Sill, W. H., Jr., and Talens, L. T. 1962. New hosts and characteristics of panicum mosaic virus. Plant Dis. Rep. 46:780–783.
Slykhuis, J. T. 1972. Poa semilatent virus from native grasses. Phytopathology 62:508–513.
Slykhuis, J. T., and Bell, W. 1966. Differentiation of agropyron mosaic, wheat streak mosaic and a hitherto unrecognized hordeum mosaic virus in Canada. Can. J. Bot. 44:1191–1208.
Slykhuis, J. T., and Paliwal, Y. C. 1972. Ryegrass mosaic virus. Descriptions of Plant Viruses, No. 86. Commonwealth Mycological Institute, Association of Applied Biologists, Kew, Surrey, England.
Sukhov, K. S., et al. 1941. Reports on Zakuklivanie (pseudorosette) of cereals in the USSR. In: Plant Virus Diseases and Their Control. Trans. Conf. Plant Virus Dis., Moscow, 1940. (R.A.M. 23:210–219)
Wellman, F. L. 1934. Infection of *Zea mays* and various other Gramineae by the celery virus in Florida. Phytopathology 24:1035–1037.

Seedborne Wheat Yellows

A seedborne wheat yellows disease of unknown etiology has been known in north China since the mid-1970s. This disease has been especially destructive in He Nan province. In susceptible cultivars, the rate of seed transmission is often 90–100%. Evidence was presented in 1986 by Xiaojie Wu, Guangrong Shuen, Suzeng Huang, and Mingqi Wang of Fudan University in Shanghai that this disease is caused by a viroid.

Viroids differ from viruses in having no protein coat. Instead, they occur as circular segments of naked, self-replicating RNA and are the smallest known agents capable of causing plant disease. Seedborne wheat yellows viroid (SWYV) apparently is the first viroid to be reported in a member of the grass family.

Symptoms

Symptoms of seedborne wheat yellows first appear as chlorotic spots on the upper and middle portions of seedling leaves. These spots gradually coalesce to form large chlorotic and then necrotic areas as the leaves age (Plate 51). Symptoms continue to appear on new leaves as they emerge.

Scientists at Fudan University reproduced the exact symptoms of this disease by inoculating healthy wheat seedlings with either sap from diseased plants or partially purified viroid RNA from plants with seedborne wheat yellows. Using two-dimensional gel electrophoresis, they found a unique band of RNA with extracts from diseased plants. Electron micrographs of the bands revealed the circular RNA typical of a viroid, whereas no viruslike particles were found in electron micrographs of diseased leaf tissues.

Corn and barley developed mild symptoms, but none of a wide array of broad-leaved (dicotyledonous) plants developed symptoms when mechanically inoculated with sap from diseased wheat leaves.

Disease Cycle

The origin of SWYV is unknown, but the viroid is thought to have been introduced with seeds of a susceptible cultivar. In addition to dispersal in seed, the viroid also is presumably spread mechanically from plant to plant in the field. Mechanical dispersal and efficient seed transmission in susceptible cultivars may account for the rapid spread and continued appearance of the disease in He Nan province.

Control

Because the disease is apparently limited to certain cultivars, breeding for disease resistance may be a practical approach to controlling seedborne wheat yellows. Using seed certified as pathogen-free can greatly reduce but may not eliminate the disease where susceptible cultivars are grown.

Diseases Caused by Parasitic Plants

Witchweeds

Witchweeds—primarily *Striga asiatica* (L.) Kuntze (syn. *S. lutea* Lour.) but also *S. hermonthica*, *S. densiflora*, and *S. aspera*—are flowering plants that parasitize wheat, barley, oats, corn, sorghum, sugarcane, rice, millet, and more than 60 species of grasses and sedges. *Striga* spp. occur most frequently in association with corn in South Africa, but they also occur in Australia, India, Southeast Asia, Indonesia, and southeastern coastal areas of the United States. They are particularly troublesome in regions with high temperatures and light, sandy soil. Witchweeds in cereal crops contributed significantly to the famine in Ethiopia in the mid-1980s.

Symptoms

Wheat associated with witchweed is stunted, mildly chlorotic, and generally low in vigor. Severe infestations of the weed compete with and kill host plants in circular patches in fields.

Although wheat and witchweed plants are unattached aboveground, their roots become interwoven and difficult to separate. Root connections are numerous and swollen. Within them a continuum of host and parasite vascular tissues can be observed in thin sections under low magnification.

Causal Organism

Witchweed parasitizes its hosts via root connections (haustoria) but otherwise resembles a self-sufficient, chlorophyllous flowering plant (Fig. 101). An erect, square-stemmed annual, it is 20–30 cm tall at maturity. Its leaves are bright green, alternate, and lanceolate and measure up to 1.5×4.7 cm. Its tubular flowers, in shades of red, yellow, or white, are 5–11 mm in diameter, with a two-lipped corolla. Seeds are numerous, striated, and brown, measure 0.15×0.22 mm, and are borne in capsules.

Disease Cycle

Witchweed survives as seed, which can remain dormant in and on soil for up to 10 years. Seeds develop into infectious germlings by producing enzymes that digest the surface cells of nearby roots. Haustoria eventually form in response to a nonspecific stimulus (quinone) from the degrading host roots.

Fig. 101. Witchweed (*Striga* sp.) parasitizing corn roots. (Courtesy USDA)

Repeated infections occur as additional roots are contacted by the developing parasite. If a suitable host root is not located within three to six days after germination, the seedlings die. However, once established on a host, the parasite emerges in four to six weeks to produce foliage and flowers. Mature seed is produced within eight to 10 weeks. Seed of *Striga* spp. is disseminated by wind and water and by agents that disperse soil and plant residues in which seeds lie dormant.

Control

Except for soil sterilization or fumigation, witchweed control is a multiyear project because of seed dormancy. Effective controls either eradicate seed or prevent their production and distribution. Herbicides best inhibit growth and prevent seed set when applied as the parasitic plants approach flowering. In some locations, ethylene is used to induce germination of soilborne witchweed seeds. In the United States and elsewhere, quarantines limit the transport of farm products, soil, and machinery from witchweed-infested areas. Research to find ways to stimulate witchweed seed germination by applying quinones to soil may offer another avenue for control.

Because the above-mentioned *Striga* spp. are specific to cereals and grasses, the use of dicotyledonous crops in rotation with wheat is encouraged. Alfalfa, soybeans, and cotton are nonhosts that also stimulate germination of witchweed seed. Vetch, clover, tobacco, and potatoes also are resistant but do not alter seed dormancy. Such crops in wheat rotations must be clean-cultivated so the parasites cannot develop on alternative grass species.

Selected References

Birch, E. B. 1963. Put a stop to the pest—The wily witchweed. Farming S. Afr. 39:29–31.

Eplee, R. E. 1975. Ethylene: A witchweed seed germination stimulant. Weed Sci. 23:433–436.

Maiti, R. K., Ramaiah, K. V., Bisen, S. S., and Chidley, V. L. 1984. A comparative study of the haustorial development of *Striga asiatica* (L.) Kuntze on sorghum cultivars. Ann. Bot. 54:447–457.

Nelson, R. R. 1958. Preliminary studies on the host range of *Striga asiatica*. Plant Dis. Rep. 42:376–382.

Parker, C. 1965. The *Striga* problem—A review. PANS 11:99–111.

Rogers, W. E., and Nelson, R. R. 1962. Penetration and nutrition of *Striga asiatica*. Phytopathology 52:1064–1070.

Vasudeva Rao, M. J. 1985. Techniques for screening sorghums for resistance to *Striga*. ICRISAT Info. Bull. 20. International Crops Research Institute for the Semi-Arid Tropics, Hyderabad, Andhra Pradesh, India. 18 pp.

Part II. Noninfectious Diseases

Noninfectious diseases are not self-sustaining. They are caused by insects and other animals and by physical, chemical, and usually inanimate variables in the environment. Many noninfectious diseases are called physiologic disorders.

Insects and Other Animal Pests

Insects

A thorough treatment of wheat-damaging insects is beyond the scope of this Compendium and could easily fill another. However, because insects are often associated with wheat symptomatology and sometimes confound the diagnosis of infectious wheat diseases, a few of the insect groups frequently observed by disease diagnosticians are briefly discussed here.

Over 100 insect species feed on wheat. Like infectious disease agents, insects usually are not a problem until environmental conditions allow one or more species to become overabundant. Plants fed upon or penetrated by insects are damaged mechanically and physiologically. Wheat plants are mechanically injured by chewing insects, are weakened by loss of sap to sucking insects, and are stressed and sometimes deformed by injected insect toxins. Also important are the many insects cited in this Compendium that act as vectors of viruses or other infectious agents and as factors predisposing plants to infection.

Chewing Injury

Many insects have chewing mouthparts and are external foragers. Insects feeding on wheat plants or seed mechanically remove tissues. Grasshoppers, cutworms, and armyworms, for example, remove conspicuous sections of leaf tissue (Fig. 102); mites and smaller insects such as leafhoppers, aphids, and thrips (Thysanoptera) have more discreet feeding patterns (see Plate 59). Feeding below ground on roots or seed, if not totally debilitating, may go unnoticed unless plants are extracted and carefully examined. Insects usually prefer wheat seedlings as food over maturing plants, and seedlings are most apt to show injury. Wheat plants in early growth stages can tolerate partial defoliation and often recover rapidly if meristem tissue remains intact.

Cutworms and armyworms. Cutworms—larvae of *Euxoa* and *Agrotis* spp.—are primarily surface feeders. They produce small, irregular holes or blank areas in unfolding leaves (Fig. 102). Leaves are chewed along the margin, and some leaves, as well as culms, may be cut off.

Cutworm larvae begin active feeding in early spring. They are full-grown by early summer, when they stop foraging and enter the soil to pupate.

Army cutworms are brown to gray-green and up to 3 cm long. The pale western cutworm is gray-white and similar in size (Plate 52). It feeds primarily below ground and often severs plants near the crown.

Armyworms—larvae of *Pseudaletia* and *Spodoptera* spp.—feed on many plant species indiscriminately and feed on maturing as well as young wheat plants. They gnaw on spikelets and seed as well as foliage.

Cereal leaf beetles. Chewing insects smaller than cutworms and armyworms produce more uniform patterns of foliar feeding. The cereal leaf beetle, *Oulema melanopa* L., for example, feeds on wheat, other small grains, and grasses (see

Fig. 102. Cutworm damage on young leaves. (Courtesy Entomology Department, University of Nebraska, Lincoln)

Cocksfoot Mottle). Adults eat narrow longitudinal holes in leaves. Larvae feed in a similar pattern but more superficially; they strip off only parenchyma cells, leaving a net of gray-white vascular tissue. Feeding is in uniform paths parallel to leaf veins (Plate 53).

First identified in Michigan in 1962, the cereal leaf beetle causes widespread damage to small-grain cereals and grasses in the eastern United States. To a lesser extent, the beetle is also reported on small grains in Europe.

Adult cereal leaf beetles are about 5 mm long and are blue-black except for a red-brown prothorax. Larvae are yellow-white and grow to 6 mm in length. They usually appear slimy black because of globular fecal mounds on their backs.

Grasshoppers. Grasshoppers are serious wheat pests in arid temperate regions (Plate 54). In the United States, five principal species use wheat as a host after moving from grassland habitats: migratory grasshoppers, *Melanoplus sanguinipes* (Fabricius); red-legged grasshoppers, *M. femurrubrum* (De Geer); two-striped grasshoppers, *M. bivittatus* (Say); clear-winged grasshoppers, *Camnula pellucida* (Scudder); and differential grasshoppers, *M. differentialis* (Thomas).

Besides the presence of nymph and adult grasshoppers on and about the plants in question, wheat damaged by grasshoppers shows various degrees of defoliation. Younger tissues are fed upon first. Heavy feeding on headed plants first leads to defoliation; then glumes and seed may also disappear, leaving an erect but barren culm (Plate 54).

Controlling grasshoppers in cropland depends on knowledge of their egg-laying habitats and their initial population in early spring. Most grasshoppers inhabit grasslands and lay their eggs in soil in late summer and autumn. Eggs are laid in clusters that are positioned 1–2 in. beneath the soil surface and contain from 15 to 80 eggs. The egg cluster is enclosed in a hardened "pod," which breaks easily as nymphs hatch in the spring. Applying insecticides in spring to their overwintering habitat can provide temporary control of grasshoppers.

Wireworms. When insects feed below ground, wheat plants may respond as if damaged by root rots or seedling blights. Damaged plants become chlorotic and flaccid and begin a slow decline. When plants are severed below ground, they become wilted, chlorotic, and necrotic and appear damped-off within one or two days. When plant decline and death are rapid, wireworms—larvae of click beetles (Coleoptera: Elateridae)—are usually responsible. Wheat is most subject to wireworm injury when grown in association with grassy weeds or in soil that supported a previous grass crop.

Wireworm damage is most evident in autumn and spring on seedlings in a short segment of drill row. Damaged plants usually appear in isolated groups; however, when infestation is severe, the damaged areas can coalesce to encompass large field areas. Plants fed upon by wireworms are torn irregularly near the crown and are predisposed to diseases like Cephalosporium stripe. The seed, radicle, coleoptile, and/or primary roots of wireworm-damaged seedlings are irregularly severed, bored, or pitted. The newest leaf on the main tiller often is wilted or necrotic and easily detached. Wireworm larvae normally are found just beneath the soil surface adjacent to the damaged plants (Plate 55).

Wireworms are milky white as they hatch from eggs in spring but soon become orange-brown. Some feed actively up to nine years before reaching adulthood. Their bodies are stiffened, sluggish, cylindrical, about 2 cm long, and wirelike. They have a distinct, usually black, head and three pairs of short legs. Adult click beetles are common in grasslands on the soil surface or low in the plant canopy.

Wireworm populations exceeding two million per hectare usually warrant control via insecticides applied to seed or soil.

Boring Injury

Some insects penetrate plants to feed, deposit eggs, establish a residence, and/or complete a portion of their life cycle. The culm and crown of wheat are suitable and usual habitats. Wheat plants invaded by boring insects are prone to lodge (see Foot Rot).

The Hessian fly (*Mayetiola destructor* Say), sawflies (*Cephus* spp.), wheat stem maggot (*Meromyza americana* Fitch), wheat jointworm (*Tetramesa tritici* Fitch), gout fly (*Chlorops pumilionis* Bjerkander), bulb fly (*Hylemya coarctata* Fallén), and wheat strawworm (*T. grandis* (Riley)) are among the major wheat-boring insects. These pests can be controlled by destroying the crop residues they inhabit. Burning residues and plowing or tilling soil interrupt their life cycles and emergence in spring.

The Hessian fly inhabits most wheat-growing countries. It is a serious wheat pest in Morocco and is potentially the most destructive wheat insect in North America. Its larvae and pupae inhabit and weaken the culm base in the spring, causing tillers to die or to develop slowly, form small heads, and/or lodge randomly. Hessian fly larvae resemble maggots and hatch from eggs laid in autumn on seedling leaves of winter cereals and grasses. They crawl beneath leaf sheaths and feed above a lower stem node, where they initially resemble grains of white rice (Plate 56). Within two weeks they develop a brown pupal skin and resemble flaxseeds. This "flaxseed" stage can overwinter and may persist for a year or more.

Some plants invaded by Hessian flies are predisposed to root rots and winter injury. Seeding wheat in autumn after the eggs are laid (flyfree date) or in early spring is the most practical control for Hessian flies. Cultivars resistant to various strains of Hessian fly are also available.

Sawflies (Hymenoptera) are not true flies (Diptera). They range across most of Europe and central North America. One important species, *Cephus cinctus* Norton, deposits eggs in wheat culms. The resultant larvae bore downward within the culm and seal themselves near a basal node with a plug of sawdustlike frass. They mine the interior of the culm base, causing it to tip over or break off near ground level (Fig. 103). Lodging is most apparent as the crop approaches maturity. Some occupied stems remain erect but are so weakened that yield is reduced or negligible.

Sawfly larvae are yellow-white, nearly legless grubs approximately 1.3 cm long, with pointed tails. Solid-stemmed cultivars have been developed to restrict their activity.

The boring activity of the wheat stem maggot is similar to that of the sawfly but is confined to the upper culm (see Loose Ear [Plate 61]).

Adults of the wheat strawworm (*T. grandis* (Riley)) and the wheat jointworm resemble winged black ants with yellow spots in the "neck" region. They deposit eggs high in the culm in summer and autumn. The resulting larvae mine the interior of

Fig. 103. Culms weakened and lodged by the wheat stem sawfly (*Cephus cinctus*) (right). (Courtesy F. H. McNeal, USDA)

the culm and sometimes destroy the primordial tissues of the spike.

Billbugs. Billbugs (*Sphenophorus* spp.) cause boring injury at soil level (on and within the first internode above the crown). Adult billbugs bore a ragged hole about 1 mm in diameter in the culm in order to deposit eggs. The resultant larvae feed on, weaken, and kill tillers, causing them to bleach gray-white. At heading, such tillers contrast sharply with the normal green color of uninvaded plants (see Plates 34 and 61). The penetration site (Plate 57) resembles a small, darkened lesion of eyespot. Dissection, however, reveals the mined culm interior with abundant frass. Larvae, if present, are short, white, legless, and curved and measure 1–1.5 × 2–4 mm.

Weevils. Weevils (*Sitophilus* spp.) are the usual cause of boring injury in stored grain. Kernels fed upon bear conspicuous holes, lack embryos, or are otherwise incomplete (Plate 58). Eggs are laid within kernels and give rise to resident larval and pupal stages. Generous quantities of grain dust and empty pupal shells of the insects among the grain are diagnostic. Adult weevils are dark brown and about 3 mm long and can fly.

Piercing-Sucking Injury

The major wheat insects that cause piercing-sucking injury are leafhoppers, planthoppers, chinch bugs, and aphids. Their mouthparts are adapted not for chewing or rasping but for penetrating plant tissue intercellularly and intracellularly. Feeding is accomplished via bidirectional pumping mechanisms that withdraw sap. Mechanical injury to host plants is less obvious and more localized (Plate 59) than with chewing or boring insects. However, damage may extend beyond parasitized cells because of phytotoxins and other products introduced during feeding (see Enanismo and Melanism). Heavy aphid feeding, for example, may weaken and even kill wheat plants.

Many piercing-sucking insects are cited in this Compendium as vectors of wheat viruses. Perhaps most significant are the aphids, which can be found the world over. Many feed on grasses and dicotyledonous hosts in addition to wheat (see Barley Yellow Dwarf). Aphids are yellow-green to black and 1–3 mm long. They migrate to wheat from alternative hosts whenever temperatures are adequate for flight. They light and feed on adjacent young wheat plants in patches or at field margins. Extensive multiplication and feeding cause wheat leaves to lose color and plants to make slow growth. The greenbug (*Schizaphis graminum* Rondani), the Russian aphid (*Diuraphis noxia* Mord.), and certain other species inject toxic saliva during feeding that causes localized discoloration of host tissues and gives some cultivars a speckled appearance (Plate 60). Aphids need not be present when such symptoms are observed because they are easily removed from plants by wind, rain, and predators. Also, warming temperatures, maturing host plants, and crowded habitats induce aphids to migrate.

Fig. 104. Seedlings with symptoms of wheat curl mite infestation. (Courtesy R. L. Forster)

In temperate zones, most aphids reproduce sexually and oviparously to overwinter as eggs. During the summer, they may reproduce parthenogenetically and viviparously. The offspring can be winged or wingless, and as many as 20 generations can be completed in one year. Mild winters and cool to moderate spring and summer temperatures favor aphid reproduction.

Chinch bugs (*Blissus leucopterus* (Say)) are more limited in distribution than aphids but occupy similar habitats. In the central United States, they overwinter on grasses and migrate to wheat when temperatures rise above 20° C. Adult chinch bugs are winged, black to dull gray, and 5–8 mm long. They fly to the base of wheat plants in spring and lay numerous eggs just below the soil surface. The nymphs suck sap from the plant base and roots before migrating to new grass habitats.

Stink bugs (Hemiptera: Pentatomidae) (*Nezara* and *Oebalus* spp.) feed on wheat heads and cause kernels to be shriveled. Flour made from grain damaged by stink bugs has undesirable odor, baking characteristics, and taste.

Several mites, including the brown wheat mite (*Petrobia latens* Müller) and the winter grain mite (*Penthaleus major* Dugés), feed on wheat sap. Heavy infestations are likely to occur in late autumn and early spring where wheat borders grasses. The wheat curl mite, *Aceria tulipae* Kiefer (syn. *Eriophyes tulipae* Kiefer), causes young leaves to be entrapped as they unfold (Fig. 104). Mites must be present in high populations to be damaging by themselves; however, they can be destructive in modest numbers when they are vectors of wheat streak mosaic virus. Spider mites (*Tetranychus* spp.) are a problem especially on wheat grown in greenhouses or in other controlled environments. Their feeding, like that of aphids, planthoppers, and leafhoppers (Plate 59) can result in numerous chlorotic-white flecks on the surface of young tissues.

Selected References

Asavanich, A. P., and Gallun, R. L. 1979. Duration of feeding by larvae of the Hessian fly and growth of susceptible wheat seedlings. Ann. Entomol. Soc. Am. 27:218–221.

Capinera, J. L., and Roltsch, W. J. 1980. Response of wheat seedlings to actual and simulated migratory grasshopper defoliation. J. Econ. Entomol. 73:258–261.

Dahms, R. G. 1967. Insects attacking wheat. Pages 411–444 in: Wheat and Wheat Improvement. K. S. Quisenberry and L. P. Reitz, eds. Agronomy Monograph 13. American Society of Agronomy, Madison, WI.

De Pew, L. J. 1980. Pale western cutworm: Chemical control and effect on yield of winter wheat in Kansas. J. Econ. Entomol. 73:139–140.

Fisher, G. C., and Weinzierl, R. 1983. The wheat stem maggot. Fact Sheet 298. Oregon State University Extension Service, Corvallis. 2 pp.

Gair, R., Jenkins, J. E. E., and Lester, E. 1972. Cereal Pests and Diseases. Farming Press Ltd., Suffolk, England. 184 pp.

Jones, F. G. W., and Jones, M. G. 1974. Pests of Field Crops. 2nd ed. St. Martin's Press, New York. 448 pp.

Martin, T. J., and Harvey, T. L. 1982. Cultivar response to wheat strawworm. Crop Sci. 22:1233–1235.

Porter, K. B., Peterson, G. L., and Vise, O. 1982. A new greenbug biotype. Crop Sci. 22:847–850.

Roberts, J. J., and Gallun, R. L. 1979. Effects of wheat leaf pubescence on the Hessian fly. J. Econ. Entomol. 72:211–214.

Storey, C. L., Sauer, D. B., Ecker, O., and Fulk, D. W. 1982. Insect infestations in wheat and corn exported from the United States. J. Econ. Entomol. 75:827–832.

Viator, H. P., Pantoja, A., and Smith, C. M. 1983. Damage to wheat seed quality and yield by stink bugs. J. Econ. Entomol. 76:1410–1413.

Watson, S. J., and Dixon, A. F. 1984. Ear structure and the resistance of cereals to aphids. Crop Prot. 3:67–76.

Webster, J. A., and Smith, D. H. 1983. Cereal leaf beetle population densities and winter wheat yield. Crop Prot. 2:431–436.

Loose (White) Ear

Isolated, loose whiteheads are uncommon but very obvious in recently headed wheat fields. The abnormal heads are more conspicuous than those bleached or white-headed from root and foot rot diseases. The head and neck are dry and whitened, while the remainder of the plant is a normal green (Plate 61). Damaged heads are easily pulled from the supporting leaf sheath because the culm is severed or injured within the sheath above the uppermost node. Attacks by the wheat stem maggot, *Meromyza americana* Fitch, are usually responsible. Aborted attacks by the wheat stem sawfly, *Cephus cinctus* Norton, and an abnormal growth rate of the stem are other possible causes.

Selected Reference

Taylor, R. E., and Schofield, E. R. 1956. An observation on loose ear of wheat. Plant Pathol. 5:94.

Birds

Ripening wheat is an accommodating food source for many seed-eating bird species. Complete shelling of the head or of random spikelets and an abundance of shattered glumes on the soil surface are indicative of bird feeding. Frequently, necks and culms are kinked, bent, or severed under the weight of feeding birds or of those attempting to light on plants (Fig. 105). With heavy feeding, field areas may appear lodged, but stem breakage does not originate near ground level as with plants lodged by wind, insects, or foot rots. Hailstorms after heading cause similar damage, except that shelled seeds, germinating seeds, and volunteer plants are more apparent with hail damage.

Birds also unearth and eat newly sown wheat seed. Grain scattered on the soil surface at seeding time is often the attractant. Partially opened furrows with lengths of missing plants when seedlings emerge are a sign of earlier bird feeding.

Chemical repellents, traps, netting, noise, and startling motion devices reduce but rarely prevent bird feeding. Birds often cause economic losses, especially along field borders and near wooded or other nesting areas.

Selected Reference

Dolbeer, R. A., and Stickley, A. R., Jr. 1979. Starling, *Sturnus vulgaris*, damage to sprouting wheat in Tennessee and Kentucky. Prot. Ecol. 1:159–169.

Fig. 105. Culms kinked and broken by the action of birds feeding on wheat kernels. (Courtesy T. H. Bowye)

Mammals

Wheat in open fields is subject to grazing and mechanical injury from wild and domestic foraging animals. Rodents, deer, and other game animals are drawn to wheat fields to forage and/or for shelter. Tracks and droppings help identify the pests.

Rabbits tend to graze along lengths of rows at random locations, whereas mice and other rodents do more damage along field borders. Mice feed on roots, foliage, sown seed, and seedlings. Leaves severed irregularly at 45° angles are typical of rodent damage. Rodents also feed on kernels in mature heads, sometimes causing significant yield losses. Fields that border hedges, woods, weedy areas, or other animal habitats are most prone to damage.

Damage from mice can be extensive when autumn-sown wheat is grown in heavy residue from a previous crop. Reduced tillage practices tend to maintain crop residues on the soil surface that offer winter protection to mice. Mice may eventually nest in the developing wheat and construct a residence of tangled culms low in the crop canopy but above ground level.

Various poisons and baits are prescribed to curb mice populations and damage. They usually are most effective when applied at field borders or within pipe or tile segments that serve as runways for the colony. Burning crop residues and clean cultivation also limit mice populations.

Disorders Caused by Environmental (Abiotic) Factors

Air Pollution Injury

Air pollutants originate from motor vehicle exhaust, coal-burning facilities, petroleum refineries, chemical and metallurgical plants, and many other sources. Gaseous pollutants such as ozone, peroxyacetyl nitrate (PAN), nitrogen oxides, sulfur dioxide, hydrogen fluoride, ethylene, and chlorine can be injurious to plants. They enter plants primarily through stomata and induce chlorosis, necrosis, and other physiologic responses.

The degree of injury to wheat and other plants depends on many factors, including cultivar, pollutant dose, and duration of exposure. Plant parts may differ in sensitivity depending on their growth stage. The effects of chlorine, sulfur dioxide, ozone, hydrogen fluoride, and ethylene on wheat have received the most attention; the effects of most other pollutants, be they gaseous or particulate, are poorly understood.

Chlorine

Petroleum refineries, glass factories, leaky storage containers, hydrochloric acid mists, and scrap-burning, sewage disposal, and water treatment plants are sources of chlorine (Cl). Normally only wheat close to such sources is injured.

Chlorine causes tip and marginal leaf necrosis; irregular necrotic spots and blotches may also occur (Plate 62). Leaves of middle age may be more sensitive than young or old leaves. Necrotic leaf areas are tan to red-brown and contrast sharply with surrounding green tissue. Epidermal and mesophyll cells can be killed by exposure for 2 hr or longer to approximately 0.1 ppm (300 $\mu g/m^3$) chlorine.

Sulfur Dioxide

Sulfur dioxide (SO_2) is released during the combustion of fossil fuels. Facilities that burn high-sulfur coal, coke, or fuel oil are major sources of sulfur dioxide. Commercial and residential heating plants, petroleum refineries, ore smelters, chemical factories, and power-generating plants emit the pollutant. Sulfur dioxide injury to vegetation is distributed in patterns that mirror the path of plume dispersal from exhaust stacks and may extend several miles downwind.

Sulfur dioxide causes leaves to appear chlorotic, tan-white, and scalded. Chlorosis develops from long-term exposure to low sulfur dioxide levels. Acute symptoms include dead spots or blotches on the leaf blade (Plate 63). Moisture in a plume containing sulfur dioxide may condense and form sulfuric acid droplets, which can induce localized leaf spotting. Chlorosis and necrosis sometimes appear as stripes between veins on the upper leaf surface. Exposure to 0.3 ppm (800 $\mu g/m^3$) sulfur dioxide for 8 hr is the approximate threshold for leaf injury; however, the presence of ozone may lower that threshold.

Sulfur dioxide reduces wheat yields, especially when toxic levels coincide with flowering. Experimental data indicate that durum wheat is more sensitive to sulfur dioxide than common wheat.

Sulfur dioxide at low levels may benefit plants because sulfur is an essential plant nutrient that can be taken up by leaves from the atmosphere as well as by roots from soil.

Ozone

Ozone (O_3) is regarded as a secondary air pollutant because it is produced from chemical reactions between nitrogen oxides and hydrocarbons in the atmosphere. Motor vehicle exhaust is the principal source of these parent gases, and sunlight provides the energy for the reactions to occur. Ozone is the most widespread gaseous pollutant and may be responsible for more injury and yield losses in wheat than all other pollutants combined. Ozone is a strong oxidant and injures a wide variety of plants in both rural and urban areas.

Leaves that have recently completed expansion are more sensitive to ozone than are younger leaves. Ozone symptoms on wheat include small, gray to white, irregular necrotic areas mainly on the upper leaf surface (Plate 64). The death of groups of subepidermal cells produces dry, gray-white flecks. Normally, oldest leaves are first to show injury. It is often difficult to distinguish ozone injury from other causes of leaf senescence.

Ozone at 0.2 ppm (390 $\mu g/m^3$) for seven days can injure wheat leaves (Plate 64). This threshold concentration for injury may be lowered when sulfur dioxide is present. As with sulfur dioxide, toxic levels of ozone are especially detrimental to yields when present at flowering. Ozone at 0.04 ppm (82 $\mu g/m^3$) for an average of 7 hr per day is sufficient to limit wheat yields.

Hydrogen Fluoride

Major sources of fluorides are fertilizer, aluminum, ceramic, glass, and ore-smelting plants. Gaseous fluorides are readily absorbed by plants and accumulate at leaf margins and tips. Some fluoride compounds are more readily absorbed when dissolved in rainwater. Conversely, rain sometimes leaches fluorides from plants.

Plant damage is distributed downwind from sources of the pollutant. Hydrogen fluoride (HF) is injurious at relatively low levels compared to other toxic gases. The chemical accumulates in plants to levels exceeding 50 ppm when as little as 0.1 ppb (0.2 $\mu g/m^3$) is in fairly constant supply.

Injury begins at leaf tips and margins, which become chlorotic, then necrotic (scorched). The injured tissue may extend down the leaf blade in irregular streaks or in a chlorotic mottle. A distinct margin usually separates healthy and dead tissue.

Wheat and other cereal crops are most sensitive to fluoride during flowering because fluoride especially inhibits pollen germination and germ tube growth.

Ethylene

Motor vehicle exhaust is the primary source of ethylene (C_2H_4), and concentrations are highest near urban centers. Wheat exposed to 50 ppb (25 $\mu g/m^3$) ethylene can be stunted, and flowering is delayed. Such damage is rare, however, and occurs only where motor vehicle density is extremely high. Sometimes ethylene accumulates in anaerobic soils to concentrations sufficient to inhibit the elongation of wheat roots.

Selected References

Abeles, F. B., and Heggestad, H. E. 1973. Ethylene: An urban air pollutant. J. Air Pollut. Control Assoc. 23:517–521.

Amundson, R. G., Kohut, R. J., Schoettle, A. W., Raba, R. M., and Reich, P. B. 1987. Correlative reductions in whole-plant photosynthesis and yield of winter wheat caused by ozone. Phytopathology 77:75–79.

Dowdell, R. J., Smith, K. A., Cress, R., and Restall, S. W. F. 1972. Field studies of ethylene in the soil atmosphere—Equipment and preliminary results. Soil Biol. Biochem. 4:325–331.

Heagle, A. S., Spencer, S., and Letchworth, M. E. 1979. Yield response of wheat to chronic doses of ozone. Can. J. Bot. 57:1999–2005.

Heggestad, H. E. 1968. Diseases of crops and ornamental plants incited by air pollutants. Phytopathology 58:1089–1097.

Hindawi, I. J. 1970. Air Pollution Injury to Vegetation. Publication AP-71. National Air Pollution Control Administration, Raleigh, NC.

Jacobson, J. S., and Hill, A. C. 1970. Recognition of Air Pollution Injury to Vegetation: A Pictorial Atlas. Air Pollution Control Association, Pittsburgh, PA.

Kohut, R. J., Amundson, R. G., Laurence, J. A., Colavito, L., van Leuken, P., and King, P. 1987. Effects of ozone and sulfur dioxide on yield of winter wheat. Phytopathology 77:71–74.

Kress, L. W., Miller, J. E., and Smith, H. J. 1985. Impact of ozone on winter wheat yield. Environ. Exp. Bot. 25:211–218.

Noggle, J. C., and Jones, H. C. 1982. Effects of air pollution on farm commodities. Pages 79–90 in: Proceedings of the Izaak Walton League Symposium. Izaak Walton League, Washington, DC.

Pack, M. R., and Shuzback, C. W. 1976. Response of plant fruiting to hydrogen fluoride fumigation. Atmos. Environ. 10:73–81.

Sechler, D., and Davis, D. R. 1964. Ozone toxicity in small grain. Plant Dis. Rep. 48:919–922.

Shannon, J. G., and Mulchi, C. L. 1974. Ozone damage to wheat varieties at anthesis. Crop Sci. 14:335–337.

Wilhour, R. G. 1978. Growth response of selected small grain and foliage crops to sulfur dioxide. CERL Report. U.S. Environmental Protection Agency, Washington, DC. 4 pp.

The Bends (Crinkle Joint)

From 1925 to 1940, an infrequent but conspicuous abnormality appeared on wheat and many grasses in the northwestern United States. Necks and culms of affected plants were weakened, bent, or kinked (Fig. 106). Repeated attempts to associate pathogens with the syndrome have been unsuccessful. The cause of the disorder remains undetermined, but the symptom is still reported to occur. Associations with tall cultivars, boring insects, frost, cold and wet growing conditions, herbicides, and hail have been postulated.

Rapidly growing but mechanically weakened stems apparently are supported by leaf sheaths when plants are young; thus, the deformity becomes visible as stems and heads emerge from leaf sheaths. The deformation apparently is the only consequence associated with this disorder because tillers with the bends tend to remain upright and produce normal heads and seed.

Selected References

Fellows, H. 1948. Falling of wheat culms due to lodging, buckling and breaking. USDA Circular 767. U.S. Department of Agriculture, Washington, DC. 16 pp.

Fischer, G. W. 1941. "Bends," a new disease of grasses and cereals. Phytopathology 31:674–676.

Chemical Injuries

Bactericides, herbicides, fumigants, fungicides, fertilizers, growth regulators, insecticides, and other chemicals applied individually or together with adjuvants and solvents play a prominent role in general crop management and in modern wheat production. However, such chemicals applied to seed, soil, and growing plants for specific, well-intentioned purposes occasionally produce undesirable or unexpected side effects. Some effects, like altered photosynthesis or inhibited respiration, may be invisible yet are critical to crop development and yield. More obvious and predictable are the consequences of applications of inappropriate chemicals or of appropriate chemicals applied in excessive amounts or at improper times. Nontarget plants may be damaged by agricultural chemicals that drift in the atmosphere, move in water, or persist in soil or on equipment. Industrial wastes, construction materials, fuels, and a vast array of other products have at times also been responsible for crop injuries.

Symptoms

Diagnosing chemical injury to wheat is difficult and requires a fairly thorough assessment of the crop, the environment, and the cultural practices employed. Symptom expression alone is often insufficient to pinpoint the cause of damage. Chemicals concentrated in particles or droplets in contact with plants may cause localized burning and desiccation of plant tissues (Plate 65) that may resemble local lesions caused by viruses, microorganisms, or other pathogens. Chlorotic mottling from difenzoquat methyl sulfate and other herbicides may resemble symptoms of virus infection (see Plate 45). Plants damaged by 2,4-D, dicamba, or other growth-regulating herbicides occasionally resemble plants damaged by root or foot rots. And deformations of heads and foliage (Fig. 107) may resemble symptoms of downy mildew.

Wheat plants emerging in herbicide-contaminated soil often have chlorotic or necrotic foliage and may be stunted, distorted, or killed. Field symptoms normally reflect the distribution of the phytotoxic chemical (Fig. 108). Symptoms of chemical injury on wheat leaves (discoloration, distortion, chlorosis, or necrosis) often reflect the mode of application and the path of the applicator in the field. Although any chemical in excess can be injurious, damage from persistent herbicides such as trifluralin, pronamide, atrazine, and triallate is especially common.

When soils contain toxic concentrations of triazine herbicides, urea, uracil herbicides, or other inhibitors of photosynthesis, the youngest leaves of wheat plants become chlorotic and then flaccid and dry at their tips (see Barley Yellow Dwarf [Plate 47]). When these chemical residues are present at sublethal concentrations, these symptoms may eventually stabilize and plants may partially recover.

When phytotoxic fungicides, fertilizers, or other chemicals are applied to seed, plant damage tends to be uniform in the field. For example, organomercury treatments applied at excessive rates or to high-moisture seed impair germination; the treated seeds either do not germinate or show appreciable thickening (hypertrophy) of the coleoptile and poor root development (Fig. 109). The eventual stand of seedlings is uniformly thin.

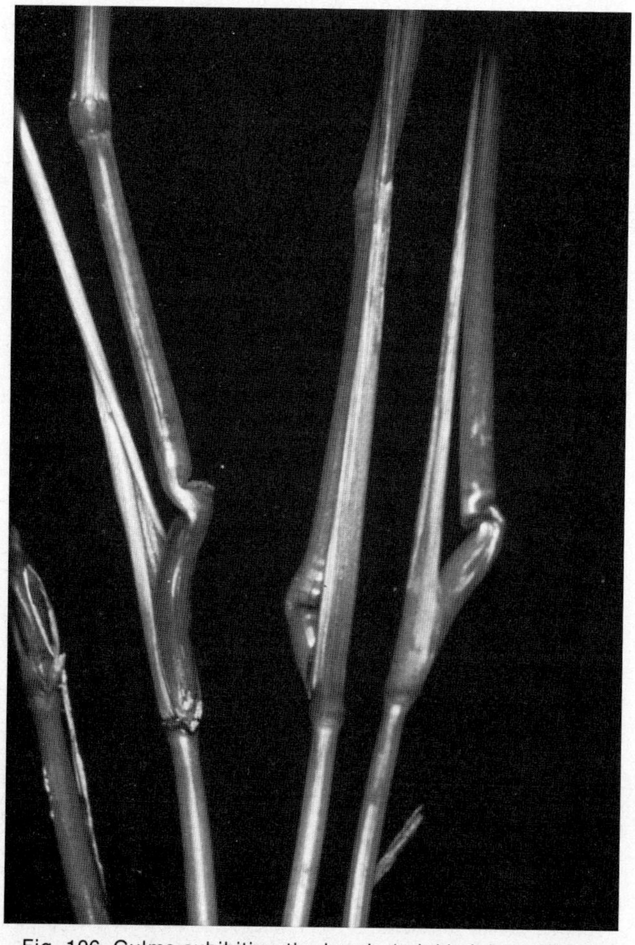

Fig. 106. Culms exhibiting the bends (crinkle joint). (Courtesy Plant Pathology Department, Washington State University)

Fig. 107. Heads distorted by growth-regulating herbicides: **A** and **B,** 2,4-D injury; **C,** spikelets stunted by glyphosate, causing head to resemble *Aegilops* sp. (A, British Crown Copyright; used by permission of the Controller of Her Britannic Majesty's Stationery Office; B and C, courtesy R. H. Callihan)

Fig. 108. Wheat seedlings showing toxic effects of insecticide banded into soil during the previous growing season. (Courtesy S. Dubetz)

Phenoxy herbicides, such as 2,4-D, MCPA, and their salt and ester formulations, may be safely applied to wheat between tillering and jointing or after the dough stage. Applied at other growth stages, they can be injurious and detrimental to yield. Benzoic acid or phenoxy herbicides in developing heads induce floret sterility and deformities such as trapped and twisted awns, branched heads, and multiple or missing spikelets (Fig. 107). These herbicides may also alter culm elongation (see The Bends [Fig. 106]).

Yield losses can be total when herbicides are applied at improper times or in excessive dosages. Symptoms of chemical damage are conditioned by formulation, dosage, environment, wheat growth stage, plant part exposed, and cultivar. Some

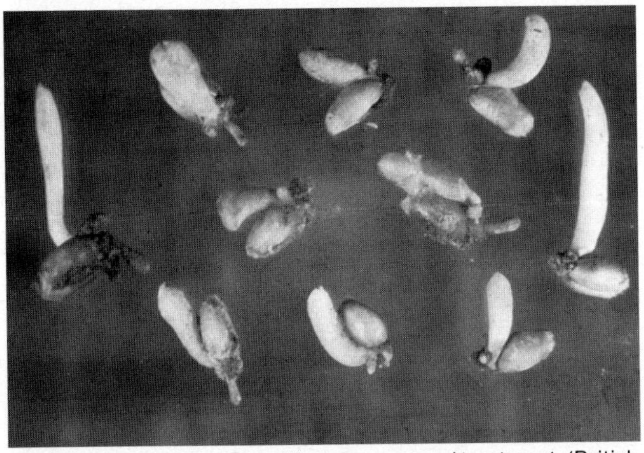

Fig. 109. Germlings distorted by mercury seed treatment. (British Crown Copyright. Used by permission of the Controller of Her Britannic Majesty's Stationery Office)

Fig. 110. Broken culms and tattered leaves caused by hail. (Courtesy A. V. Ravenscroft)

cultivars that mature early under chemical stress may support secondary "sooty" molds (see Plate 9).

Evidence is increasing that many crop and weed species are sources of toxic organic acids, alkaloids, and other products. Via allelopathy or other mechanisms, wheat plants may be stunted or even killed in areas where natural plant toxicants accumulate.

Selected References

Ashton, F. M., and Crafts, A. S. 1981. Mode of Action of Herbicides. 2nd ed. John Wiley & Sons, New York. 525 pp.

Audus, L. J., ed. 1976. Herbicides: Physiology, Biochemistry, Ecology. 2 vols. Academic Press, New York. 608 and 564 pp.

Buchenau, G. S. 1963. Winter survival of wheat in chloropicrin-treated soil. Plant Dis. Rep. 47:602–604.

Eagle, D. J., Caverly, D. J., and Holly, K. 1981. Diagnosis of Herbicide Damage to Crops. Chemical Publishing Co., New York. 70 pp.

Hageman, L. H., and Behrens, R. 1981. Response of small grain cultivars to chlorsulfuron. Weed Sci. 29:414–420.

Huber, D. M., Seely, C. I., and Watson, R. D. 1966. Effects of the herbicide diuron on foot rot of winter wheat. Plant Dis. Rep. 50:852–854.

Miller, S. D., and Nalewaja, J. D. 1980. Herbicide antidotes with triallate. Agron. J. 72:662–664.

Nilsson, H. E. 1973. Influence of herbicides on take-all and eyespot diseases of winter wheat in a field trial. Swed. J. Agric. Res. 3:115–118.

North American Symposium on Allelopathy. 1983. J. Chem. Ecol. 9:935–1292.

Prusky, D., Dinoor, A., and Eshel, Y. 1974. Symptomless effects of experimental fungicides on wheat. Phytopathology 64:812–813.

Rovira, A. D., and McDonald, H. J. 1986. Effects of the herbicide chlorsulfuron on Rhizoctonia bare patch and take-all of barley and wheat. Plant Dis. 70:879–882.

Weed Science Society of America. 1979. Herbicide Handbook of the Weed Science Society of America. 4th ed. Weed Science Society of America, Champaign, IL. 479 pp.

Color Banding

Color banding is an uncommon but conspicuous abnormality on emerging seedlings of wheat and other small grains. It results from rapid and wide fluctuations in temperatures at the soil surface. The abnormality, called "rugby stocking" or "rugby banding" in Europe, is temporary and intermittent.

Color banding appears most frequently on wheat that is deep-seeded in compacted soils. Banded seedlings have a series of one or more distinct, yellow-white, transverse bands on coleoptiles and seedling leaves. The bands are up to 2 cm wide and alternate with bands of normal green tissue (Plate 66).

Banding sometimes results when water-saturated or frosted leaves are dried rapidly by surface soil heat or by bright sun (heat banding). Because emergence normally is complete within three or four days, affected plants have three or four bands but are otherwise normal. Tissue sensitivity to temperature extremes declines rapidly with age and reduced water content. The severity of color banding is cultivar-related.

Selected References

Richards, B. L. 1934. Frost-banding of cereal seedlings. Proc. Utah Acad. Sci. Arts Lett. 11:3–5.

Vanterpool, T. C. 1949. Chlorotic banding of cereal seedlings. Sci. Agric. 29:334–339.

Wilcox, H. J. 1959. Color-banding of cereal seedlings. Plant Pathol. 8:34–35.

Hail Damage

Wheat plants sustain random mechanical injury when pelted by hail (Fig. 110). When the crop is mature, hail and bird damage are similar except that hail damage is more uniformly distributed across the field and produces more evidence of shattered grain, germinating grain, volunteer plants, and localized injury on culms. Hail-induced volunteer stands are important epidemiologically and influence the incidence of diseases like wheat streak mosaic and powdery mildew. At any stage of plant growth, hail randomly kinks and severs plant parts. Tattered plant parts will be distributed on or sometimes driven into the soil. Unlike lodging from root and foot rots, lodging from hail tends to originate at random points well above ground rather than at the plant base. Lodging associated with hail tends to be directional because of accompanying winds, whereas lodging from bird feeding, take-all, and foot rots tends to be nondirectional. Random and localized drying and bleaching of immature tissues are late symptoms of hail injury, as are broken whiteheads and localized stem lesions.

Frost and Winter Injury

Climate is the most important variable regulating wheat growth and production. Wheat is best adapted to temperate zones, but inconsistencies in weather patterns and temperature extremes sometimes exceed its tolerance. In North America, for example, the frequency of winter damage on autumn-sown crops defines the boundary between winter and spring wheat production. At this boundary, winter wheat is preferred to spring wheat because of its greater yield potential. However, many winter-damaged fields (Fig. 111) are abandoned each year or replaced by an alternative spring-sown crop. Losses to frost and freezing among cereals in the United States are estimated at 3–4% annually. In Brazil and other parts of the Southern Hemisphere, frosts reduce wheat production by nearly 5% annually.

In general, snow insulates wheat against desiccation and low temperatures. However, persistent snow, and especially ice, is stressful. Patches of dead plants in early spring can be the direct result of waterlogging, wind desiccation, extreme temperature, or a combination of temperature, moisture, and disease stress (see, for example, Snow Molds, Foot Rot, Cephalosporium Stripe, Pythium Root Rot, and Soilborne Wheat Mosaic). The effects of cold temperatures independent of disease stress are discussed here.

Frost

Transient freezing temperatures are most damaging to wheat when they coincide with tiller elongation, heading, or flowering. Frosts during tiller elongation cause the collapse and death of meristematic (growing) tissues immediately above and below nodes. Such damage may lead to early tiller maturity, whiteheads, and new shoot formation at the crown (Fig. 112). Frosts during heading cause floret sterility and abortion. Awns and newly extruded anthers are easily destroyed or distorted (Plate 67).

Fig. 111. Irregular spring stand of autumn-sown wheat injured by winter freezing. (Courtesy M. V. Wiese)

Autumn frosts tend to be less consequential than spring frosts because normal cooling temperatures tend to "harden" seedlings and months of growing time lie ahead for damaged plants to recover. In the spring, seedlings tend to lose their cold-hardiness when exposed to warm temperatures.

Wheat tissues sustain three principal types of frost injury. The first, apparently most prevalent in plants that have not hardened or have lost their cold-hardiness, involves the mechanical disruption of cell protoplasts by ice crystals that enlarge within and between cells. This injury commonly results from spring frosts. Damage patterns in the field reflect the topographic pattern of cold air flow and settling. Field areas with lush vegetative growth and plant parts near the surface of the crop canopy are most prone to injury. Seedlings may be severed or banded at the soil line (see Color Banding [Plate 66]). The collapsed internodal tissues of jointing and headed plants often appear bleached and/or water-soaked initially but eventually are darkened by secondary microorganisms (Fig. 112). Culm tissues discolored by frost may resemble culms damaged by foot rots (see Plate 32). When actively growing tissues freeze, they rapidly become dry, bleached, and shrunken (Plate 68). Heads appear tip-blighted or bleached and may emerge variously distorted, with bands of sterile florets (Plate 67). Dull blue-gray streaks on glumes are indicative of frost damage on some cultivars.

A second and more common effect of freezing injury to wheat is the formation of ice between cells and the concomitant diffusion of water from cells to the developing ice lattice. This process, which normally occurs in hardened tissues, causes cellular and tissue death by dehydration and severe plasmolysis. The extent of tissue damage depends on water content. Crown tissue with 55–65% moisture can withstand temperatures to −20° C. At higher or lower moisture contents, crowns are more susceptible to freezing injury and may be damaged at −10° C. This dehydration is often responsible for the "winterkill" (Fig. 111) of hardened winter wheat.

A third form of injury occurs in roots and culms, where quick-frozen columns of ice form in vascular tissues. Local

Fig. 112. Culms weakened and darkened at nodes by frost. (Courtesy R. J. Cook, USDA)

vascular cells are mechanically damaged, and adjacent living cells become dehydrated. Tissue hardening and cold-temperature tolerance are related to the kinds and quantity of solutes, especially carbohydrates, in cell sap; higher levels of solutes in cell sap tend to lower its freezing point.

Suffocation

Where precipitation is heavy during winter months, soils become waterlogged and wheat plants receive inadequate aeration. Smothering occurs especially in flooded or wet fields that freeze over. Damage, though unpredictable, is related to the amount of oxygen available to plants and possibly also to accumulation of ethylene in the flooded soil. Plants that are growing (rather than dormant) can die from suffocation within a few days.

Soil Heaving

Soil heaving is a temperature-associated phenomenon that lifts wheat plants and exposes their crowns and roots to desiccation by sun and wind. The effect is lethal unless new supportive roots can be generated. Wet soils exposed to freezing temperatures are especially prone to heaving. Crop damage can be severe when alternate freezing and thawing repeatedly lift plants (Plate 69).

Heaving results when horizontal lenses of ice form beneath the soil surface. Plants on top of or bound in the growing lenses are lifted sometimes as much as 5 cm. The powerful freezing forces easily exceed root extensibility, and roots that are not extracted are severed. If new supportive roots can be generated, plants may totally recover. Most lifted plants, however, remain underdeveloped through maturity. Soil heaving and associated root damage appear to predispose wheat to infection by *Cephalosporium gramineum* and may provide infection courts for other soilborne pathogens.

Heaving is inconsequential in well-drained soils and beneath snow and can be reduced when foliage or crop residues insulate the soil surface. Seeding should be timed so that the top growth of overwintering wheat is substantial (a well-developed crown and three seedling leaves is generally optimal) but not so great as to increase the risk of matted foliage and poor aeration at soil level.

Selected References

Afanasiev, M. M. 1966. Frost injury to wheat. Plant Dis. Rep. 50:928–930.

Andrews, C. J., and Pomeroy, M. K. 1979. Toxicity of anaerobic metabolites accumulating in winter wheat seedlings during ice encasement. Plant Physiol. 64:120–125.

Chen, T. H. H., Gusta, L. V., and Fowler, D. B. 1983. Freezing injury and root development in winter cereals. Plant Physiol. 73:773–777.

Fowler, D. B., and Carles, R. J. 1979. Growth, development and cold tolerance of fall acclimated cereal grains. Crop Sci. 19:915–922.

Limin, A. L., and Fowler, D. B. 1985. Cold hardiness response of sequential winter wheat tissue segments to differing temperature regimes. Crop Sci. 25:838–843.

Livingston, J. E., and Swinbank, J. C. 1947. Two types of late spring frost injury to winter wheat. Agron. J. 39:536–544.

Mayland, H. F., and Cary, J. W. 1970. Frost and chilling injury to growing plants. Adv. Agron. 22:203–234.

Metcalf, E. L., Cress, C. E., Olien, C. R., and Everson, E. H. 1970. Relationship between moisture content and killing temperatures for three wheat and three barley cultivars. Crop Sci. 10:362–365.

Olien, C. R. 1967. Freezing stresses and survival. Annu. Rev. Plant Physiol. 18:387–408.

Paulsen, G. M., and Heyne, E. G. 1983. Grain production of winter wheat after spring freeze injury. Agron. J. 75:706–707.

Shaw, M. W., and Blackett, G. A. 1966. Unusual frost injury to winter sown wheat and barley. Plant Pathol. 15:119–120.

Singh, J. 1981. Isolation and freezing tolerances of mesophyll cells from cold-hardened and nonhardened wheat. Plant Physiol. 67:906–909.

Single, W. V., and Marcellos, H. 1981. Ice formation and freezing injury in actively growing cereals. Pages 17–33 in: Analysis and Improvement of Plant Cold Hardiness. C. R. Olien and M. N. Smith, eds. CRC Press, Boca Raton, FL.

Melanism (Brown Necrosis)

Wheat cultivars differ in the production and expression of melanoid pigments in cells and tissues. In certain cultivars the pigments appear as brown-black streaks or blotches, mostly in response to nonspecific stresses. Natural plant maturation can increase their visibility, as can certain types of mechanical injury or localized physiologic stress (see Chemical Injuries). Some cultivars develop streaks or blotches on glumes (pseudo black chaff) that resemble symptoms of black chaff (Plate 4) or glume blotch (Plate 29). Darkened areas on peduncles and culms also occur (see Basal Glume Rot [Fig. 9]).

The term "brown necrosis" was originally associated with wheat lines containing a gene linked to the Hope gene for resistance to stem rust. In these lines, rust infections and various other diseases, stresses, and injuries induced the darkening response in epidermal cells. In some cultivars, the intensity of the pigmentation is enhanced by low light, low temperature, and spike sterility. A recessive mutant on chromosome 7B in durum wheat causes the spike, awns, and sometimes the entire plant to turn deep chocolate brown at maturity.

Physiologic Leaf Spots

Leaf spotting in wheat that cannot be attributed to a biological pathogen is often assumed to be physiologic or genetic in nature. Many leaf spots that are unexplained initially become less mysterious when thoroughly researched. Some eventually are found to be associated with specific pathogens, and others may be attributed to mineral imbalances (see Nutrient Imbalances [manganese]) or other environmental stresses. Some disorders, such as bacterial leaf blight and aster yellows, were originally described incorrectly as "physiologic" or "genetic" abnormalities.

On the plains of North America, particularly the Canadian prairies, a leaf spot called "splotch" is reported on durum wheats. The spots are tan-white and numerous. Most are irregular in shape and less than 20 mm long. Their margins are diffuse initially and later become more distinct. The spots appear as plants approach heading and increase in size and number toward the top of the plant. The spotting can encompass major areas of upper leaves and resembles bacterial leaf blight or white blotch. The disease appears to be a nutritional disorder perhaps associated with marginal nitrogen deficiency. Spotted plants otherwise are normal and do not show usual nitrogen deficiency symptoms.

Some cultivars develop leaf spots and blotches when hot, sunny weather follows a cool, cloudy, moist period. Apparently the rapid loss of water from water-saturated leaf cells can induce leaf spotting in addition to wilt. *Pseudomonas syringae* pv. *syringae* (see Plate 6) and *Bacillus megaterium* (see White Blotch) may influence the spotting but are not consistently present nor apparently necessary for the disease.

Another unexplained malady causes flecks and spots about 2 mm in diameter to develop on the upper leaves of some wheat genotypes. Microbial associations have been investigated, but none appear causative or influential.

Selected References

Atkinson, T. G., and Grant, M. N. 1967. A nonparasitic leaf blotch of winter wheat. Proc. Can. Phytopathol. Soc. 34:15.

Chester, K. S. 1944. A cause of "physiological leaf spot" of cereals. Plant Dis. Rep. 28:497–499.

Sallans, B. J., and Tinline, R. D. 1962. Splotch, a physiological leaf disease of wheat. I. Nutritional studies. Can. J. Plant Sci. 42:403–414.

Sprouting

The dormancy of newly mature wheat seed depends on genotype and environment. Some genotypes produce seed that is dormant (incapable of germination) for several weeks; with others, mature seed is immediately germinable. Where little or no dormancy is exhibited, ambient moisture may induce seeds to germinate (sprout) within the spike (Plate 70).

Cultivars without seed dormancy may begin sprouting when rains fall on ripened grain. Wet periods that delay harvest not only encourage sprouting but also promote the activity of many resident fungi (see Black Head Molds). Sprouting often is an additional consequence in lodged or hail-damaged fields. In storage, high-moisture seed, whether germinating or not, tends to heat, cake, or otherwise deteriorate (see Storage Molds). When high-moisture seeds eventually dry, test weights usually are significantly reduced.

White wheats are generally more prone to preharvest sprouting than red wheats. Cultivars of both white and red wheats that resist sprouting in the field are in use in parts of the United States, Australia, and Europe.

Selected Reference

Hagemann, M. G., and Ciha, A. J. 1984. Evolution of methods used in testing winter wheat susceptibility to preharvest sprouting. Crop Sci. 24:249–254.

Water Stress (Flooding and Drought)

Patches of dead or dying plants in lowlands often are the result of excessive moisture. Even hardened and winter-dormant plants may drown in shallow pools of standing water. Wet or flooded soil interferes with root aeration, nitrification, and the availability and utilization of most soil nutrients. In some instances phytotoxic products accumulate in wet soils. Excessive moisture keeps soil temperatures low and elevates the relative humidity in the crop canopy. It favors many infectious diseases and complicates harvest. Sprouting is promoted, and seed quality and storability are reduced (see Black Head Molds and Storage Molds). Resistance to excess moisture is not available, but some improved cultivars better withstand disease and physiologic disorders associated with wet conditions.

When moisture is sparse, germination, tiller formation, and growth are slowed (Fig. 113), but viability is usually maintained. Secondary roots are often limited in size and number. With restricted root systems between tillering and maturity, drought-stressed plants may appear dark blue-green. Tillering is reduced or prevented, and florets may be aborted. The isolated kernels that remain tend to retain moisture at the expense of other plant parts but become shriveled by maturity. Drought-stressed leaves at first are more erect and rolled toward the upper midrib. In severe cases, leaves may become ashen to bronze in color, hardened, and dry. Such "firing" progresses from leaf tip to base and from lower to upper leaves. Drought stress after heading causes plants to wilt slightly, lose color, cease growth, and ripen prematurely. Plants deficient in moisture also are more susceptible to damage by Fusarium root rot (see Common Root and Foot Rot).

Fig. 113. Irregular emergence and development of wheat sown in autumn in moisture-deficient soil. (Courtesy M. V. Wiese)

Water is utilized primarily in transpiration; only marginal amounts are retained in the plant. Under drought conditions, stomata close, transpiration decreases, and respiration and leaf temperature increase. Drought stress is increased by low relative humidity, high temperatures, wind, and high light intensity. Drought-stressed and drought-hardy plants, like cold-hardy plants, typically have high salt and sugar contents (osmotic values). Small cell size also appears to impart drought resistance.

Selected References

Jackson, R. D., Idso, S. B., Reginato, R. J., and Pinter, D. J. 1981. Canopy temperature as a crop water stress indicator. Water Resources Res. 4:1133–1138.

Papendick, R. I., and Campbell, G. S. 1975. Water potential in the rhizosphere and plant and methods of measurement and experimental control. Pages 39–49 in: Biology and Control of Soil-Borne Plant Pathogens. G. W. Bruehl, ed. American Phytopathological Society, St. Paul, MN.

Pomeroy, M. K., and Andrews, C. J. 1979. Metabolic and structural changes associated with flooding at low temperature in winter wheat and barley. Plant Physiol. 64:635–639.

Wind Damage (Lodging)

Winds, often in combination with heavy rain, cause plants to lean, become kinked, or break off near their base (see Hail and Birds). Wind-lodged plants tend to lie in one direction (see Foot Rot [Fig. 60], Birds [Fig. 105], and Hail [Fig. 110]). Immature plants often partially recover, but mature plants remain in dense prostrate layers and are difficult to harvest. Lodged plants are prone to infection by field fungi (see Black Head Molds) and sprouting because heads tend to remain moist in the dense crop canopy. Although wind contributes directly to lodging, wheat is frequently predisposed to lodge by other factors such as culm, root, and foot diseases or a rank growth habit from excessive soil fertility.

Winds also tatter or shred leaves, induce leaf tip necrosis, increase transpiration, and shatter ripened grain. Plants in the path of windblown soil or other particulates become stippled and flecked. Wind speeds greater than 3.6 m/sec increase plant respiration as much as 40%. Such increased utilization of food and moisture reserves in the plant may detract from yield.

Selected References

Patterson, F. L. 1957. Lodging by node-bending in wheat and barley. Agron. J. 49:518–519.

Todd, G. W., Chadwick, D. L., and Tsai, S. D. 1972. Effect of wind on plant respiration. Physiol. Plant. 27:342–346.

Yellow Berry

The term yellow berry originated in the United States, where wheat-grading standards take into account the hardness or vitreous nature of kernels. Vitreous, amber kernels normally contain more protein than faded, yellow seed (yellow berries) (Plate 71). Wholly or partially yellow-white seed has a higher proportion of horny starch, an undesirable character in some market and special-use areas.

Yellow berry results when metabolism and chemical translocations are slowed within maturing plants before kernels are completely filled. The cheek area of seed is last to fill and thus may remain starchy and lighter in color. Stresses from deficient nitrogen and warm temperatures during heading and seed maturation are associated with yellow berry. Excessive moisture during this period also increases the proportion of yellow kernels, probably by leaching nitrogen from the root zone or by slowing plant maturation.

The incidence of yellow berry differs among cultivars and is perhaps most easily diagnosed in red wheats.

Selected References

Dennis, R. E., Thompson, R. K., Openshaw, M. D., Gardner, B. R., and Halderman, A. D. 1975. Minimizing yellow berry in durum wheat. Ext. Agri-File 255.26. University of Arizona, Tucson.

Henson, J. F., and Waines, W. G. 1983. Nitrogen metabolism and yellow berry of two bread wheat cultivars. Crop Sci. 23:20–22.

Pryor, A. 1981. Sampling for yellow berry control. Calif. Farmer 254:32-G.

Robinson, F. E., and Cudney, D. W. 1979. Nitrate fertilizer timing, irrigation, protein, and yellow berry in durum wheat. Agron. J. 71:304–308.

Nutrient Imbalances

Intensive cultivation of high-yielding wheat cultivars requires an abundant supply of plant nutrients. Because all agricultural crops remove nutrients from the soil, without replacement nutrient depletion is inevitable. Thus, nutrient supplementation, especially through the use of commercial fertilizers and organic materials such as manure, is commonplace and often necessary to attain economic yields. Nutrition also influences the reaction of wheat cultivars to disease. Nutrients can be managed to enhance the biological control of wheat pathogens or to increase the crop's capacity to tolerate or resist disease stress.

Wheat requires at least 16 essential nutrients. Of these, nitrogen, potassium, and phosphorus are the major and most utilized nutrients for plant growth. Secondary nutrients such as calcium, magnesium, and sulfur and micronutrients such as manganese, zinc, and copper are required in moderate or minute quantities. Many elements are interrelated, so that adjustments to one may affect the requirement for or availability of others. The availability of all nutrients is influenced by moisture, temperature, pH, and other chemical and physical properties of soil. When present in excess amounts, some nutrients are toxic.

Because different crops utilize different amounts of nutrient elements, cropping frequency and sequence have an important effect on nutrient supplies. Wheat grown in rotation with corn or sorghum, for example, is likely to be more nutrient-deficient than wheat grown in fallow soil or following legume crops.

Symptoms of nutritional disorders in wheat are vague and not easily diagnosed. Stunted or uneven growth (Fig. 114), loss of green color, lack of tillering and vigor, low yields, and shriveled grain are common symptoms of nutrient deficiency. Deficiency symptoms are most apparent between tillering and heading, when plant growth is rapid and the demand for nutrients is high. Likewise, nutrient toxicity symptoms are best observed on immature plants. The diagnosis of nutritional disorders often requires chemical analysis of the plants and soil in question as well as observation of plant and field symptoms.

Nutrient deficiencies are corrected by application of fertilizer. Nutrient toxicities are less easily corrected; treatment may involve leaching, extraction, degradation, or inactivation of the toxic element.

Nitrogen

Nitrogen (N) is an important element in proteins and is the nutrient most frequently deficient in plants. In wheat, the demand for nitrogen is greatest during periods of rapid growth (jointing through heading) and declines gradually toward maturity. Most plant nitrogen in wheat eventually is translocated to the seed (see Yellow Berry).

Wheat takes up nitrogen primarily as nitrate (NO_3^-) and to a lesser extent as ammonium (NH_4^+). Organic nitrogen is unavailable until converted to inorganic forms by microbial activity. Nitrogen is often deficient in early spring when plants begin growth before soil temperatures warm sufficiently for microbial activity to make nitrogen available.

Nitrogen deficiency in wheat is first expressed as a loss of green color and then as stunted growth. Because nitrogen is translocated to physiologically active tissues, chlorosis proceeds from lower to upper leaves and begins at leaf tips. Extreme shortages cause premature death of older leaves.

Nitrogen deficiency is corrected by spring or autumn applications of nitrogen fertilizers such as urea, ammonium sulfate, ammonium nitrate, or aqueous ammonia. Nitrogen applied to foliage just before heading may not influence yield but can significantly increase the protein content of seed.

Excess nitrogen promotes lush, rank vegetative growth and delays maturity. Thus, lodging, frost injury, and diseases favored within a lush, moist crop canopy may be accentuated.

Phosphorus

Soils that are calcareous (pH 7.5–8.2), strongly acidic ($pH<5.5$), or low in organic matter are most likely to be deficient in phosphorus (P). Phosphorus does not move appreciably in soil and is less available to seedlings with small root systems than to larger plants. Absorbed mainly as orthophosphate ($H_2PO_4^-$), phosphorus is vital for cell division and for photosynthetic and respiratory reactions.

Wheat plants deficient in phosphorus are physiologically stressed and are prone to develop diseases such as Pythium root rot. Phosphorus-deficient plants maintain their green color but grow slowly and mature late. Leaf tips die back when shortages are severe. The red, brown, or purple discoloration of leaves, indicative of phosphorus shortages in corn and certain other plants, is less common in wheat.

Phosphorus deficiency tends to be associated with cold, wet soils high in calcium and exchangeable aluminum. Phosphate fertilizer used to correct deficiencies must be placed near plant roots because the element is relatively immobile in soil.

Potassium

Potassium (K) is likely to be deficient in sandy, intensively cropped, and highly weathered soils. Potassium, like nitrogen, is translocated to growing tissues, so shortages first appear in older leaves. Leaves lose color from tip to base and may be streaked with yellow. Severe deficiencies give plants a scorched or blighted appearance. Typically, plants deficient in potassium develop weak straw and are prone to lodge. They may be

Fig. 114. Wheat in nutrient-deficient soil, showing growth responses to excrement from grazing livestock. (Courtesy M. V. Wiese)

predisposed to diseases like powdery mildew. In contrast to nitrogen, only about 25% of the potassium taken up by wheat plants is translocated to the seed. Thus, straw removal has a greater impact on potassium status in soil than grain removal.

Sulfur

Symptoms of sulfur (S) deficiency in wheat closely resemble those described for nitrogen deficiency. Sulfur deficiency occurs when soil organic matter becomes depleted. Heavy applications of nitrogen and harvest of most crop biomass lead to the eventual depletion of both organic matter and sulfur in soil. Wheat grown in sulfur-deficient soil may appear brightly chlorotic and yield poorly. Sulfur-deficient flour is lower in milling and baking quality than normal flour. Sulfur deficiencies are corrected by applications of sulfate salts or elemental sulfur.

Manganese

Manganese (Mn) tends to be deficient or unavailable in highly organic soils and is occasionally deficient in mineral soils where pH exceeds 6.0. Shortages are accentuated by cool, dry weather.

Gray-white spots on old and young leaves are a symptom of manganese deficiency in wheat. This condition, known as "gray speck," may be accompanied by chlorotic or gray-white streaks. When the spots coalesce, leaves may kink or droop at the base of the blade. Affected plants also are lighter green than healthy plants and are slow to mature. Some wheat cultivars are reported to be tolerant or resistant to low manganese levels.

In some soils, as acidity increases ($pH<5$), manganese solubility can increase to toxic levels. Manganese toxicity is rare and is not easily diagnosed because toxicity symptoms somewhat resemble deficiency symptoms in wheat. Plants develop marginal chlorosis, curled young leaves, and necrotic spots on older leaves.

Magnesium

Magnesium (Mg) shortages are usually confined to acid soils. Because magnesium is an integral part of chlorophyll, a deficiency can limit photosynthetic activities. Fields deficient in magnesium produce pale, stunted plants. Yellowing is most apparent on older leaves because the element is translocated to younger, growing tissues.

Deficiencies are corrected by foliar sprays of magnesium sulfate. Limestone containing magnesium (dolomite) is also therapeutic, but plant response is much slower.

Copper

Organic soils in parts of the United States, Australia, and Western Europe provide insufficient copper (Cu) for good wheat growth. Although required only in minute quantities, the element is yield-limiting.

Copper-deficient plants develop light green leaves that become dry at their tips (Plate 72). Chlorosis and bleaching are followed by death, curling, and twisting of leaf tips and margins. This condition results from a lack of calcium translocation from old to young tissues, a process that depends on copper. Roots of affected plants are stunted, excessively branched, and rosetted. Heads are bleached (whiteheads), poorly filled, and sometimes incompletely emerged. Occasionally, darkened nodes or areas on the neck are part of the deficiency syndrome (see Chemical Injuries).

Copper sulfate applied to soil and copper oxychloride or copper oxide applied as foliar sprays at late tillering reduce damage. Excess phosphorus may reduce copper availability.

Iron

Iron (Fe) deficiencies occur in calcareous soils and cause a vague yellowing in wheat. Some plants are uniformly chlorotic, and others develop interveinal, yellow leaf stripes (Plate 73). Younger leaves are affected first and may turn ivory in color. Elements like copper, manganese, aluminum, and phosphorus may form complexes with iron and decrease its availability in soil and its utilization in plants.

Chloride

Chloride (Cl) deficiency in wheat has not been reported under field conditions. However, positive yield responses to chloride applications are not unusual. The nature of the response is not completely understood but apparently involves factors beyond nutrition. High levels of plant chloride are associated with reduced disease (take-all, stripe rust, leaf rust, tan spot, Septoria leaf spot) in wheat. Chloride supplementation is most likely to be beneficial where soils are leached by heavy rainfall or where potassium chloride has not previously been applied.

Zinc

Zinc (Zn) deficiencies are most apt to occur in calcareous soils and in soils rich in phosphorus. Plants deficient in zinc are stunted, with short internodes and few tillers. Leaves are chlorotic, especially between the margin and midrib. Severe deficiencies cause leaves to turn gray-white and die. Foliar applications of zinc can correct the deficiency in the existing crop, but soil applications are necessary to restore minimal zinc levels in soil.

Aluminum

Aluminum (Al) is not an essential plant nutrient. However, aluminum may be growth-limiting in highly weathered, acid soils ($pH<5.4$) because of its toxicity rather than its deficiency. Wheat cultivars differ in tolerance to high-aluminum soils, apparently because they differ in uptake and accumulation of the element. Susceptible cultivars have reduced root and top growth. A shallow root system often results because aluminum concentration increases with depth in soils developed from acidic parent material. Aluminum, like copper, decreases calcium transport. Applications of lime or chelating agents sometimes relieve aluminum toxicity.

Selected References

Brown, J. C., Ambler, J. E., Chaney, R. L., and Foy, C. D. 1972. Differential responses of plant genotypes to micronutrients. Pages 389–418 in: Micronutrients in Agriculture. J. J. Mortvedt, P. M. Giordano, and W. L. Lindsey, eds. Soil Science Society of America, Madison, WI.

Christensen, N. W., and Brett, M. 1985. Chloride and liming effects on soil nitrogen form and take-all of wheat. Agron. J. 77:157–163.

Fleming, A. L. 1983. Ammonium uptake of wheat varieties differing in aluminum tolerance. Agron. J. 75:726–730.

Foy, C. D., Fleming, A. L., and Schwartz, J. W. 1973. Opposite aluminum and manganese tolerances in two wheat cultivars. Agron. J. 65:123–126.

Huber, D. M. 1981. The use of fertilizers and organic amendments in the control of plant diseases. Pages 357–394 in: Handbook of Pest Management in Agriculture. D. Pimentel, ed. CRC Press, Boca Raton, FL.

Krantz, B. A., and Melsted, S. W. 1964. Nutrient deficiencies in corn, sorghum and small grains. Pages 25–58 in: Hunger Signs in Crops. H. B. Sprague, ed. David McKay Co., New York.

Lamb, C. A. 1967. The nutrient elements. Pages 207–217 in: Wheat and Wheat Improvement. K. S. Quisenberry and L. P. Reitz, eds. Agronomy Monograph 13. American Society of Agronomy, Madison, WI.

Ohki, K. 1984. Manganese deficiency and toxicity effects on growth, development and nutrient composition in wheat. Agron. J. 76:213–218.

Papastylianou, I., Graham, R. D., and Puckridge, D. W. 1982. The diagnosis of nitrogen deficiency in wheat by means of a critical nitrate concentration in stem bases. Commun. Soil Sci. Plant Anal. 13:473–485.

Taylor, R. G., Jackson, T. L., Powelson, R. L., and Christensen, N. W. 1983. Chloride, nitrogen form, lime, and planting date effects on take-all root rot of winter wheat. Plant Dis. 67:1116–1120.

Ward, R. D., Whitney, D. A., and Westfall, D. G. 1973. Plant analysis as an aid in fertilizing small grains. Pages 329–348 in: Soil Testing and

Plant Analyses. L. M. Walsh and J. D. Beaton, eds. Soil Science Society of America, Madison, WI.
Wrigley, C. W., du Cros, D. L., Moss, H. J., Randall, P. J., and Fullerton, J. G. 1984. Effect of sulfur deficiency on wheat quality. Sulfur Agric. 8:2–7.

Soil Compaction

Soil compaction is a process by which the soil bulk is compressed into a smaller volume. As soil bulk density increases, longitudinal root growth normally slows down (Fig. 115). Soil texture influences the specific bulk density at which wheat root growth is impaired. In most soils, bulk density values greater than 1.3 g/cm^3 limit wheat growth. Normally, the risk of compaction is greatest in fine-textured soils and in soils low in organic matter.

Reduced top growth and yield are usual consequences of impaired root systems. Wheel tracks from heavy machinery may remain visible throughout the growing season because the compacted soil supports fewer and shorter plants with fewer tillers. Compacted soils have lower water-holding capacities and less air space and erode more easily. Layers of compacted soil sometimes form just beneath plow or tillage depth or at field gateways and headlands. Increasing organic matter content and deep (subsoil) cultivation are partial remedies.

In contrast, soil with too much air space may dry quickly. As a consequence, plants become moisture-stressed, seedling emergence is retarded, and winterkill is accentuated.

Soil pH

Wheat is adapted for growth in soils with pH between 5.5 and 7.5. The pH optimum for vegetative and reproductive growth lies between 6.2 and 6.8. Excessive soil acidity or alkalinity damages wheat indirectly via effects on nutrient availability (Plate 74), aluminum toxicity, nitrogen form, and pathogens.

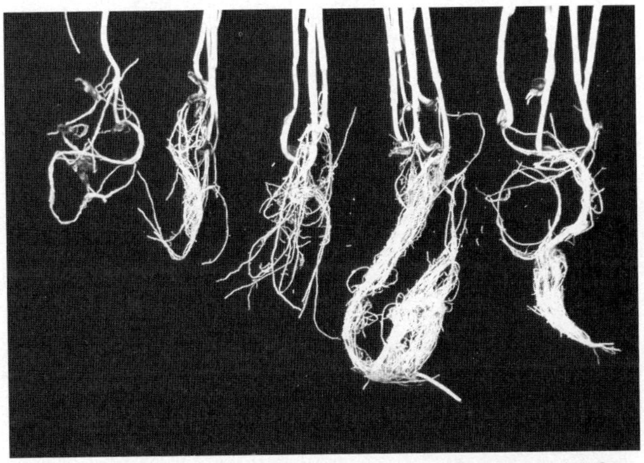

Fig. 115. Root development relative to soil bulk density of (left to right) 1.75, 1.55, 1.4, 1.25, and 1.1 g/cm^3. (Courtesy E. D. Kerr)

Continued use of commercial fertilizers, such as ammonium and elemental sulfur, tends to increase soil acidity. Soil acidity increases primarily via biological oxidation of ammonium and elemental sulfur and the production of protons (H^+), which replace other cations at exchange sites. Crushed limestone (calcium carbonate) is used to increase soil pH. On a weight basis, approximately three units of calcium carbonate are needed to neutralize the acidity contributed by four units of nitrogen. Elemental sulfur is used occasionally to reduce the pH of calcareous or alkaline soils.

Selected References

Kamprath, E. J., and Foy, C. D. 1971. Lime-fertilizer-plant interactions in acid soils. Pages 105–151 in: Fertilizer Technology and Use. 2nd ed. Soil Science Society of America, Madison, WI.
Smiley, R. W. 1975. Forms of nitrogen and the pH in the root zone and their importance to root infections. Pages 55–62 in: Biology and Control of Soil-Borne Plant Pathogens. G. W. Bruehl, ed. American Phytopathological Society, St. Paul, MN.

Glossary

C—centigrade or Celsius
cm—centimeter (1 cm = 0.01 m = 0.39 in.)
g—gram (1 g = 0.35 ounce; 453.6 g = 1 pound)
ha—hectare (1 ha = 2.47 acres)
in.—inch (1 in. = 2.54 cm)
m—meter (1 m = 39.37 in.)
min—minute
mm—millimeter (1 mm = 0.001 m = 0.04 in.)
μg—microgram (1 μg = 10^{-6} g)
μm—micrometer (1 μm = 10^{-6} m)
nm—nanometer (1 nm = 10^{-9} m)
ppb—parts per billion
ppm—parts per million
sec—second
syn(s).—synonym(s)

acervulus (pl. acervuli)—erumpent mycelial cushion bearing conidiophores and conidia
acrogenous—developing from an apex or terminus rather than a base
acropetal—toward the tip, apex, or growing point
acute—pertaining to symptoms that develop suddenly (as opposed to chronic)
aeciospore—dehiscent, dikaryotic spore from an aecium
aecium (pl. aecia)—cuplike fruiting body of rust fungi
aerial—in the air; aboveground
aerobic—in the presence of oxygen
aflatoxin—chemical by-product from *Aspergillus flavus* harmful to people and animals
agar—gelatin from seaweed; growth medium
alkaline—having basic (nonacidic) properties
alkaloid—organic chemical with alkaline (basic) properties
allelopathy—competition between plant species based on emission of natural toxicants
alternate host—an additional plant species required by some pathogens such as rust fungi to complete their life cycle; host of lesser economic importance
alternative host—another host of a given pathogen
anaerobic—not using air or oxygen
anamorph—asexual or imperfect form of a fungus
anastomosis—fusion, as of hyphal strands, and combination of their contents
annual—plant that develops from seed, matures, and sets seed in one growing season
annulate—having rings
annule—ring or collarlike fold
anterior—toward the front or head (as opposed to posterior)
anther—pollen-producing organ of flowering plants
antheridium (pl. antheridia)—male sexual organ of some fungal and lower plant species
anthesis—growth stage at which flowers open and shed pollen
anthracnose—disease caused by fungi that form acervuli
antibiotic—chemical (usually of microbial origin) that inhibits or kills other organisms
antibody—a product of the immune system in animals; a specific protein (globulin) produced to combine with, and counter the effects of, an antigen
antigen—any foreign chemical (normally a protein) that induces antibody formation in animals
apex (adj. apical)—tip
aplerotic—characterizing an oospore that does not completely fill the oogonium
apothecium (pl. apothecia)—saucer-shaped ascocarp
appressorium (pl. appressoria)—swollen or flattened portion of a germ tube or hypha that functions in host penetration
arginine dihydrolase—enzyme of qualitative significance in some bacteria
arthrospore—spore formed from cells of a fragmented hypha
ascocarp—sexual fruiting body of an ascomycete
ascogenous—developing or originating from an ascus
ascomycete—any of a class of fungi that produce sexual spores (ascospores) within an ascus
ascospore—spore produced within an ascus
ascus (pl. asci)—saclike structure within an ascocarp in which ascospores are borne
aseptate—without cross walls
asexual—vegetative; without sex organs, gametes, or sexual spores; imperfect
auricle—earlike structure at juncture of leaf sheath and blade
autoclave—sterilize with pressurized steam
avirulent—without capacity to cause disease; nonpathogenic
awn—bristlelike structure at the terminus of a glume

bacilliform—shaped like a blunt, thick rod
bacteriocin—antibiotic produced by some bacteria active against strains of the same or related bacterial species
ballistospore—a spore that is forcibly discharged from its sporophore
bar—unit of pressure used to express water potential (1 atmosphere = 1.013 bars)
basidiocarp—sexual fruiting body of a basidiomycete
basidiomycete—fungus that forms sexual spores (basidiospores) on a basidium
basidiospore—haploid sexual spore produced on a basidium
basidium (pl. basidia)—short, club-shaped, haploid promycelium produced by basidiomycetes
basipetal—toward the bottom
biflagellate—having two flagella
binucleate—having two nuclei
biological control—disease or pest control through counterbalance of microorganisms or other natural components of the environment
biotrophic—nourished by organic foods or substrates
biotype—subspecies group of organisms that differs from other members of the species in biochemical, physiologic, or behavioral properties
bitubular—having a tube within a tube
blade—flat portion of a grass leaf above the sheath
blastospore—asexual spore developed by budding or extrusion from an existing spore or cell
blight—general and rapid death of plants or plant parts
blotch—usually superficial blot or spot
boot—sheath of the uppermost leaf on a wheat or grass plant that encloses the head
bract—modified leaf subtending a flower or inflorescence
bromovirus—an isometric plant virus of the brome mosaic virus type
bunt—a covered smut disease
bunt ball—smut sorus that displaces a wheat kernel
bursa—lateral cuticular extension at the posterior of male nematodes

calcareous—rich in calcium carbonate (lime)
cap—structural part of higher basidiomycetes that supports the hymenium
capsid—protein coat of a virion
carbohydrate—food compound composed of carbon, hydrogen, and

oxygen
catalase—enzyme of qualitative significance in some bacteria
cereal—grass grown for its edible seed
chaff—nonseed portion of a mature wheat head
cheek—rounded surface or face of a wheat kernel
chitin—carbohydrate in fungal cell walls
chlamydospore—thick-walled, asexual spore formed by modification of a hyphal or conidial cell
chlorophyll—green, light-sensitive pigment in higher plants that participates in photosynthesis
chlorosis (adj. chlorotic)—fading of green plant color to light green or yellow
chromosome—aggregate of genes in a replicating, rodlike body
chronic—developing slowly; persistent
circulative—pertaining to viruses that accumulate within or pass through the gut of insect vectors before being transmitted to plants
cirrhus (pl. cirrhi)—column of mucus laden with spores
citriform—lemon-shaped
clamp connection—hyphal bypass in basidiomycetes formed at cell division and connecting the newly divided cells
clavate—club-shaped
cleistothecium (pl. cleistothecia)—closed, usually spherical ascocarp
closterovirus—a flexuous rod plant virus of the beet yellows virus type
clypeus (pl. clypei)—shieldlike stromatic growth over one or more perithecia
coalesce—to run together; overlap; merge
coccoid—spherical
coenocytic—nonseptate; having nuclei not separated by cross walls
coleoptile—ephemeral, nonpigmented tissue sheathing the first true leaf of a grass seedling
conidiophore—hypha differentiated to bear conidia
conidium (pl. conidia)—any asexual spore except a sporangiospore or chlamydospore
convex—having a surface that curves outward
corolla—collectively, the petals of a flower
cortex (adj. cortical)—tissue between the epidermis and phloem in culms and roots
coryneform—club-shaped
cotyledon—seed leaf; primary embryonic leaf
crazy top—disease symptom characterized by twisting, proliferation, and distortion of upper plant parts
cross-protective—pertaining to a pathogen established in a host that prevents or inhibits the establishment of a second pathogen in the same host
crown—in wheat, the compacted series of nodes from which culms and secondary roots arise
culm—stem of a grass plant
cultivar—cultivated variety
cuticle (adj. cuticular)—outer sheath or membrane
cyst—swollen, egg-laden carcass of a dead female nematode of the genus *Heterodera* and other cyst-forming genera
cystosorus (pl. cystosori)—resting spore
cytoplasm—living contents of a cell, excluding the nucleus

damping-off—damage near soil level to germinating seed or seedlings that results in a rapid, lethal decline before or soon after emergence
degree-day—unit of heat measurement
dehiscent—opening by breaking into parts
denticle—indentation; pore
desiccate—to dry
dextrose—glucose; sugar
diclinous—having an oogonium and antheridium on different hyphae
dicotyledonous—having two seed leaves
dictyospore—spore with longitudinal and lateral septa
digitate—fingerlike
dikaryon (adj. dikaryotic)—fungal cell having two sexually compatible nuclei ($n + n$ chromosomes) per cell
diploid—having a double set of chromosomes ($2n$) per cell
diurnal—daily; during the day (as opposed to nocturnal)
dormant—resting; living in a state of reduced physiologic activity
dorsal—pertaining to the back or top (as opposed to ventral)

ear—spike or head of wheat
echinulate—having spines or other sharp projections
ectoparasite—parasitic organism that lives outside its host
effuse—flat and spreading
ELISA—*See* enzyme-linked immunosorbent assay
elutriation—a specific mechanical and gravimetric process for separating nematodes from soil
embryo—seed germ; the rudimentary plant within a seed
enation—small swelling or gall
encyst—to form a cyst or encasement and become dormant
endogenous—coming or developing from within
endoparasite—parasitic organism that lives within its host
enzyme—protein that catalyzes a specific biochemical reaction
enzyme-linked immunosorbent assay (ELISA)—a color-enzyme enhanced antigen-antibody reaction
epiblast—superficial swelling
epidemic—widespread outbreak of disease
epidemiology—study of disease initiation, development, and spread
epidermis—superficial layer of cells
epiphyte (adj. epiphytic)—plant growing on the surface of its host or substrate
eradicate—destroy; eliminate
ergotism—disease of people and animals resulting from ingestion of ergot alkaloids
erumpent—bursting or erupting through the surface
etiology—science or study of the causes or origins of a disease or abnormality
exogenous—coming or developing from outside
extrude—to push out; emit to the outside
exudate—that which is excreted or discharged; ooze

facultative parasite—saprophytic organism capable of parasitism
fairy ring—confined circular pattern of growth of toadstools and other gill fungi
fermentation—growth or metabolism under anaerobic conditions
filamentous—threadlike
filiform—long, needlelike
fission—form of cell division; splitting
flaccid—limp, as opposed to turgid
flagellum (pl. flagella)—hairlike, whiplike, or tinsel-like appendage on bacterial cells and zoospores that provides locomotion
flag leaf—leaf originating from the first culm node below the rachis; uppermost leaf
fleck—small spot
flexuous—having turns or bends
floret—small flower, as in a spikelet
flower primordia—meristematic tissues differentiable into flower parts
foliar—pertaining to leaves
foot—basal area of a culm
foot cell—basal cell, as of a spore or conidiophore
frass—excrement of an insect
fructification—fruiting body
fruiting body—complex sporulating fungal structure
fumigant—chemical pesticide that is active in the vapor state
fungicide—agent, usually a chemical, that kills or inhibits fungi
furovirus—fungusborne, rod-shaped plant virus of the soilborne wheat mosaic virus type
fusoid—narrowing toward the ends

gall—swelling; localized overgrowth
gametangium (pl. gametangia)—cell or organ that produces gametes
gamete—male or female sex cell
geminate—paired
geminivirus—a geminate plant virus of the maize streak virus type
gene—smallest functional unit of genetic material on a chromosome; unit of inheritance
genome—set or group of chromosomes
genotype—genetic constitution of an individual or group; class or group of individuals sharing a particular genetic makeup
genus (pl. genera)—taxonomic group above species and below family
germinate—begin growth (as of a seed or spore)
germling—initial thallus or body produced by germination
germ pore—thin area within a spore wall through which a germ tube can emerge
germ tube—hypha resulting from an outgrowth of the spore wall or cytoplasm
gills—platelike hymenial arrangement in the cap of higher basidiomycetes
girdle—to encircle
globose—nearly spherical
gluten—a cereal protein
glume—empty bract at the base of a spikelet
Gramineae—taxonomic family name of grasses
gram-positive (gram-negative)—pertaining to bacteria that retain (release) the violet dye in the Gram stain procedure
gregarious—tending to exist in groups or clusters

haploid—having a single complete set of chromosomes (n)
hardiness—ability to withstand stress

haustorium (pl. haustoria)—specialized hypha within penetrated host cells that probably functions in food absorption
headland—roadway or field margin
herbaceous—developing primary but not woody tissue
herbicide—chemical that limits the growth of or kills herbaceous plants
heterogeneous—containing dissimilar genes
heterothallic—having sexes separated in different mycelia
hexaploid—having six sets of chromosomes ($6n$)
homothallic—having both sexes in the same body or mycelium
honeydew—ooze or exudate of sweetish fluid
hordeivirus—a rigid rod plant virus of the barley stripe mosaic virus type
humus—an organic soil component
hyaline—colorless; transparent
hybrid (v. hybridize)—offspring of two individuals of different genetic character
hymenium (adj. hymenial)—spore-bearing (sporogenous) layer of a fungal fruiting body
hyperparasitism (v. hyperparasitize)—parasitism of one microorganism by another
hyperplasia (adj. hyperplastic)—process of enlargement by excessive cell division
hypertrophy (adj. hypertrophic)—process of enlargement by excessive increase in cell size
hypha (pl. hyphae)—tubular filament of a fungal thallus or mycelium
hyphopodium (pl. hyphopodia)—usually flattened hypha of an epiphytic fungus specialized for host attachment or penetration

immune—not affected by or responsive to disease
imperfect stage—asexual portion of a life cycle
inclusion body—subcellular structure induced to form in virus-infected cells
indigenous—native
infect—invade or penetrate as an initial phase of disease
infection court—site in or on host plant where infection can occur
infection type—gross appearance or size of disease symptom
infectious—capable of infecting and spreading from plant to plant
infest—to contaminate, as with organisms
inflorescence—flower cluster
inoculate—place inoculum in an infection court
inoculum—pathogen or its parts accessible to or brought into contact with a host
insecticide—chemical used to control or kill insects
intercellular—between or among cells
internode—culm area between two adjacent nodes
intracellular—through or within a cell or cells
in vitro—in glass or in an artificial environment
isolate (n.)—separated or confined spore or microbial culture
isometric—having equal sides or dimensions

jointing—growth stage of rapid culm elongation between tillering and heading

kernel—a seed of a cereal plant

labium (pl. labia)—lip
lanceolate—like a lance or long spear
larva (pl. larvae)—juvenile; growth stage between the embryo and adult
latent—present but not manifested or visible; symptomless
lesion—wound or delimited diseased area
ligule—outgrowth at the inner juncture of a leaf sheath and blade
lobate—lobed
lodging—lying down, by which culms approach horizontal rather than vertical
lumen—hollow center of a culm
luteovirus—an isometric plant virus of the barley yellow dwarf virus type
lux—unit of illumination equivalent to 0.093 footcandles

macroconidium (pl. macroconidia)—long or large conidium relative to microconidium
male-sterile—having nonfunctional or no male sex organs
manual transmission or inoculation—dissemination or introduction of inoculum to infection courts by hand manipulation
mechanical transmission or inoculation—dissemination or introduction of inoculum to infection courts (especially wounds) created by physical disruption of host tissues
meiosis—process by which a zygote divides and produces haploid gametes
melanin (adj. melanoid)—brown-black pigment
meristem (adj. meristematic)—plant tissues that function principally in cell division and differentiation
mesophyll—central internal tissues of a leaf
metabolite—chemical component of a reaction series or process
microconidium (pl. microconidia)—small conidium relative to a macroconidium; microspore
microflora—composite of microscopic plants
microspore—small spore relative to macrospore
midrib—central, thickened vein of grass leaves
mitosis—process of nuclear division where chromosomes duplicate, divide, and produce daughter nuclei of the same genetic constitution as the parent nucleus
Mollicutes—group of prokaryotic organisms bounded by flexuous membranes
molt—to shed a cuticle or body encasement during a growth phase
monoclinous—having an antheridium and an oogonium on the same hypha
monoclonal—having identity with a type individual; pertaining to antibodies developed through replication of an antigen-producing cell and a myeloma (cancerous) cell
monocotyledonous—having one seed leaf
morphology—study of the form and structure of organisms
mosaic—disease symptom characterized by mixed green, light green, and yellow patches
mottle—disease symptom characterized by light and dark areas in an irregular pattern
mucus—slime; bacterial exudate
multigenic—containing or controlled by a number of genes
multiline—cultivar with a mixed population of plant genotypes
multinucleate—having more than one nucleus
multiple infection—invasion by more than one parasite
multiseptate—having many septa
mutation (n. mutant)—abrupt heritable change in an individual
mutually exclusive—describing two or more parasites or pathogens that cannot exist together in the same host cell or tissue
mycelium (pl. mycelia)—mass of hyphae constituting the thallus or body of a fungus
mycoplasma—prokaryotic organism, smaller than conventional bacteria, without rigid cell walls, and variable in shape
mycorrhiza (pl. mycorrhizae, adj. mycorrhizal)—association or resting structure of a synergistic or nonpathogenic fungus in roots of higher plants
mycotoxin—toxic chemical compound produced by a fungus

neck—culm area between the head and the uppermost leaf of wheat; peduncle
necrosis (adj. necrotic)—death of tissue, usually accompanied by darkening or discoloration
nematicide—agent, usually a chemical, that kills or inhibits nematodes
nitrification—the biological oxidation of ammonium (NH_4^+) to nitrate (NO_3^-)
node—joint of a culm or rachis; site of attachment of leaves and spikelets
nonpersistent—dissipating; describing viruses that are infectious within insect vectors for short periods and are transmissible without a latent period and without prior multiplication and translocation within the vector
nonseptate—without cross walls
nymph—juvenile insect that resembles the adult

obligate parasite—organism that can survive only on or in living tissues
onchiostyle—stiff, slender, grooved tooth of stubby root nematodes
oogonium (pl. oogonia)—female structure (gametangium) in fungi that produce oospores
oospore—thick-walled, sexually or asexually derived resting spore of phycomycetes
organelle—delimited, membranous structure within a cell having a specialized function
ostiole (adj. ostiolate)—pore; opening
outcrossing—cross-fertilization, as opposed to self-fertilization
ovary—female reproductive structure that contains an ovule or egg
oversummer—to survive over the summer
overwinter—to survive over the winter
oviparous—egg-laying; reproducing from eggs
ovoid—egg-shaped
oxidase—enzyme of qualitative significance in some bacteria

papilla (adj. papillate)—small, blunt projection
parasexual—pertaining to the recombination of genetic characters without sexual process

parasite—organism that lives with, in, or on another organism for its own advantage
parenchyma—physiologically active plant tissue that is less important structurally
parthenogenesis (adj. parthenogenetic)—process of asexual reproduction in some animals whereby offspring develop directly from female gametes
particulate—having minute separate particles
pathogen—agent that causes disease
pathogenicity—ability to cause disease
pathotype—variant of a given pathogen; pathogenic reaction on a given host
pathovar (abbrev. pv.)—strain of a bacterial species
pedicellate—having a stalk or pedicel
peduncle—*See* neck
perennial—persisting for more than two years or growing seasons
perfect (of fungi)—sexual; capable of sexual reproduction
pericarp—outer layer of a seed or fruit
perithecium (pl. perithecia)—flasklike ascocarp with an ostiolelike opening
peritrichous—having flagella distributed along a circumference
persistent—pertaining to viruses that are infectious within and transmissible by insect vectors for long periods
phasmid—posterior sensory pore on certain nematodes
phlallde—conidiophore of fixed length with one or more open ends through which a basipetal succession of conidia develop
phialospore—conidium borne by a phialide
phloem—food-conducting tissues in plants
photosynthesis—process of manufacturing carbohydrates from carbon dioxide, water, chlorophyll, and light
phycomycete—member of a group of fungi with or without a true mycelium and, if a mycelium is present, having multinucleate hyphal protoplasm and occasionally but not regularly septate hyphae
physiologic form—subspecies group that differs from other physiologic forms of the species in behavior or other characteristics but not in morphology
physiologic race—*See* race
phytopathogenic—causing disease in plants
phytotoxin (adj. phytotoxic)—substance that is harmful to plants
pinwheel—shape of some virus inclusion bodies seen in thin section
plasmalemma—membrane that surrounds a protoplast
plasmodium (pl. plasmodia, adj. plasmodiophoraceous)—naked mass of protoplasm containing nuclei
plasmolysis—shrinking of a protoplast caused by water loss
polar—at the ends or poles
pollen—male sex cells produced by the anthers of flowering plants
polyhedron—form with many sides or faces
polyploid—having more than two sets of chromosomes
porospore—spore developed by expansion through an opening of fixed dimension
potyvirus—a flexuous rod plant virus of the potato virus Y type
predispose—to make prone to infection and disease
primary inoculum—inoculum that initiates rather than spreads or magnifies disease
primary root—root developed from a seed
primordium—*See* flower primordia, seed primordia
prokaryotic (n. prokaryote)—without a nucleus or membrane-bound organelles
promycelium—basidium; initial structure produced when teliospores germinate
propagule—any part of an organism capable of independent growth
protectant—describing an agent, usually a chemical, that prevents or inhibits infection
prothorax—structure between the thorax and head of an insect
protoplasm—living contents of a cell
protoplast—plant cell exclusive of its wall
pseudothecium (pl. pseudothecia)—ascocarp similar to a perithecium and having a dispersed rather than an organized hymenium
punctate (n. punctation)—with spots, pits, or hollows
pupa (pl. pupae)—an immature, inactive stage in insect metamorphosis
pupate—to be inactive, as with certain insects in a cocoon or case
pustule—blisterlike, usually erumpent spot or sorus
pycnidiospore—conidium produced within a pycnidium
pycnidium (pl. pycnidia)—asexual, globose or flask-shaped fruiting body produced by fungi
pycniospore—haploid, sexually derived spore formed in a pycnium
pycnium (pl. pycnia)—haploid, pycnidiumlike fruiting body produced by some rust fungi

quarantine—legislative control of the transport of plants, plant parts, or other materials to prevent disease spread
quiescent—quiet; dormant

race—group of individuals within a variety or species distinguished by behavior but not by morphology
rachis—axis of the wheat head
receptive hyphae—specialized hyphae that protrude from a pycnium and function as female gametes
reductive division—stage of meiosis where two daughter nuclei each receive half the chromosomes of the parent nucleus
reovirus—a double-shelled, isometric plant virus of the wound tumor virus type
resistance (adj. resistant)—property of hosts that prevents or impedes disease development
respiration—series of chemical reactions that produce energy at the expense of oxygen, carbohydrate, and fat
resting spore—temporarily dormant, usually thick-walled spore
reticulate—netlike; with netlike ridges
rhabdovirus—a bacilliform (enveloped) plant virus of the lettuce necrotic yellows virus type
rhizoid—stiff, rootlike hyphal strand
root cap—protective tissue at the root apex
root hair—threadlike, single-celled outgrowth from a root epidermal cell
rosette—disease symptom characterized by short, bunchy growth habit resulting from restricted elongation of internodes
rotation—sequence and duration of crop cultivation in a given area
rugose—wrinkled
runner hyphae—thickened hyphal strands

saprophyte (adj. saprophytic, saprophagous)—organism that feeds on nonliving organic matter
sclerotium (pl. sclerotia)—hard, usually darkened and rounded mass of dormant hyphae
scutellum (adj. scutellar)—cotyledon of a grass embryo
secondary inoculum—inoculum resulting from primary infections
secondary root—root developed from a crown or node, as opposed to a seed root
sedentary—remaining in a fixed location
seed primordia—meristematic tissues differentiable into seed
seminal root—root derived from a seed
senescence—decline, as with maturation, age, or disease stress
septum (pl. septa, adj. septate)—cross wall
serology—study, detection, and identification of antigens, antibodies, and their reaction
seta (pl. setae)—stiff, hairlike appendage
sheath—lower part of a grass leaf that clasps the culm; membranous cover
sign—indication of disease from direct visibility of the pathogen or its parts
sobemovirus—an isometric plant virus of the southern bean mosaic virus type
solute—compounds or ions dissolved in a solvent
sorus (pl. sori)—compact fruiting structure of rust and smut fungi
species (abbrev. sp., pl. spp.)—taxonomic category ranking below a genus but above a race, strain, or variety; a genus name followed by sp. indicates that the particular species is undetermined; spp. after a genus name indicates that two or more species are being referred to without being named individually
spicule—copulatory organ of male nematodes
spike—inflorescence with florets and/or spikelets on an axis; wheat head
spikelet—spike appendage composed of glumes and florets
sporangiophore—differentiated hypha that bears a sporangium
sporangiospore—spore that develops in a sporangium
sporangium (pl. sporangia)—flasklike fungal structure whose contents differentiate into asexual spores
spore—one- to many-celled reproductive body in fungi and lower plants
spore ball—teliospore and accompanying sterile cells of flag smut fungi
sporidium (pl. sporidia)—basidiospore
sporodochium (pl. sporodochia)—superficial, cushion-shaped, asexual fruiting body
sporophore—any structure that bears spores
sporulate—to produce spores
spring wheat—wheat sown in spring to mature in the same growing season
sprout—wheat germling; to begin to germinate, as a seed
stellate—pointed, as a star
sterigma (pl. sterigmata)—small, usually pointed protruberance or projection

sterile—infertile; free from contaminant organisms; nonsporing, anamorphic
stigma—portion of a flower that receives pollen
stomate (pl. stomata)—regulated opening in the plant epidermis for passage of gases and water vapor
strain—descendants of an isolated organism; biotype; race
striate (n. striation)—marked with lines, grooves, or ridges
stroma (pl. stromata)—compact mass of mycelium (with or without host tissue) that supports fruiting bodies
stubble—culm bases and crowns of harvested wheat still rooted in soil
stylet—stiff, slender, hollow feeding organ in plant-parasitic nematodes
subcrown internode—short, culmlike connection between the crown and seed roots of wheat
susceptible—not immune; lacking resistance; prone to infection
symptom—indication of disease by reaction of the host
symptomatology—study of disease symptoms
syncytium (pl. syncytia)—multinucleate mass of protoplasm resulting from the fusion of protoplasts and surrounded by a common cell wall
synergistic—pertaining to a host response to concurrent pathogens that exceeds the sum of the separate responses to each pathogen
synnema (pl. synnemata)—compact or fused, generally upright group of conidiophores
systemic—pertaining to chemicals or pathogens that spread throughout plants, as opposed to remaining localized

teleomorph—sexual or perfect form of a fungus
teliospore—thick-walled resting spore of rust and smut fungi that germinates to form a basidium
telium (pl. telia)—sorus that produces teliospores
tetraploid—having four sets of chromosomes ($4n$)
thallus (pl. thalli)—fungus body
therapeutic—able to relieve symptoms and cure established disease
thorax—insect body part between the head and abdomen
thrifty—growing vigorously
tiller—shoot, culm, or stalk arising from a crown bud
tissue—group of cells of similar structure and function
tobamovirus—a rigid rod plant virus of the tobacco mosaic virus type
tolerant—sustaining disease without serious damage or yield loss
toruloid—yeastlike
toxin—poison
translocation—movement of water, minerals, food, or pathogens within a plant
translucent—intermediate between clear and opaque relative to light passage
transmission—spread of virus or other pathogen from plant to plant or by a vector
transovarial—through ovaries and eggs
transpiration—loss of water vapor through leaves
triticale—hybrid cereal from a wheat (*Triticum*) and rye (*Secale*) cross
truncate—ending abruptly

ubiquitous—omnipresent
unicellular—having one cell
uniflagellate—having one flagellum
uninucleate—having one nucleus
unitunicate—having one definable wall or cover
urediospore—binucleate, dikaryotic, asexual rust spore
uredium (pl. uredia)—fruiting body (sorus) of rust fungi that produces urediospores

vascular—pertaining to conductive (xylem and phloem) tissues
vector—agent that transmits inoculum
ventral—pertaining to the front or belly (as opposed to dorsal)
vermiform—worm-shaped
vernalize—to expose plants to cold temperatures to change their growth habit from vegetative to reproductive
vesicle—subcellular membranous enclosure
virion—complete virus particle
virulence (adj. virulent)—degree of pathogenicity; capacity to cause disease
viruliferous—virus-laden
vitreous—glasslike
viviparous—bearing live young, as opposed to laying eggs
volunteer—self-set plant; plant seeded by chance

water-soaked—describing diseased tissues that appear wet, dark, and usually sunken and translucent
whitehead—bleached, dry, immature spike
whorl—circle of leaves, flowers, or roots arising from a single node or point
winterkill—tissue and plant death caused by injurious winter environment
winter wheat—wheat sown in autumn to overwinter and set seed in the next growing season

xylem—water-conducting tissue in plants

zoosporangium (pl. zoosporangia)—structure that produces and contains zoospores
zoospore—flagellated fungal spore capable of locomotion in water
zygote—diploid cell arising from the fusion of two haploid cells or gametes

Selected References

Agrios, G. N. 1978. Plant Pathology. 2nd ed. Academic Press, New York. 703 pp.

Ainsworth, G. C. 1971. Ainsworth & Bisby's Dictionary of the Fungi. 6th ed. Commonwealth Mycological Institute, Kew, Surrey, England.

Federation of British Pathologists (Terminology Sub-Committee). 1973. A guide to the use of terms in plant pathology. Phytopathological Papers, No. 17. Commonwealth Mycological Institute, Kew, Surrey, England.

Robinson, R. A. 1969. Disease resistance terminology. Rev. Appl. Mycol. 48:593–606.

Stern, W. T. 1973. Botanical Latin. David and Charles, Ltd., London. 566 pp.

Index